D0621220

MATHQUEST THREE

Brendan Kelly
Co-ordinator of Mathematics
Halton Board of Education

Chester Carlow
Associate Professor, Department of Curriculum
Ontario Institute for Studies in Education

J Symington
Special Education Consultant, Mathematics/Science
Peel Board of Education

Joan Worth
Professor, Elementary Education
University of Alberta

Addison-Wesley Publishers Limited
Don Mills, Ontario • Reading, Massachusetts
Menlo Park, California • Wokingham, Berkshire
Amsterdam • Sydney • Singapore • Tokyo
Mexico City • Bogota • Santiago • San Juan

EDITORIAL
Dianne Goffin
Philip Gebhardt

DESIGN, ART DIRECTION AND ILLUSTRATION
Pronk & Associates

Gord Pronk, Mary Pronk, Walter Augustowitsch, Alan Barnard, Bruce Bond, Ralph Oesterreich, David Peden, Steve Pilcher, Margo Stahl, Gay Wieringa

Illustrators
Raffi Anderian, Graham Bardell, Alan Barnard, David Bathurst, Thach Bui, Scott Caple, Ian Carr, Frank Esch, Rick Fischer, Chuck Gammage, Peter Grau, Robert Hughes, Robert Johannsen, Danielle Jones, Bernadette Lau, Paul McCusker, Marilyn Mets, David Partington, Bill Payne, Steve Pilcher, Paul Rivoche, Greg Ruhl, Mark Summers, Carol Watson

Photographers
Jeremy Jones, Teri Kelly

Cover Illustration
Graham Bardell

Canadian Cataloguing in Publication Data

Kelly, B. (Brendan), 1943–
MathQuest 3
For use in grade 3.
Includes index.

ISBN 0-201-19300-0

1. Arithmetic – 1961 – I. Title.

QA108.K46 1985 513 C85-098688-5

Printed and bound in Canada.

ISBN 0-201-19300-0

G H -BP- 93 92

ACKNOWLEDGEMENTS

The authors and publishers would like to express their appreciation for the invaluable advice and encouragement received from educators across Canada during the development of this program. We particularly wish to thank the following people: Dianne Brow, Janis Cleugh, Jan Cornwall, Sheila Fitzgerald, Sandra Folk, David Glebe, Gordon Jeffery, Ann Maher, Eileen Mansfield, Elizabeth McNaught, Bill Nimigon, Alexander Norrie, and Katherine Willson. Special thanks to the students at Percy Williams Jr. Public School.

Grateful acknowledgement is made for permission to reprint copyrighted materials.

"Toucans Two" from *Zoo Doings* by Jack Prelutsky. Copyright © 1970, 1983 Jack Prelutsky. Reprinted by permission of Greenwillow Books (a division of William Morrow & Company).

"Homework Machine" and "Shapes" from *A Light in the Attic* by Shel Silverstein. Copyright © 1981 Snake Eye Music Inc. "Two Boxes" and "Me and My Giant" from *Where the Sidewalk Ends* by Shel Silverstein. Copyright © 1974 Snake Eye Music Inc. Reprinted by permission of Harper & Row, Publishers Inc.

"The Puzzle" from *Jelly Belly* by Dennis Lee. "Alligator Pie" and "The Fishes of Kempenfelt Bay" from *Alligator Pie* by Dennis Lee. Reprinted by permission of Macmillan of Canada, a Division of Canadian Publishing Corporation.

"Canada Is" by Steve Hyde. Reprinted by permission of Tembo Music Canada.

"Biking" by Margaret Hillert. Copyright © 1969 Margaret Hillert. Reprinted by permission of the author.

"Lengths of Time" by Phyllis McGinley. Text copyright © 1965, 1966 by Phyllis McGinley. Reprinted by permission of Curtis Brown, Ltd.

"Sugaring Off" from *Nunny Bag 3* by Helen Guiton. Copyright © 1964 W.J. Gage Ltd. Used by permission of Gage Educational Publishing Company (a Division of Canada Publishing Corporation).

"Fog" from *Chicago Poems* by Carl Sandburg. Copyright © Holt, Rinehart and Winston, Inc., renewed 1944 by Carl Sandburg. Reprinted by permission of Harcourt Brace Jovanovich, Inc.

Table of Contents

UNIT 1

Addition and Subtraction Facts to 18

UNIT 2

Place Value to 999

UNIT 3

Addition of 2-Digit Numbers

UNIT 4

Subtraction of 2-Digit Numbers

UNIT 5

Geometric Solids

UNIT 6

Linear Measurement

UNIT 7

Place Value to 9999

UNIT 8

Addition and Subtraction of 3-Digit Numbers

UNIT 9

Geometric Figures

UNIT 10

Multiplication Facts to 9 × 5

UNIT 11

Division Facts to 45 ÷ 5

UNIT 12

More Measurement

UNIT 13

More Multiplication

UNIT 14

More Division

UNIT 15

Fractions and Decimals

Sometimes you will read me.

Sometimes you will solve problems from me.

Sometimes you will copy questions from me...

and answer them in your notebook.

Here's a problem for you to solve.

How many monkeys are shown here?
Try to find out without counting every one.

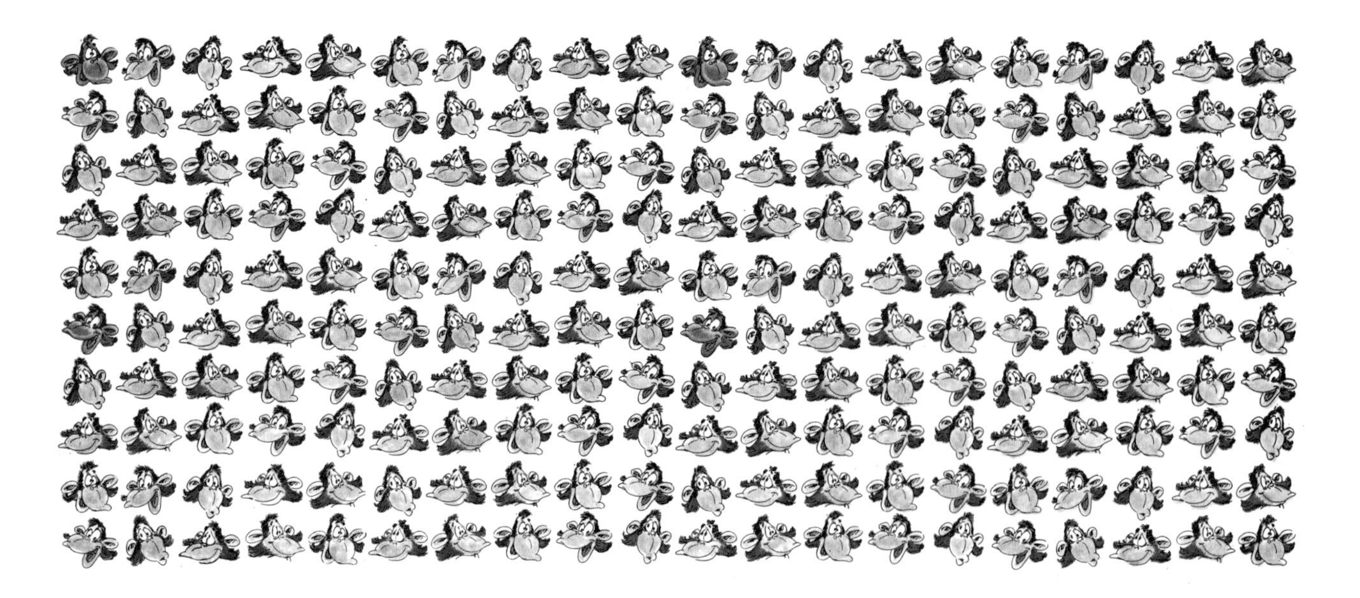

To copy from me, just follow these easy steps.

1. Write the MathQuest page number.
2. Write the date.

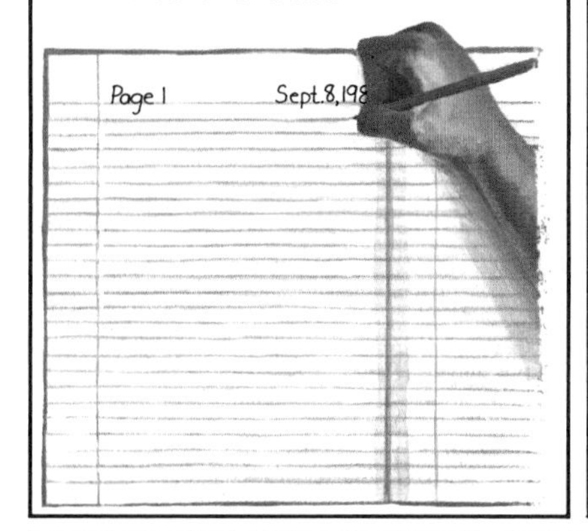

3. Write the question number.
4. Then copy the question.

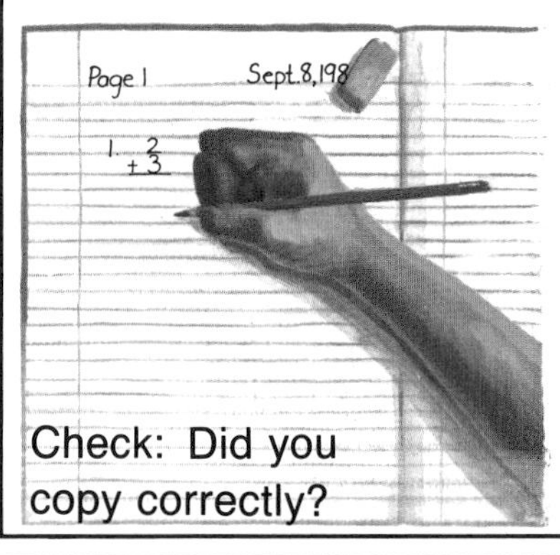

Check: Did you copy correctly?

5. Now answer the question. Use counters if you wish.

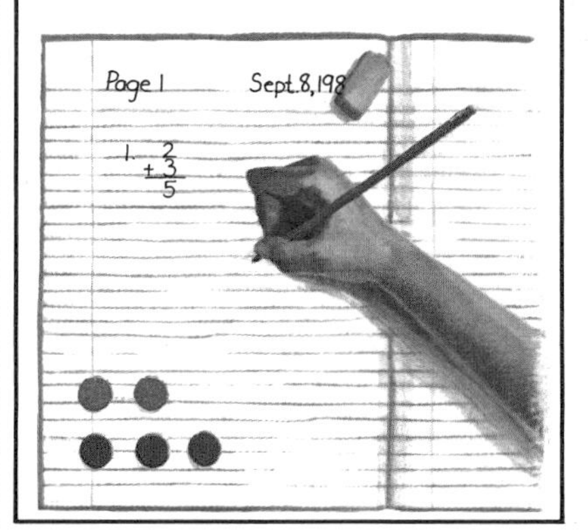

Here are some questions to copy and complete.

Add.

1. $\begin{array}{r} 2 \\ +3 \\ \hline \end{array}$

2. $\begin{array}{r} 5 \\ +1 \\ \hline \end{array}$

3. $\begin{array}{r} 3 \\ +4 \\ \hline \end{array}$

4. $\begin{array}{r} 0 \\ +2 \\ \hline \end{array}$

5. $\begin{array}{r} 5 \\ +2 \\ \hline \end{array}$

6. $\begin{array}{r} 4 \\ +1 \\ \hline \end{array}$

7. $\begin{array}{r} 3 \\ +5 \\ \hline \end{array}$

8. $\begin{array}{r} 6 \\ +1 \\ \hline \end{array}$

9. $\begin{array}{r} 4 \\ +4 \\ \hline \end{array}$

10. $\begin{array}{r} 7 \\ +2 \\ \hline \end{array}$

11. $\begin{array}{r} 4 \\ +0 \\ \hline \end{array}$

12. $\begin{array}{r} 1 \\ +8 \\ \hline \end{array}$

ADDITION AND SUBTRACTION TO 18

1

TOUCANS TWO

whatever one toucan can do
is sooner done by toucans two
and three toucans it's very true
can do much more than two can do
and toucans numbering two plus two can
manage more than all the zoo can
in fact there is no toucan who can
do what four or three or two can.

—Jack Prelutsky

Addition Facts to 10

How many toucans are there in all?

$$\begin{array}{r} 2 \\ +3 \\ \hline 5 \end{array}$$

Add. Use counters if you wish.

1. $4 + 2$
2. $4 + 3$
3. $6 + 1$
4. $3 + 2$
5. $4 + 3$
6. $3 + 6$
7. $2 + 4$
8. $3 + 3$
9. $5 + 4$
10. $1 + 7$
11. $0 + 8$
12. $1 + 9$

Write the missing numbers.

2 children running
+ 2 children walking

4 children in all

1.

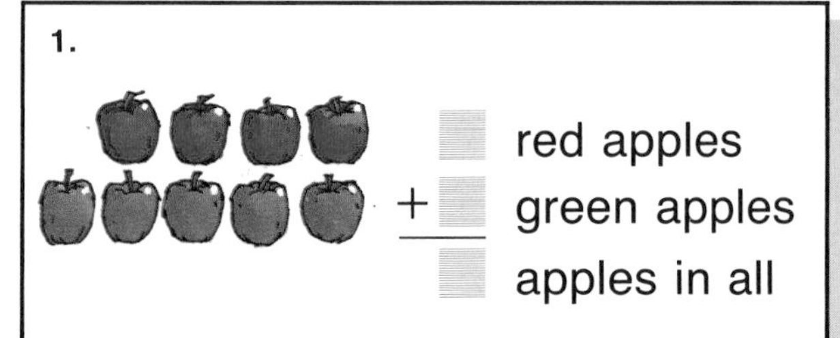

red apples
+ green apples

apples in all

2.

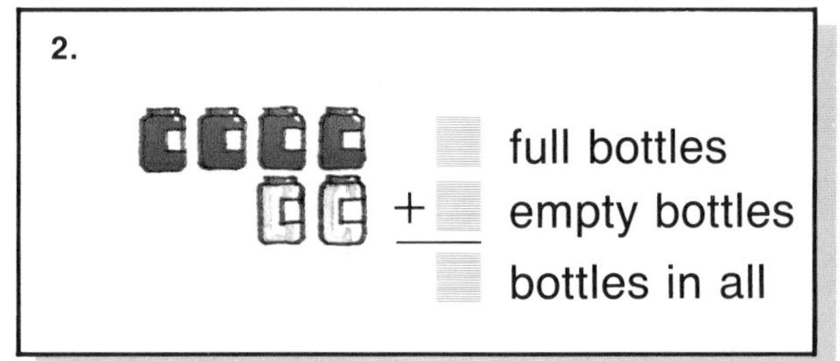

full bottles
+ empty bottles

bottles in all

3.

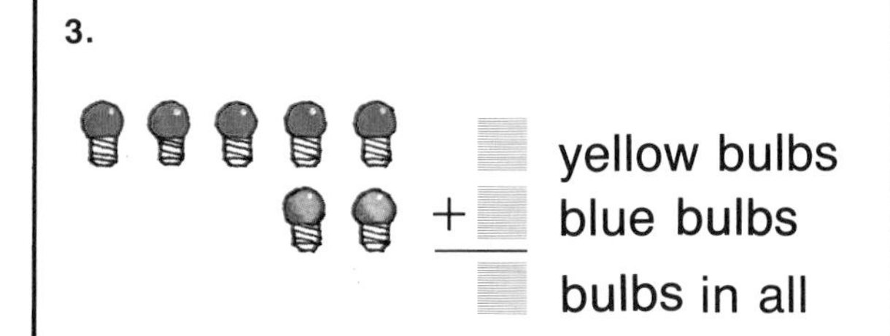

yellow bulbs
+ blue bulbs

bulbs in all

4.

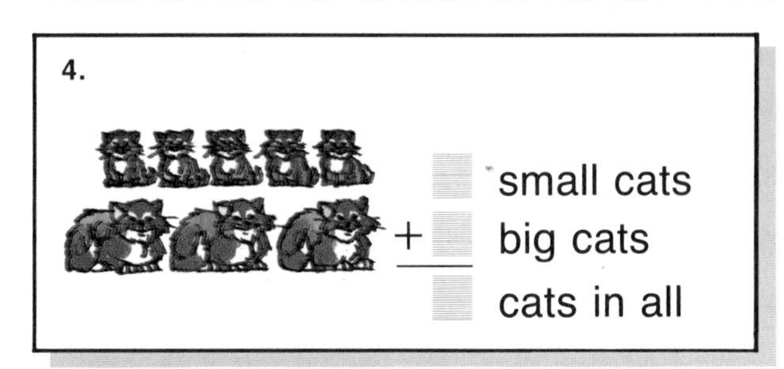

small cats
+ big cats

cats in all

Add.

5. $3 + 3 =$

6. $6 + 0 =$

7. $5 + 5 =$

8. $2 + 5 =$

9. $1 + 0 =$

10. $1 + 9 =$

11. $2 + 6 =$

12. $3 + 7 =$

Subtraction Facts to 10

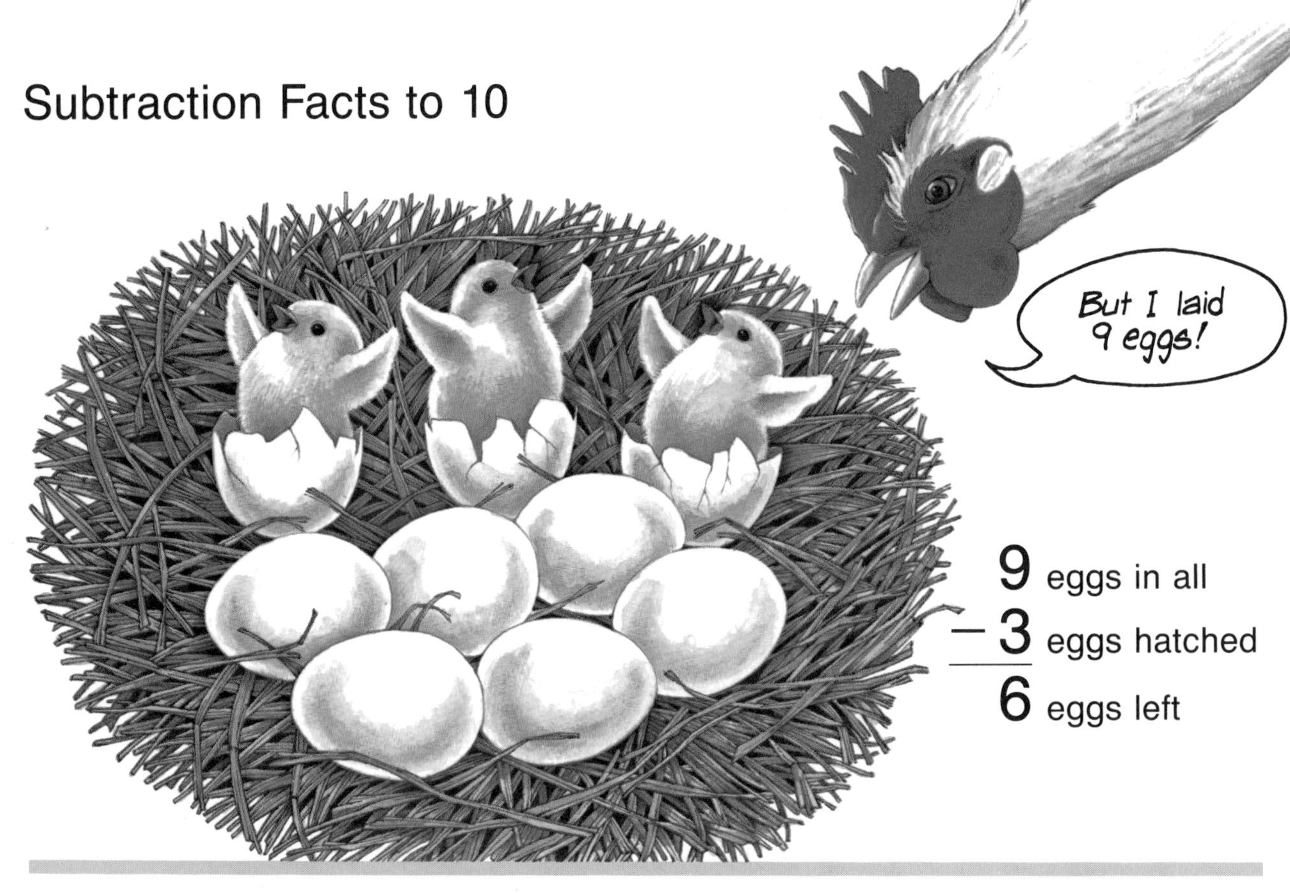

$$\begin{array}{rl} 9 & \text{eggs in all} \\ -\,3 & \text{eggs hatched} \\ \hline 6 & \text{eggs left} \end{array}$$

Subtract. Use counters if you wish.

1. $\begin{array}{r} 7 \\ -3 \\ \hline \end{array}$
2. $\begin{array}{r} 8 \\ -5 \\ \hline \end{array}$
3. $\begin{array}{r} 7 \\ -6 \\ \hline \end{array}$
4. $\begin{array}{r} 8 \\ -6 \\ \hline \end{array}$
5. $\begin{array}{r} 9 \\ -3 \\ \hline \end{array}$
6. $\begin{array}{r} 8 \\ -3 \\ \hline \end{array}$
7. $\begin{array}{r} 10 \\ -\ 3 \\ \hline \end{array}$
8. $\begin{array}{r} 10 \\ -\ 6 \\ \hline \end{array}$
9. $\begin{array}{r} 10 \\ -\ 8 \\ \hline \end{array}$
10. $\begin{array}{r} 9 \\ -5 \\ \hline \end{array}$
11. $\begin{array}{r} 10 \\ -\ 9 \\ \hline \end{array}$
12. $\begin{array}{r} 10 \\ -\ 5 \\ \hline \end{array}$

Write the missing numbers.

	10	balloons in all
$-$	3	balloons lost
	7	balloons left

1.

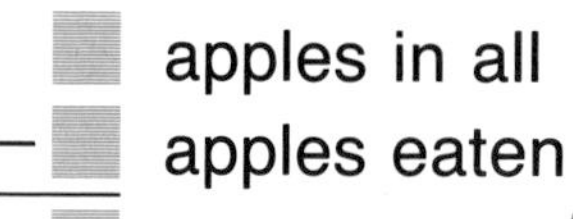

apples in all

$-$ apples eaten

apples left

2.

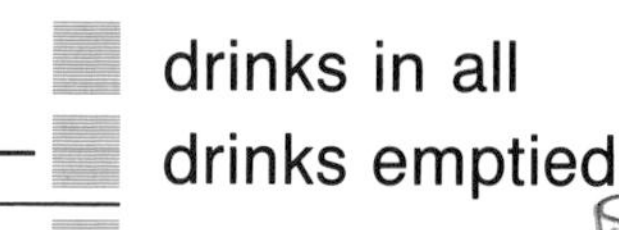

drinks in all

$-$ drinks emptied

drinks left

3.

flowers in all

$-$ flowers picked

flowers left

4.

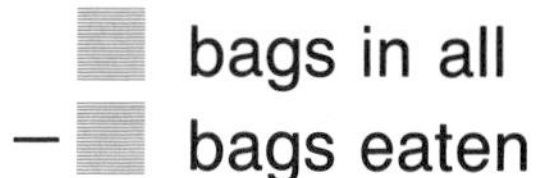

bags in all

$-$ bags eaten

bags left

Subtract.

5. $8 - 4 =$ ▒

6. $10 - 9 =$ ▒

7. $6 - 4 =$ ▒

8. $9 - 4 =$ ▒

9. $8 - 6 =$ ▒

10. $10 - 4 =$ ▒

11. $9 - 6 =$ ▒

12. $8 - 3 =$ ▒

PROBLEM SOLVING

Choosing the Appropriate Operation

Think of a story. Then write a number sentence.

1.

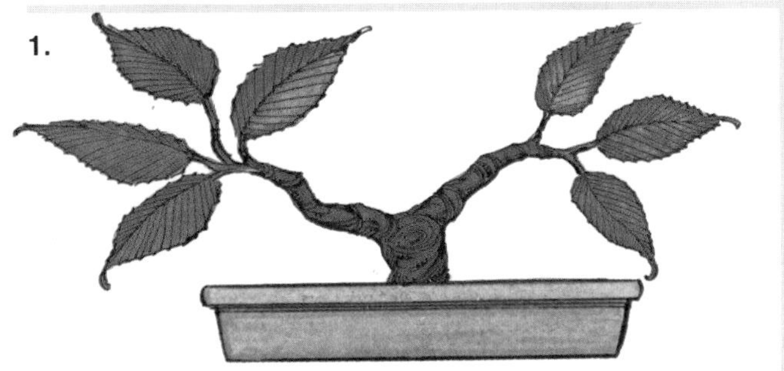

2.

3.

4.

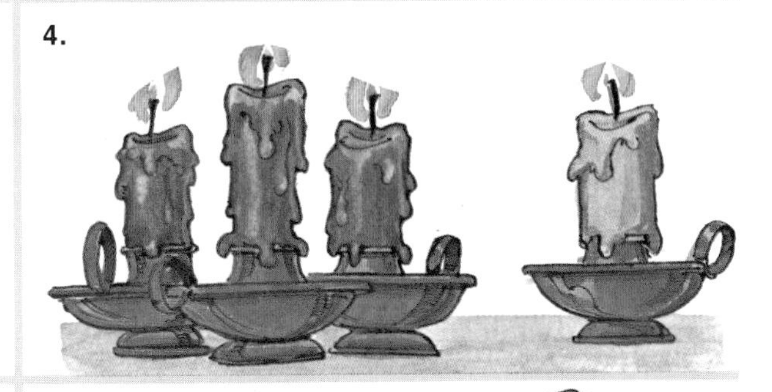

5.

6.

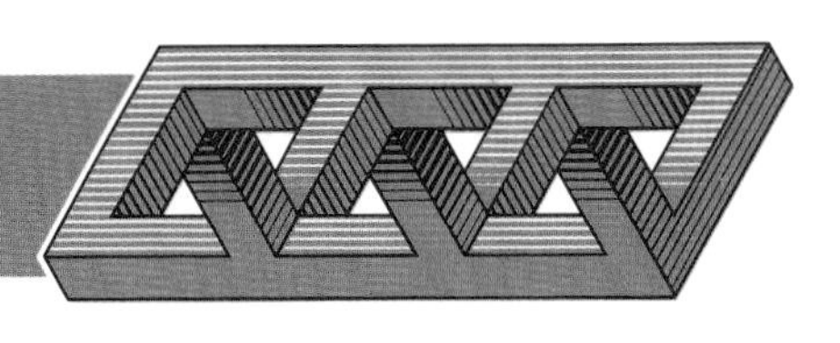

Think of a story. Then write a number sentence.

1.

2.

3.

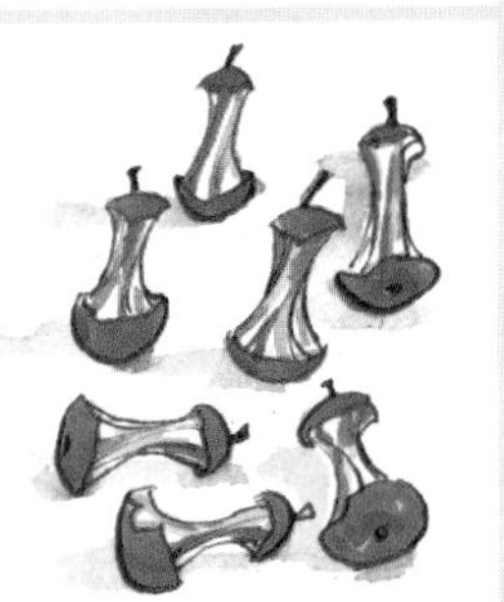

4.

5.

6.

7. Write an addition sentence.
   Draw a picture for it.

8. Write a subtraction sentence.
   Draw a picture for it.

Addition Facts to 12

$$8 + 3 = 11$$

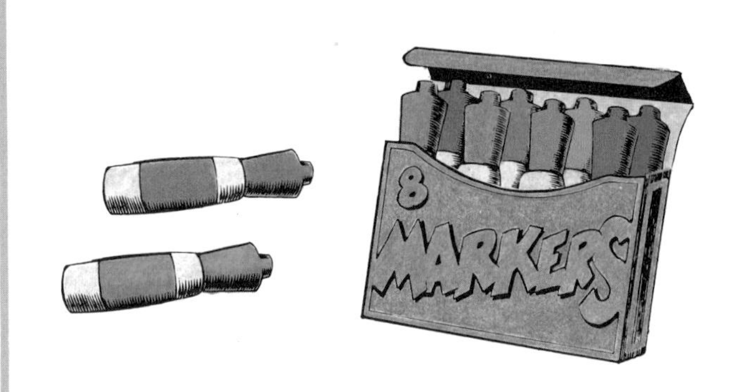

$$2 + 8 = 10$$

Add. Use counters if you wish.

1. $3 + 7$
2. $2 + 9$
3. $9 + 1$
4. $8 + 2$
5. $9 + 0$
6. $3 + 8$
7. $8 + 3$
8. $2 + 8$
9. $3 + 9$
10. $7 + 3$
11. $1 + 9$
12. $9 + 2$

13. $9 + 3 =$ ▢

14. $3 + 7 =$ ▢

15. $8 + 2 =$ ▢

16. $0 + 7 =$ ▢

Addition Facts to 18

$8 + 8 = 16$

$8 + 9 = 17$

Add. Use counters if you wish.

1. $\begin{array}{r} 7 \\ +7 \\ \hline \end{array}$

2. $\begin{array}{r} 5 \\ +6 \\ \hline \end{array}$

3. $\begin{array}{r} 9 \\ +8 \\ \hline \end{array}$

4. $\begin{array}{r} 6 \\ +7 \\ \hline \end{array}$

5. $\begin{array}{r} 8 \\ +8 \\ \hline \end{array}$

6. $\begin{array}{r} 7 \\ +8 \\ \hline \end{array}$

7. $\begin{array}{r} 5 \\ +5 \\ \hline \end{array}$

8. $\begin{array}{r} 8 \\ +9 \\ \hline \end{array}$

9. Chung has 7 crayons.
   His sister has 8 crayons.
   How many crayons do they have?

10. Anna-Maria used 6 of her crayons.
    She has 5 she did not use.
    How many crayons does she have altogether?

More Addition Facts to 18

How many steps above the ground is the top of each ladder?

$$\begin{array}{r} 6 \\ +9 \\ \hline 15 \end{array}$$

$$\begin{array}{r} 4 \\ +9 \\ \hline 13 \end{array}$$

Add.
Use counters if you wish.

1. $\begin{array}{r} 9 \\ +4 \\ \hline \end{array}$
2. $\begin{array}{r} 9 \\ +9 \\ \hline \end{array}$
3. $\begin{array}{r} 7 \\ +9 \\ \hline \end{array}$
4. $\begin{array}{r} 6 \\ +9 \\ \hline \end{array}$
5. $\begin{array}{r} 5 \\ +9 \\ \hline \end{array}$
6. $\begin{array}{r} 9 \\ +8 \\ \hline \end{array}$
7. $\begin{array}{r} 8 \\ +4 \\ \hline \end{array}$
8. $\begin{array}{r} 6 \\ +8 \\ \hline \end{array}$
9. $\begin{array}{r} 7 \\ +5 \\ \hline \end{array}$
10. $\begin{array}{r} 7 \\ +4 \\ \hline \end{array}$
11. $\begin{array}{r} 8 \\ +5 \\ \hline \end{array}$
12. $\begin{array}{r} 4 \\ +7 \\ \hline \end{array}$

JUST FOR FUN

RIDDLE:

Why do firefighters wear red suspenders?

Add.

H. 5 + 6	L. 2 + 3	A. 8 + 4
B. 7 + 3	S. 6 + 9	K. 8 + 8
O. 6 + 1	C. 4 + 4	R. 0 + 1
I. 2 + 4	M. 4 + 5	T. 6 + 7
P. 1 + 1	N. 9 + 9	V. 2 + 1
X. 1 + 3	E. 8 + 9	U. 7 + 7

To answer the riddle, write the letters of the questions with these answers.

13 7   16 17 17 2   13 11 17 6 1

2 12 18 13 15   14 2

Related Addition Facts

We can write 2 facts for this picture.

6 tulips +5 daisies 11 flowers	5 daisies +6 tulips 11 flowers

We can write the same facts this way:

$$6 + 5 = 11$$
$$5 + 6 = 11$$

Write 2 addition facts for each picture.

1\.

2\.

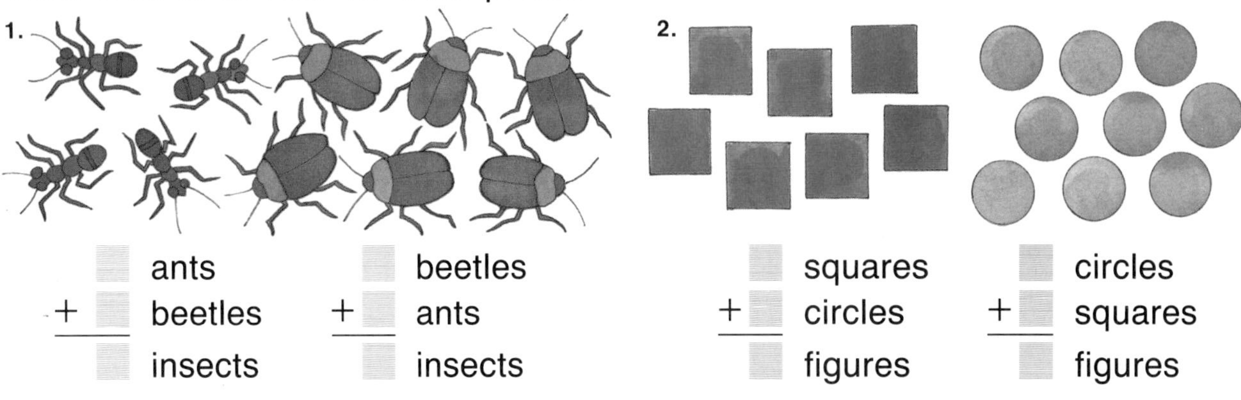

1\.

☐ ants + ☐ beetles ☐ insects	☐ beetles + ☐ ants ☐ insects

2\.

☐ squares + ☐ circles ☐ figures	☐ circles + ☐ squares ☐ figures

Complete each addition sentence.
Then write the related addition fact.

3\. $2 + 9 =$ ☐

4\. $8 + 5 =$ ☐

5\. $4 + 7 =$ ☐

6\. $7 + 6 =$ ☐

7\. $9 + 4 =$ ☐

8\. $6 + 8 =$ ☐

Finding Missing Numbers

How many puzzle pieces are in the box?

We see:

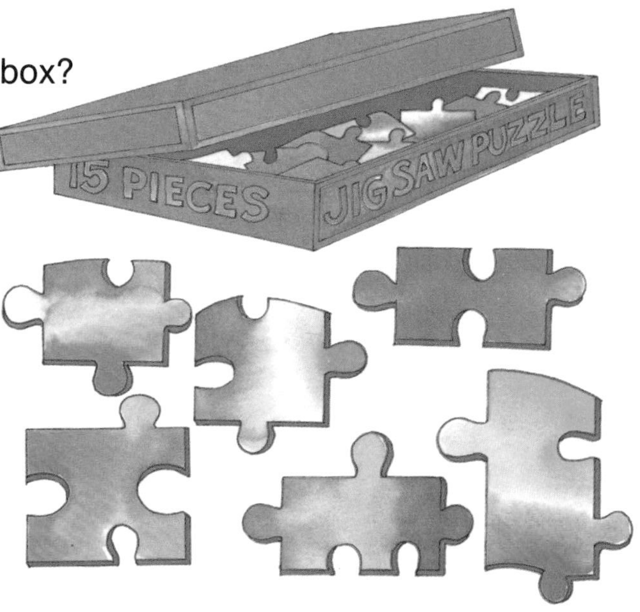

We say: 6 pieces plus the hidden pieces equal 15 pieces.

We write: $6 + \square = 15$

or $6 + 9 = 15$

There are 9 pieces in the box.

Copy and complete.

1. $6 + \square = 9$
2. $4 + \square = 11$
3. $\square + 8 = 17$
4. $9 + \square = 18$
5. $\square + 7 = 13$
6. $5 + \square = 14$
7. $\begin{array}{r} 5 \\ + \square \\ \hline 8 \end{array}$
8. $\begin{array}{r} \square \\ + 7 \\ \hline 9 \end{array}$
9. $\begin{array}{r} 6 \\ + \square \\ \hline 10 \end{array}$
10. $\begin{array}{r} \square \\ + 9 \\ \hline 9 \end{array}$
11. $\begin{array}{r} 2 \\ + \square \\ \hline 7 \end{array}$
12. $\begin{array}{r} \square \\ + 3 \\ \hline 4 \end{array}$
13. $\begin{array}{r} 9 \\ + \square \\ \hline 14 \end{array}$
14. $\begin{array}{r} \square \\ + 6 \\ \hline 12 \end{array}$

Subtraction Facts to 18

13 people in all
− 4 people floating away
9 people left

Subtract. Use counters if you wish.

1. $11 - 2$
2. $14 - 6$
3. $13 - 8$
4. $10 - 8$
5. $14 - 7$
6. $18 - 9$
7. $11 - 8$
8. $15 - 9$
9. $13 - 7$
10. $14 - 8$
11. $16 - 9$
12. $12 - 8$
13. $11 - 9$
14. $12 - 3$
15. $17 - 9$
16. $15 - 7$

Subtract. Use counters if you wish.

1. $10 - 8 = \square$

2. $11 - 7 = \square$

3. $14 - 5 = \square$

4. $12 - 6 = \square$

5. $11 - 0 = \square$

6. $13 - 8 = \square$

7. $14 - 9 = \square$

8. $15 - 7 = \square$

9. $18 - 9 = \square$

10. $13 - 6 = \square$

11. There are 14 children.
9 of them are playing ball.
How many are not playing ball?

12. There are 14 children.
There are 6 adults.
How many more children than adults are there?

PROBLEM SOLVING

There are 15 balloons.
6 balloons are red.
The rest are yellow.
How many more yellow balloons than red balloons are there?

JUST FOR FUN

RIDDLE: What has 4 wheels and flies?

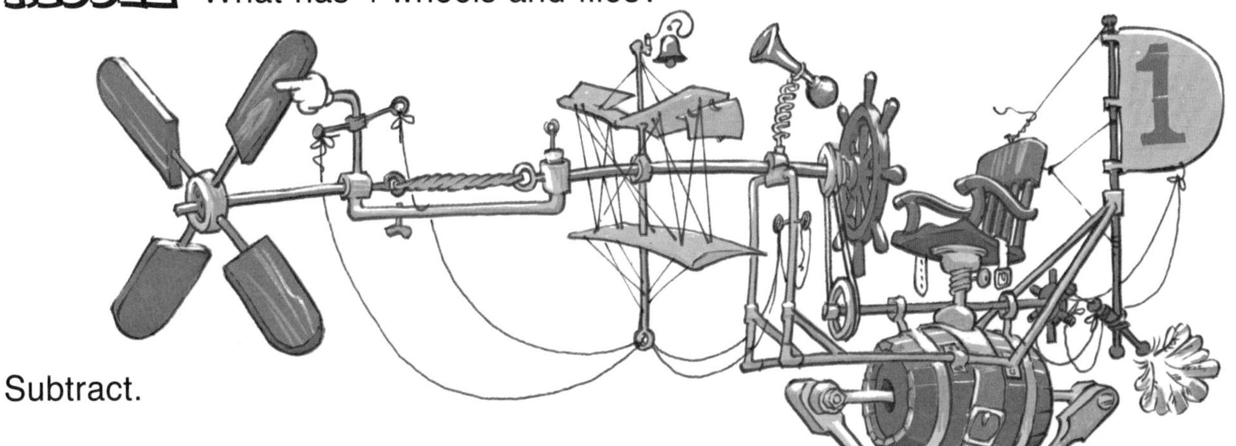

Subtract.

T. $16 - 7$

A. $13 - 8$

K. $15 - 9$

B. $11 - 8$

C. $5 - 4$

R. $12 - 8$

G. $14 - 6$

E. $11 - 9$

U. $13 - 6$

To answer the riddle, write the letters of the questions with these answers.

5 8 5 4 3 5 8 2 9 4 7 1 6

Related Subtraction Facts

We can write 2 facts for this picture.

11	leaves in all	11	leaves in all
− 6	red leaves	− 5	yellow leaves
5	yellow leaves	6	red leaves

We can write the same facts this way:

$$11 - 6 = 5$$
$$11 - 5 = 6$$

Write 2 subtraction facts for each picture.

1.

▢	pets in all	▢	pets in all
− ▢	dogs	− ▢	cats
▢	cats	▢	dogs

2.

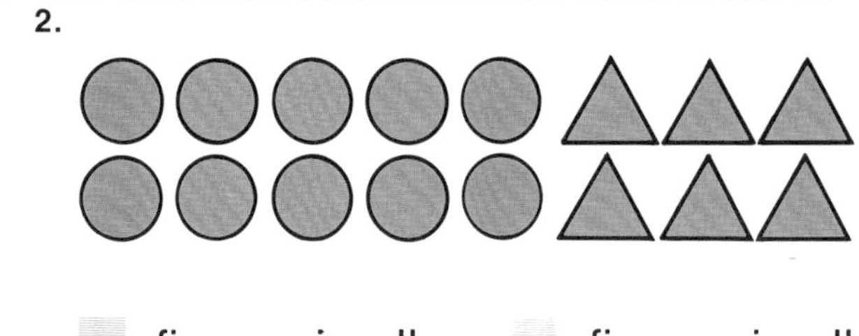

▢	figures in all	▢	figures in all
− ▢	circles	− ▢	triangles
▢	triangles	▢	circles

Complete each subtraction sentence.
Then write the related subtraction fact.

3. $11 - 4 = ▢$
4. $13 - 8 = ▢$
5. $11 - 2 = ▢$
6. $10 - 3 = ▢$
7. $16 - 9 = ▢$
8. $14 - 8 = ▢$

Sums and Differences to 18

Watch the signs!

Add or subtract.

1. $\begin{array}{r} 8 \\ +3 \\ \hline \end{array}$
2. $\begin{array}{r} 14 \\ -\ 6 \\ \hline \end{array}$
3. $\begin{array}{r} 13 \\ -\ 8 \\ \hline \end{array}$
4. $\begin{array}{r} 9 \\ +7 \\ \hline \end{array}$
5. $\begin{array}{r} 6 \\ +7 \\ \hline \end{array}$
6. $\begin{array}{r} 13 \\ -\ 5 \\ \hline \end{array}$
7. $\begin{array}{r} 9 \\ +5 \\ \hline \end{array}$
8. $\begin{array}{r} 17 \\ -\ 9 \\ \hline \end{array}$
9. $\begin{array}{r} 12 \\ -\ 7 \\ \hline \end{array}$
10. $\begin{array}{r} 8 \\ -0 \\ \hline \end{array}$
11. $\begin{array}{r} 9 \\ +6 \\ \hline \end{array}$
12. $\begin{array}{r} 15 \\ -\ 9 \\ \hline \end{array}$
13. $\begin{array}{r} 6 \\ +6 \\ \hline \end{array}$
14. $\begin{array}{r} 15 \\ -\ 6 \\ \hline \end{array}$
15. $\begin{array}{r} 9 \\ +9 \\ \hline \end{array}$
16. $\begin{array}{r} 14 \\ -\ 9 \\ \hline \end{array}$

17. Kelly had 8 shiny stones.
She found 6 more.
How many does she have now?

18. Jacques had 17 baseball cards.
He gave 9 to a friend.
How many does he have now?

JUST FOR FUN
I am Marvin Horse
Of course,
And no one knows
My age.
To find out just how old
I am.
Please copy and complete
This page.
How old is Marvin?
Start
5 + 2 = A
+ 8
B − 6 = C
+ 3
D − 5 = E
+ 9
F − 7 = G
− 7
H + 6 = I
Marvin's Age

PROBLEM SOLVING

Looking for Possibilities

Show these numbers in many ways.

1. eight

2. twelve

3. Choose your own number.
   Show it in many ways.
   How many ways did you find?

PROBLEM SOLVING

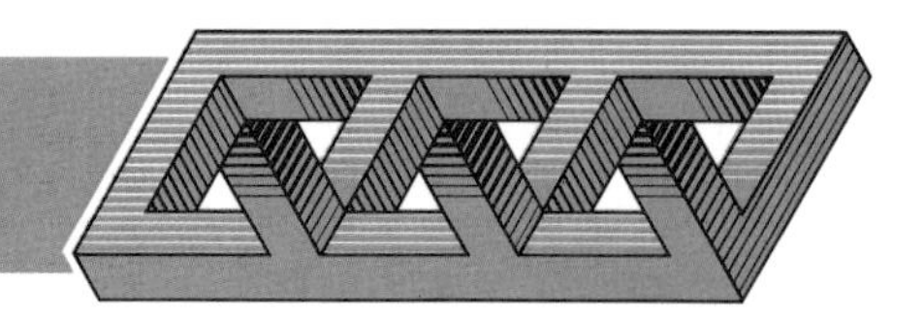

Performing a Sequence of Operations

Mrs. Carson picks up 13 children at the first bus stop.

9 children get off at Elm School.

7 children get on at Lake Street.

8 children get off at Maple School.

How many children are still on the bus?

Write a bus story problem of your own.

Fact Families

We can write 4 facts
for this family of kittens.

$$4 + 2 = 6$$

$$2 + 4 = 6$$

$$6 - 2 = 4$$

$$6 - 4 = 2$$

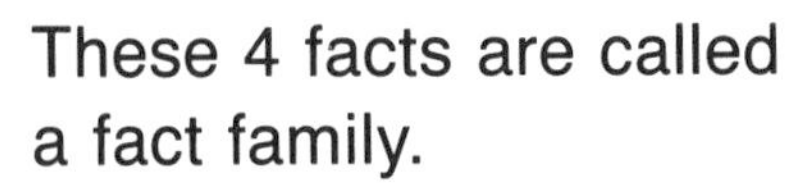

These 4 facts are called
a fact family.

Write 4 facts for each picture.

1.

2.

3.

4.

Writing Fact Families

Each fact in a family uses the same 3 numbers.

$5 + 8 = 13$

$8 + 5 = 13$

$13 - 8 = 5$

$13 - 5 = 8$

Write 4 facts for each family of numbers.

1. 2. 3.

Estimate: How many kittens are there?

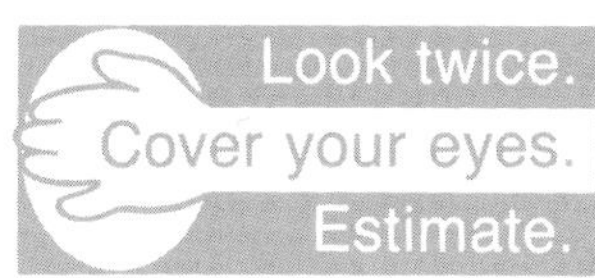

Rows and Columns

Choose the word "first", "second" or "third" to complete each sentence.

1. The Grey family is in the ▒ column.
2. The Miller family is in the ▒ column.
3. All the girls are in the ▒ row.
4. All the boys named Pat are in the ▒ row.
5. Mike Wong is in the ▒ row and ▒ column.

Write the name of each person.

Addition Tables

To fill in a square on an addition table add the number beside that row to the number at the top of the column.

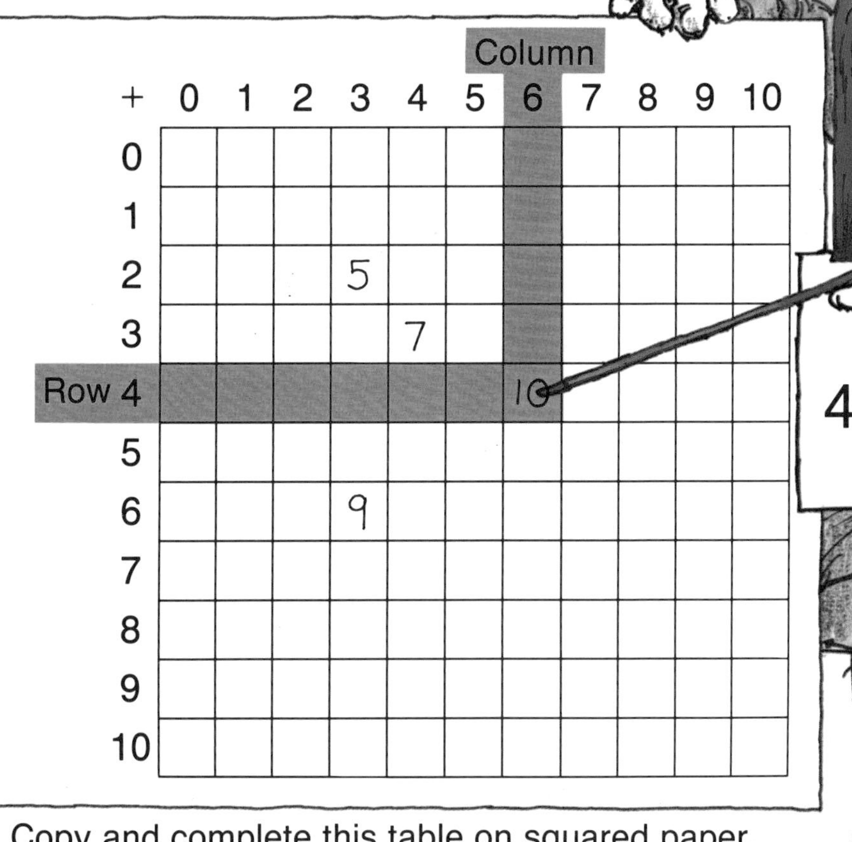

+	0	1	2	3	4	5	6	7	8	9	10
0											
1											
2				5							
3					7						
4							10				
5											
6				9							
7											
8											
9											
10											

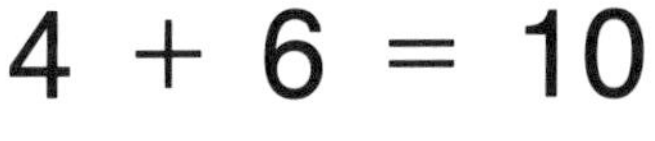

1. Copy and complete this table on squared paper.

2. Shade in all the squares which contain 10.
   What pattern do the squares make?

Add. Use the table to check.

3. $6 + 9 =$ ▒ 4. $7 + 8 =$ ▒

5. $5 + 5 =$ ▒ 6. $9 + 2 =$ ▒

PROBLEM SOLVING

Interpreting a Pictograph

Janet feeds the birds every morning. The graph shows how many birds she saw in one day.

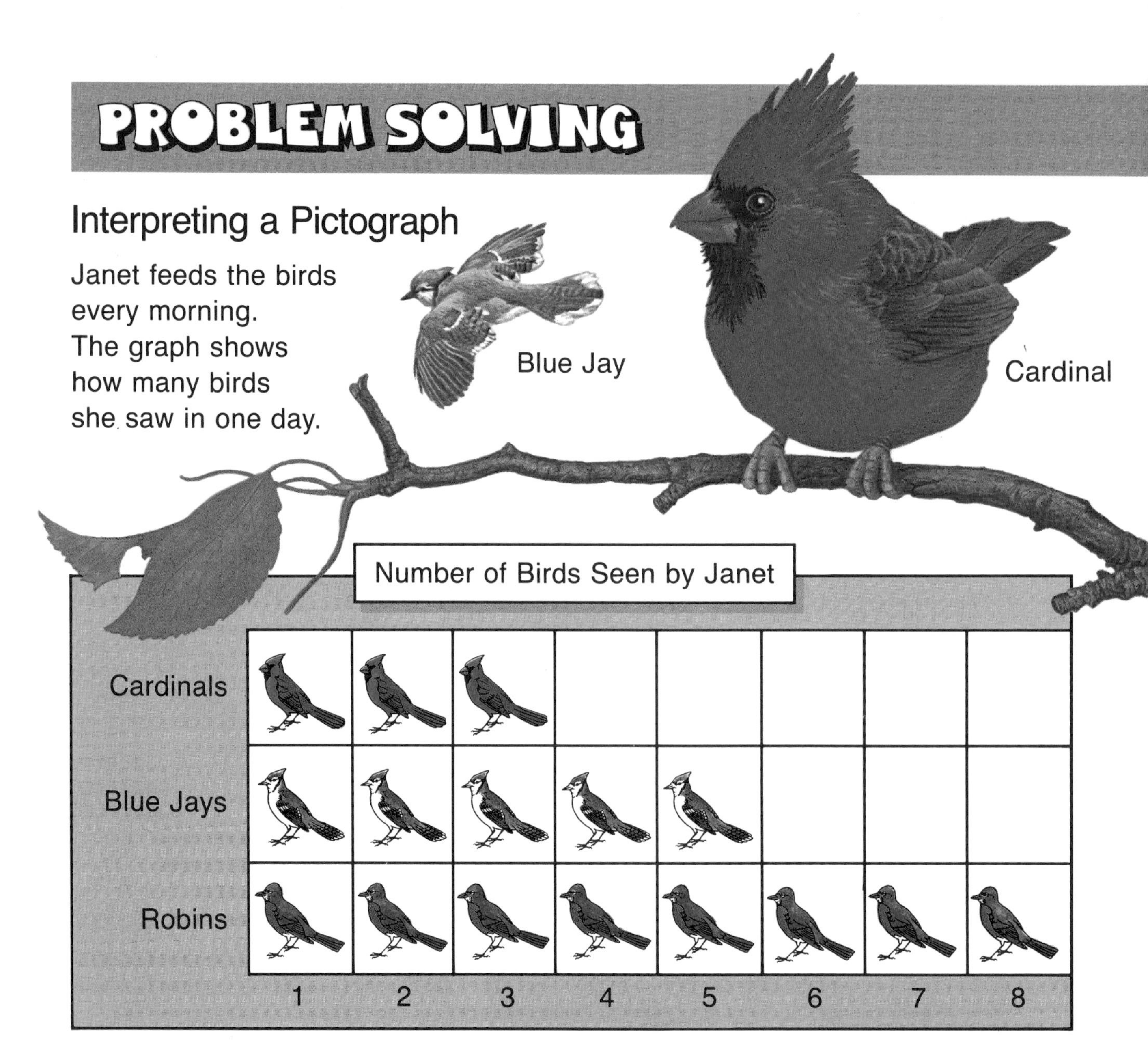

1. Which kind of bird did Janet see most often?
2. Did she see more cardinals or more blue jays?
3. How many cardinals did she see?
4. How many more blue jays than cardinals did she see?
5. How many more robins than cardinals did Janet see?
6. How many birds did she see in all?

Jason also feeds the birds.
The graph shows how many
birds he saw.

1. How many robins did Jason see?
2. How many more robins than cardinals did he see?
3. How many birds did Jason see?
4. How many more cardinals did Jason see than Janet?
5. How many more birds did Jason see than Janet?
6. Write a question of your own about this graph.

CHECK-UP

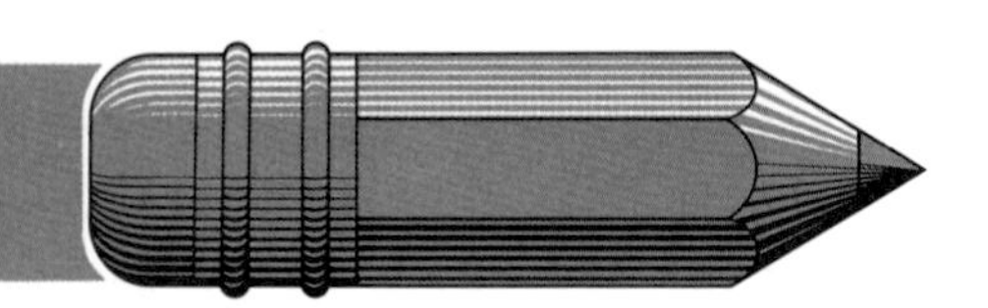

Write the missing numbers.

1.

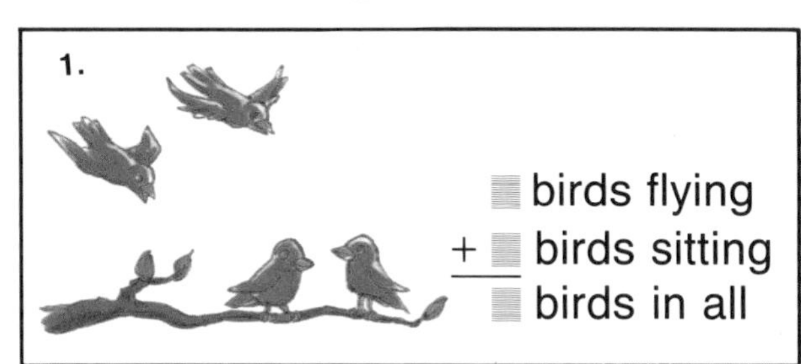

▒ birds flying
+ ▒ birds sitting
▒ birds in all

2.

▒ birds in all
− ▒ birds flying
▒ birds left

3. $\begin{array}{r} 9 \\ -2 \\ \hline \end{array}$

4. $\begin{array}{r} 3 \\ +6 \\ \hline \end{array}$

5. $\begin{array}{r} 5 \\ +4 \\ \hline \end{array}$

6. $\begin{array}{r} 8 \\ -3 \\ \hline \end{array}$

7. $\begin{array}{r} 8 \\ +6 \\ \hline \end{array}$

8. $\begin{array}{r} 15 \\ -\ 7 \\ \hline \end{array}$

9. $\begin{array}{r} 10 \\ -\ 8 \\ \hline \end{array}$

10. $\begin{array}{r} 9 \\ +4 \\ \hline \end{array}$

11. $\begin{array}{r} 9 \\ +9 \\ \hline \end{array}$

12. $\begin{array}{r} 17 \\ -\ 9 \\ \hline \end{array}$

13. $\begin{array}{r} 15 \\ -\ 8 \\ \hline \end{array}$

14. $\begin{array}{r} 8 \\ +8 \\ \hline \end{array}$

15. $\begin{array}{r} 14 \\ -\ 8 \\ \hline \end{array}$

16. $\begin{array}{r} 9 \\ +8 \\ \hline \end{array}$

17. $\begin{array}{r} 9 \\ +7 \\ \hline \end{array}$

18. $\begin{array}{r} 16 \\ -\ 9 \\ \hline \end{array}$

19. $17 - 8 =$ ▒

20. $8 + 7 =$ ▒

21. Write 4 facts for this picture.

22. Susan has 8 cards.
Jack has 5 cards.
How many cards are there in all?

23. Sarah saved 14 stickers.
She gave 8 to Dan.
How many stickers did she have left?

Homework Machine

The Homework Machine, oh the Homework Machine,
Most perfect contraption that's ever been seen.
Just put in your homework, then drop in a dime,
Snap on the switch, and in ten seconds' time,
Your homework comes out, quick and clean as can be.
Here it is—"nine plus four?" and the answer is "three."
Three?
Oh me...
I guess it's not as perfect
As I thought it would be.

—Shel Silverstein

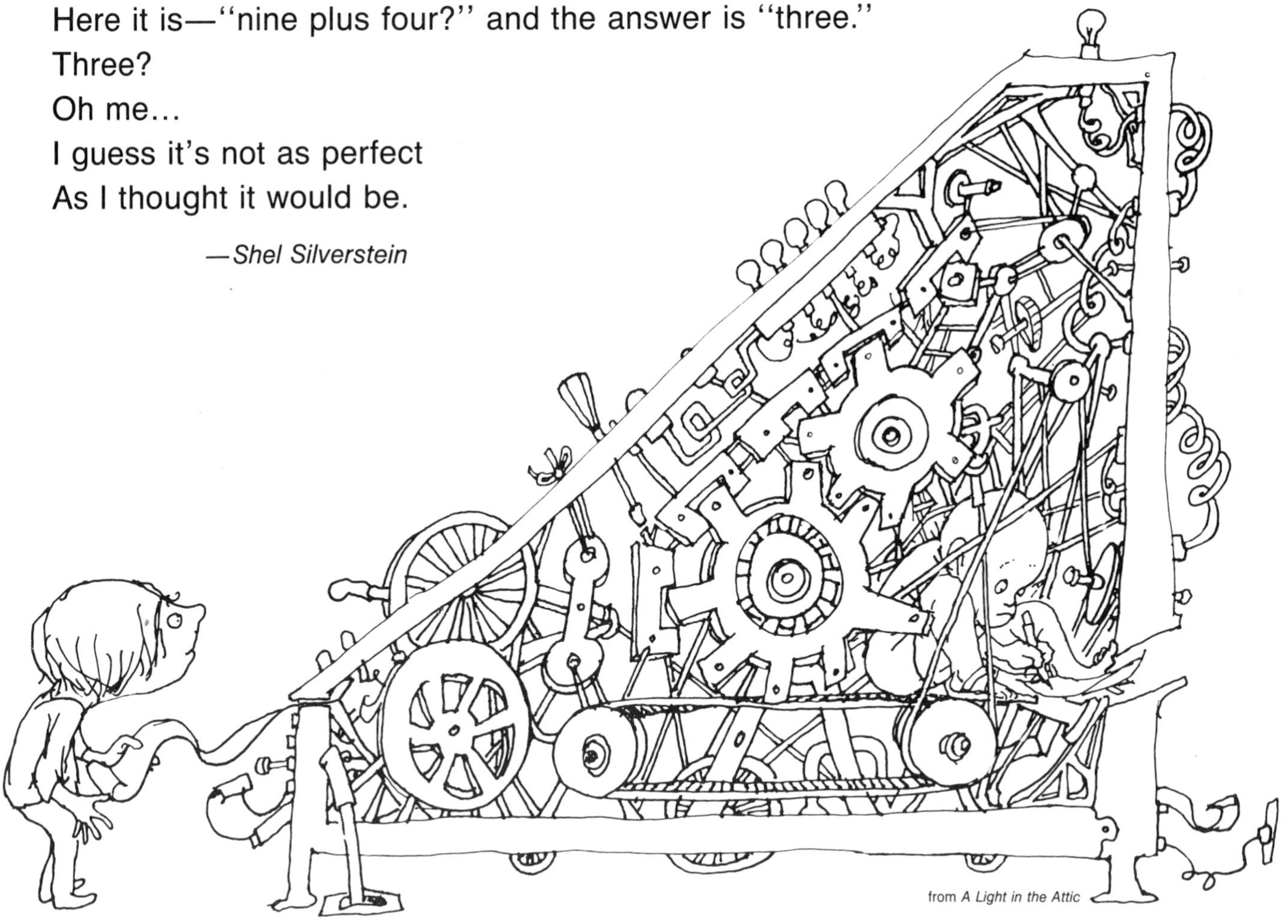

from *A Light in the Attic*

Groups of 10

The graph shows how much money each child has saved.

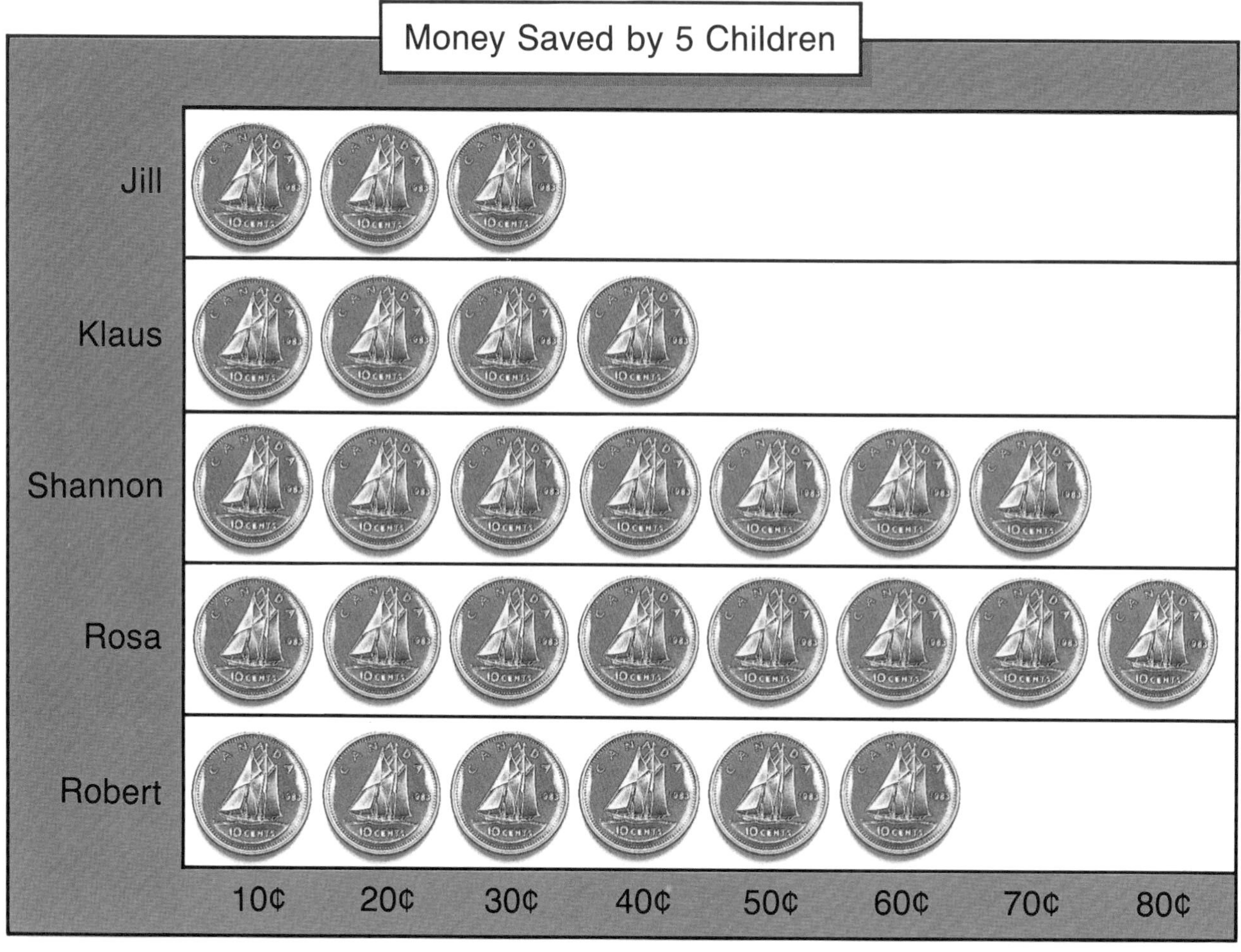

How much has each child saved?

1. Jill
2. Klaus
3. Shannon
4. Rosa
5. Robert

Tens and Ones

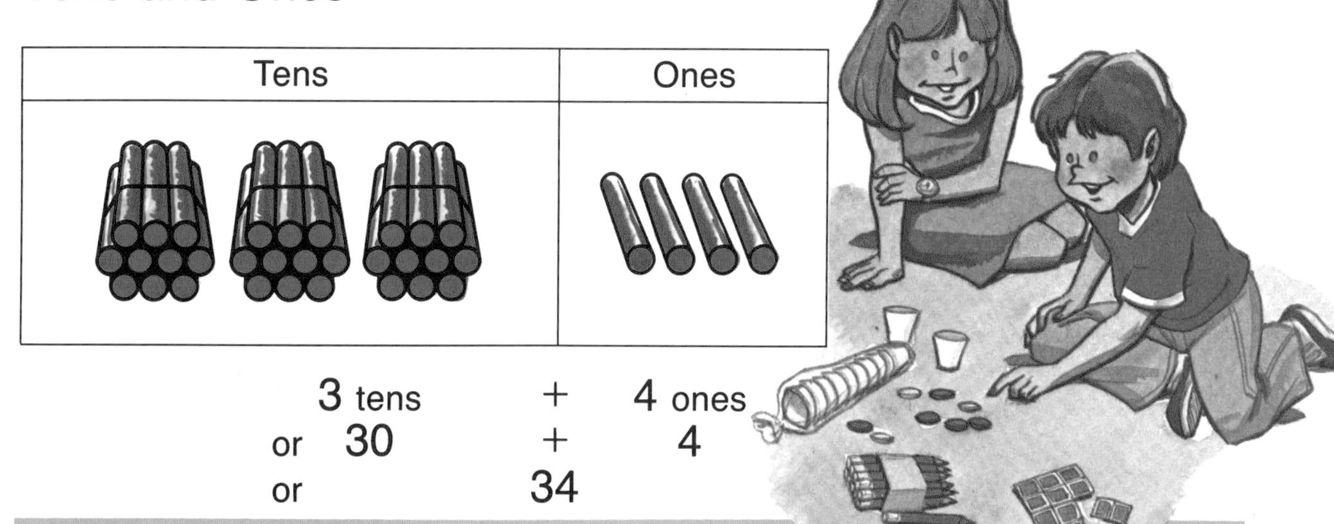

3 tens + 4 ones
or 30 + 4
or 34

Write the answer in 3 ways.

1. How many pencils are there?

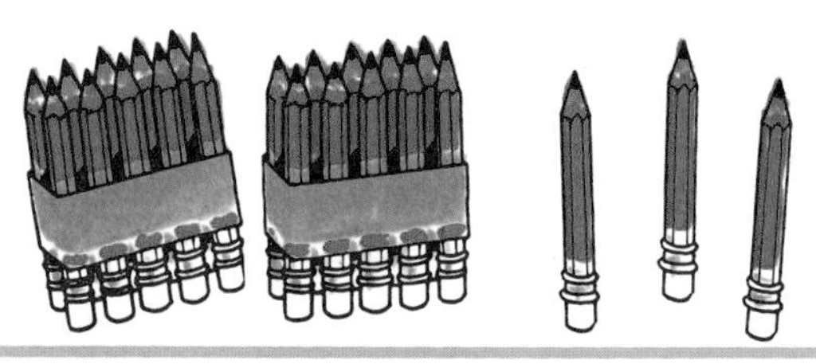

2. How many cups are there altogether?

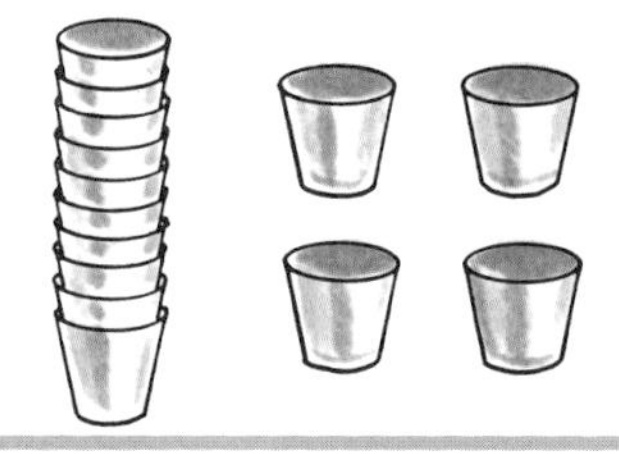

3. How many cents are there?

4. How many cents are there?

5. How many stamps are there altogether?

Tens and Ones

How old is Kim?

How old is each person in Kim's family?

Each [dark candle] stands for 10 years.

Each [light candle] stands for 1 year.

Numbers to 99 in Expanded Form

Copy and complete this addition table.
Use squared paper.

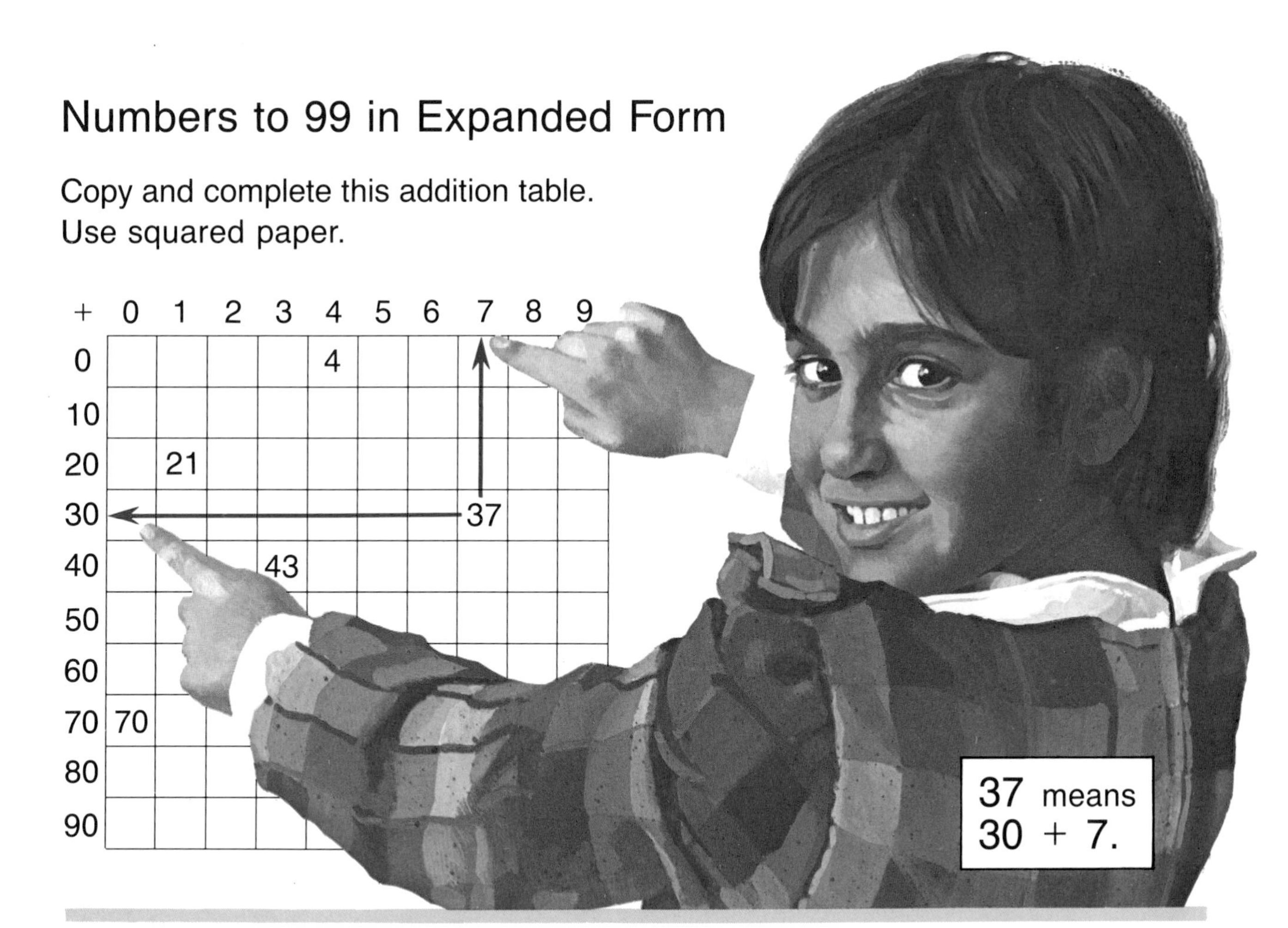

Copy and complete.

1. 40 + 5 = ▒
2. 60 + 7 = ▒
3. 70 + 9 = ▒
4. 10 + 6 = ▒
5. 50 + 9 = ▒
6. 90 + 8 = ▒
7. 39 = 30 + ▒
8. 47 = 40 + ▒
9. 92 = 90 + ▒

Write the meaning of each coloured digit.

10. 42

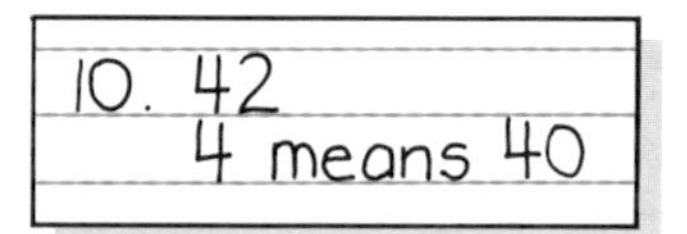

11. 67
12. 78
13. 92
14. 53
15. 14

Counting Patterns

This is a 100-chart.

1	2	3	4	5	6	7	8	9	10
11	12	13	14	15	16	17	18	19	20
21	22	23	24	25	26	27	28	29	30
31	32	33	34	35	36	37	38	39	40
41	42	43	44	45	46	47	48	49	50
51	52	53	54	55	56	57	58	59	60
61	62	63	64	65	66	67	68	69	70
71	72	73	74	75	76	77	78	79	80
81	82	83	84	85	86	87	88	89	90
91	92	93	94	95	96	97	98	99	100

Look at a row.

What is the counting pattern?

Look at a column.

What is the counting pattern?

Write the missing numbers.

1. 10, 11, 12, 17
2. 73, 74, 75, 80
3. 37, 38, 39, 45
4. 42, 44, 46, 56
5. 24, 34, 44, 94
6. 19, 29, 39, 89
7. 45, 50, 55, 80
8. 29, 28, 27, 22
9. 46, 45, 44, 39
10. 64, 63, 62, 57
11. 92, 82, 72, 22
12. 40, 38, 36, 28

Find other patterns in the 100-chart.

JUST FOR FUN

1	2	3	4	5	6	7	8	9	10
11	12	13	14	15	16	17	18	19	20
21	22	23	24	25	26	27	28	29	30
31	32	33	34	35	36	37	38	39	40
41	42	43	44	45	46	47	48	49	50
51	52	53	54	55	56	57	58	59	60
61	62	63	64	65	66	67	68	69	70
71	72	73	74	75	76	77	78	79	80
81	82	83	84	85	86	87	88	89	90
91	92	93	94	95	96	97	98	99	100

Use a completed 100-chart. Shade all squares which contain these numbers.

1. 4 tens and 5 ones
2. 5 tens and 8 ones
3. 20 + 3
4. 10 greater than 22
5. 1 greater than 29
6. 1 less than 27
7. 10 greater than 28
8. 10 less than 62
9. between 68 and 70
10. ten less than sixty-five
11. 20 greater than 50
12. 20 greater than 29
13. between 61 and 63
14. 10 greater than 15
15. 10 less than 52
16. 60 + 3
17. 30 greater than 35
18. 5 greater than 61
19. two less than seventy
20. 11 greater than 10
21. 12 greater than 10
22. between twenty-eight and thirty
23. 4 greater than 31
24. 2 greater than 59
25. 20 greater than 8
26. 20 less than 68

Comparing Numbers to 99

Estimate: Which page has more stamps?

Check: Count the stamps.

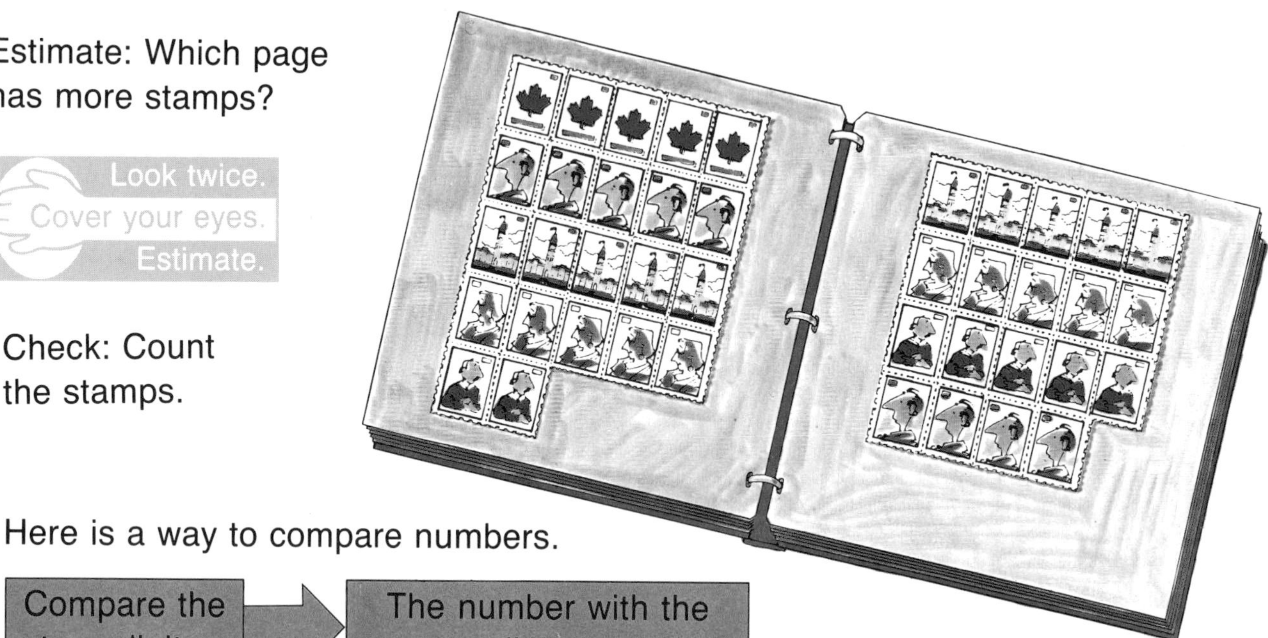

Here is a way to compare numbers.

Compare the tens digits.	→	The number with the greater digit is greater.

22
19   2 is greater than 1.

We say: 22 is greater than 19 or 19 is less than 22.

We write: $22 > 19$ or $19 < 22$

The left page has more stamps.

Write the number that is greater.

1. 47 / 32
2. 19 / 52
3. 65 / 17

Write the number that is less.

4. 47 / 32
5. 95 / 64
6. 17 / 71

Write $>$ or $<$ for each ▒ .

7. 68 ▒ 73
8. 92 ▒ 51
9. 80 ▒ 30
10. 44 ▒ 12

Ordering Numbers to 99

Which class has the most students?

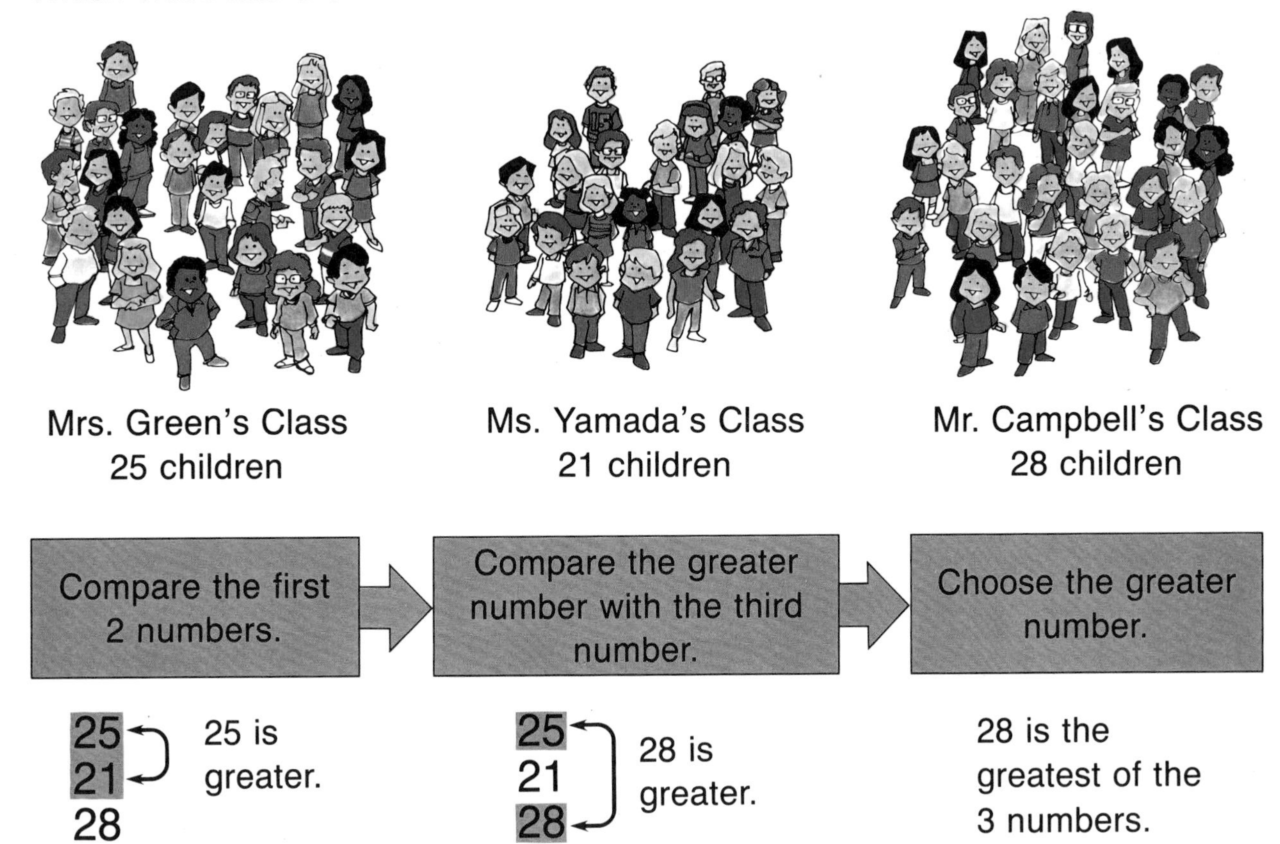

Mr. Campbell's class has the most students.

Write the greatest of the 3 numbers.

1. 57, 52, 55
2. 83, 81, 85
3. 69, 72, 70

Write the least of the 3 numbers.

4. 92, 36, 60
5. 52, 62, 72
6. 79, 86, 78

Write the numbers from least to greatest.

7. 28, 19, 17
8. 88, 89, 90
9. 50, 32, 71

PROBLEM SOLVING

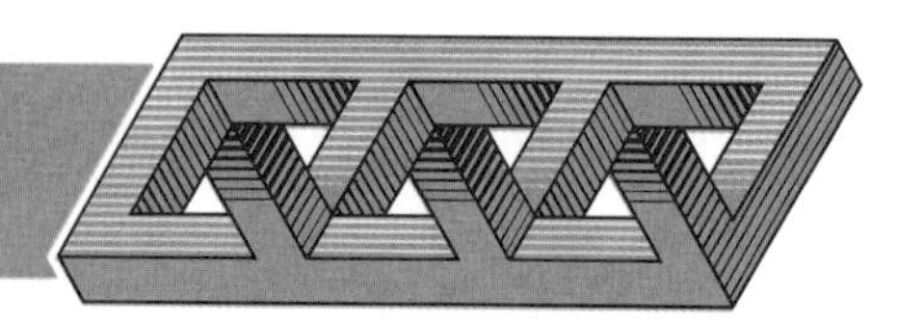

Looking for a Pattern

Some of the numbers on this 100-chart are hidden.

Find the numbers which belong in the blue squares.

Write them in a table like this.

A	B	C	D	E	F	G	H	I	J
26									

EXTENSIONS

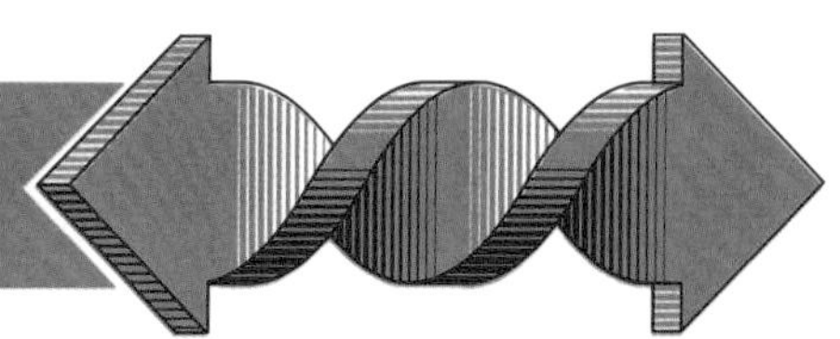

Ordinals to 31st

What animal is:

1. first in line?
2. eighth in line?
3. eighteenth in line?
4. twentieth in line?
5. thirtieth in line?

In what position is:

6. the last bear?
7. the first cougar?
8. the last cougar?
9. the third duck?
10. the turtle?

Estimating Number

Estimate: Are there more squares or more circles?

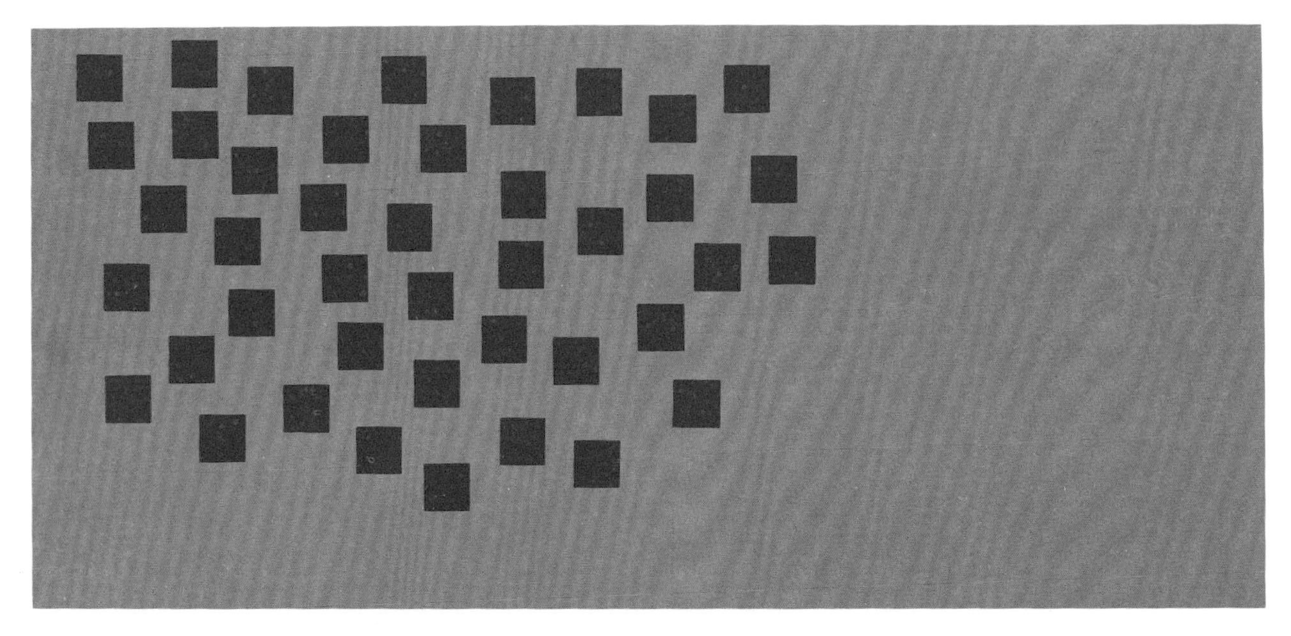

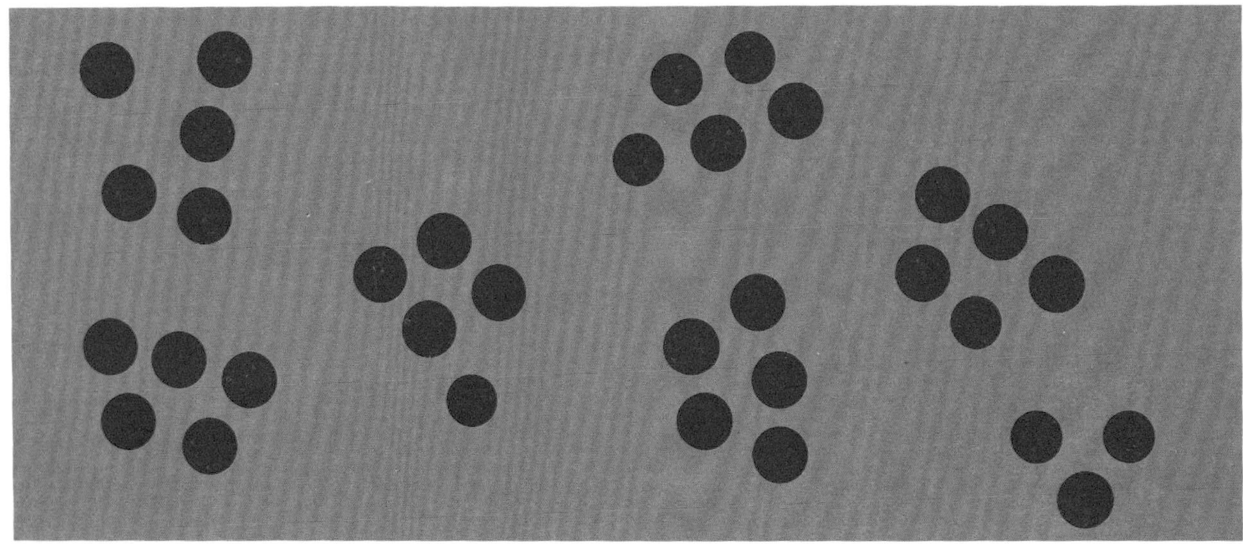

Count to check.

Estimate:
How many oats
are there?
Look twice.
Cover your eyes.
Estimate.
Work with a partner.
Find a good way to count the oats.
Do not mark the page!

Hundreds, Tens and Ones

This shows 1 cube.

This shows 10 cubes.

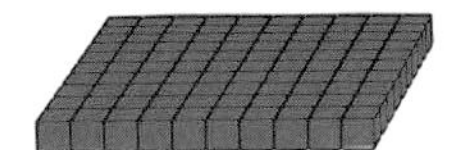

This shows 100 cubes.

Hundreds	Tens	Ones

3 hundreds + 5 tens + 2 ones
or 300 + 50 + 2
or 352

What number is shown?
Write the answer in 3 ways.

1.

Hundreds	Tens	Ones

2.

Hundreds	Tens	Ones

3.

Hundreds	Tens	Ones

4.

Hundreds	Tens	Ones

How many cubes are there?
Write the numeral.

1.

Hundreds	Tens	Ones

2.

Hundreds	Tens	Ones

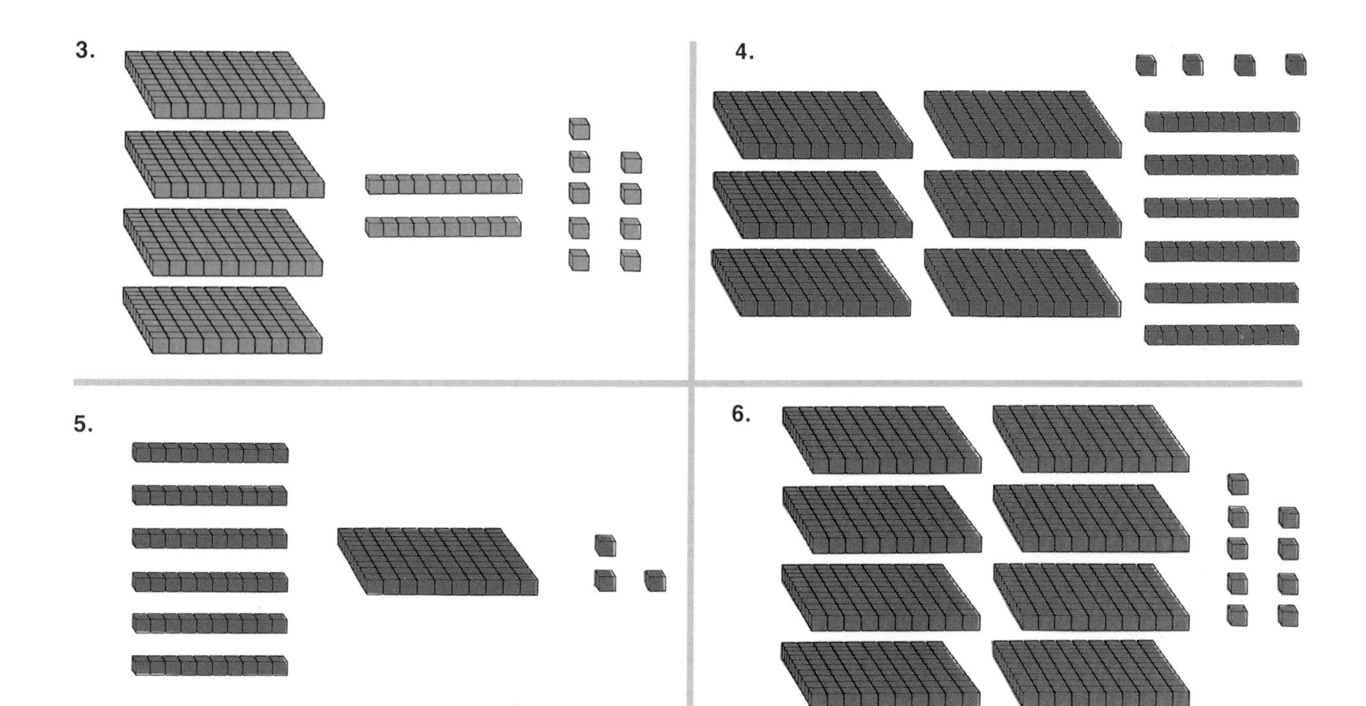

Write the numeral.

7. 8 hundreds, 5 tens, 7 ones

8. 6 hundreds, 3 ones, 9 tens

9. 2 hundreds, 0 tens, 3 ones

10. 2 ones, 5 tens, 4 hundreds

Money Amounts to $999

Here are some Canadian bills and their values.

We see:

We think: $300 + $20 + $4

We write: $324

How much money is shown?

1.     

2.     

3.      

4.      

Numbers to 999 in Expanded Form

This is an abacus.

 shows 100.

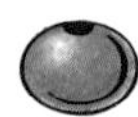 shows 10.

 shows 1.

The abacus shows: 200 + 60 + 3
or 263

What number is shown? Write the answer in 2 ways.

1. 

2. 

Write the meaning of the coloured digit.

3. 7**5**4
4. **2**38
5. 40**6**
6. **6**15
7. 42**2**
8. **9**63
9. 63**0**
10. 3**9**7

Copy and complete.

11. 300 + 20 + 4 = ▒
12. 600 + 50 + 3 = ▒
13. 846 = 800 + ▒ + ▒
14. 592 = ▒ + 90 + ▒
15. 453 = ▒ + ▒ + ▒
16. 716 = ▒ + ▒ + ▒

Counting Patterns

Look at the chart.

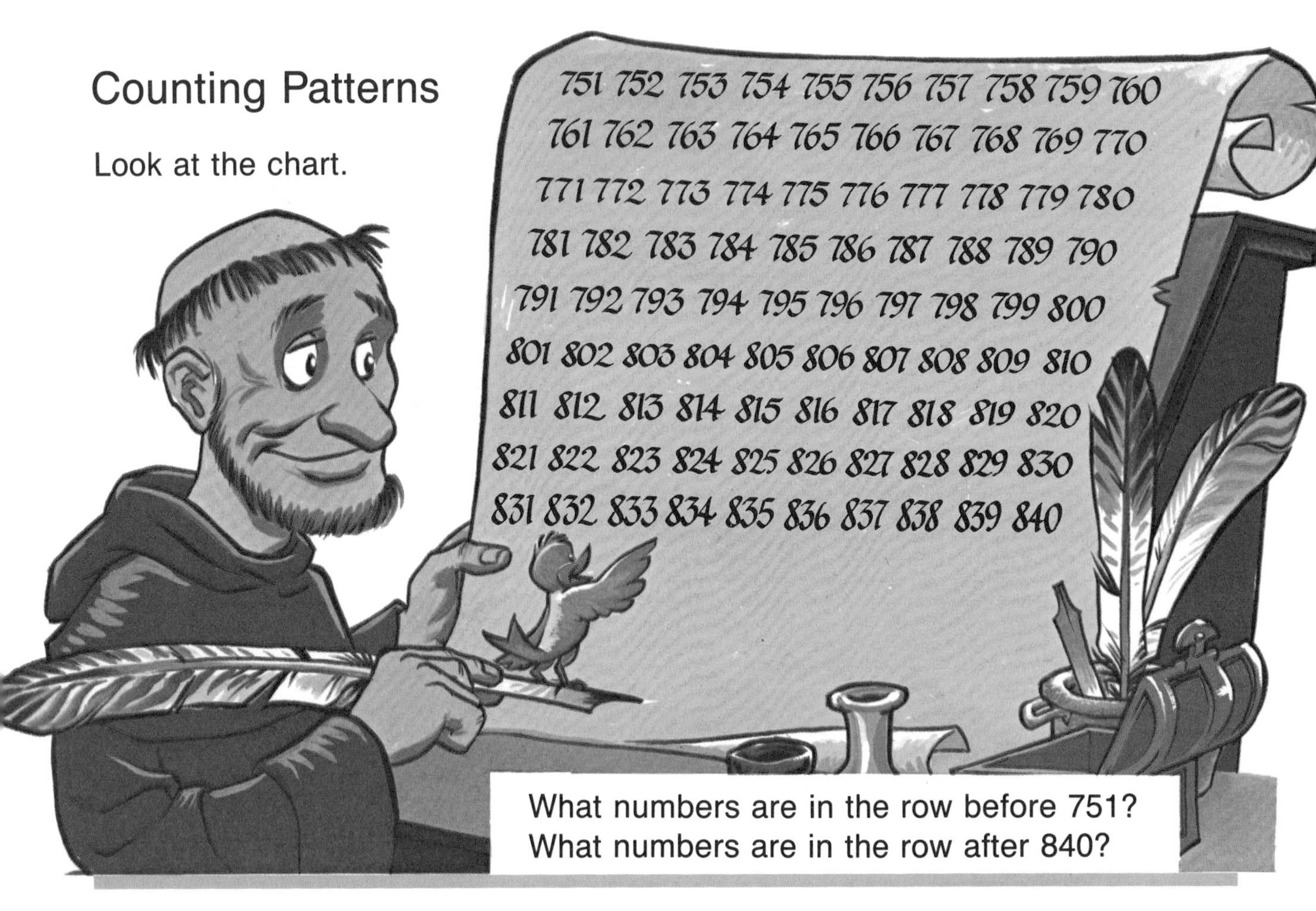

Write the next 4 numbers in each pattern.

1. 758, 759, 760,...
2. 837, 838, 839,...
3. 620, 621, 622,...
4. 584, 585, 586,...
5. 740, 742, 744,...
6. 165, 170, 175,...
7. 202, 201, 200,...
8. 355, 455, 555,...
9. 990, 980, 970,...
10. 515, 513, 511,...
11. 645, 646, 647,...
12. 823, 833, 843,...
13. 896, 796, 696,...
14. 125, 120, 115,...

What are the missing numbers?

Copy each pattern. Fill in the missing numbers.

1. 8, 10, ▒ , 14, ▒ , 18
2. 925, 930, 935, ▒ , 945, 950,
3. 146, 246, ▒ , ▒ , 546, 646
4. 549, 547, ▒ , 543, ▒ , 539
5. Make up some number patterns of your own.

BITS AND BYTES

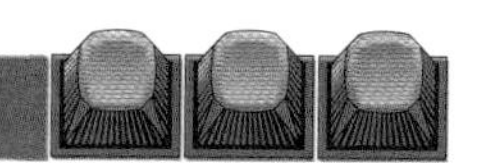

You can use a calculator to make patterns.

To count by tens, Kathy pushes

1 0 + 1 0 + 1 0 + 1 0 ...

Use your calculator to count by 25s to 200.

Push 2 5 + 2 5 + 2 5 + 2 5 ...

Continue the pattern until you reach 200.

Use your calculator to make your own pattern.

EXTENSIONS

Roman Numerals

The picture shows Roman numerals for numbers one to ten.

Write a Roman numeral for each of these.

1. 5
2. 1 less than 5
3. 1 more than 5
4. 10
5. 1 less than 10
6. 1 more than 10
7. 2
8. 2 more than 5
9. 3 more than 5
10. 2 more than 10
11. 3 more than 10
12. thirteen

13. Copy and complete.

I	II	▒	IV	V	▒	▒	▒	IX	▒
XI	▒	XIII	XIV	XV	XVI	▒	▒	XIX	XX
XXI	▒	▒	▒	▒	▒	XXVII	▒	▒	▒

Write numbers for these.

1. XV
2. XX
3. XIV
4. XIX
5. XXIX
6. IX
7. XXV
8. XII
9. XXVII
10. XXIV

Write the number that is greater.

11. VI
IV
12. V
IV
13. XVI
XV
14. XIX
XIV
15. XVIII
XIX
16. XVI
XX
17. XXIV
XXVI
18. IX
XI
19. XVI
XV
20. XI
X

21. Copy and complete this addition table.

+	I	II	III	IV	V	VI	VII	VIII	IX
X	XI	▒	▒	▒	▒	▒	▒	▒	▒
XX	▒	▒	▒	▒	XXV	▒	▒	▒	▒
XXX	▒	XXXII	▒	XXXIV	▒	▒	▒	▒	▒

Comparing Numbers to 999

Laura saved 578 stickers.
Aaron saved 583 stickers.
Who saved more?

Compare hundreds digits. → Compare tens digits. → The number with the greater digit is greater.

578 583	578 583	583 is greater than 578.
same	8 is greater than 7	

We write: $583 > 578$
Aaron saved more than Laura.

Write the number that is greater.

1. 684
   529
2. 327
   527
3. 872
   839
4. 709
   706
5. 198
   201

Write the number that is less.

6. 756
   387
7. 541
   620
8. 927
   963
9. 846
   446
10. 202
    197

Copy each sentence. Write > or < for each ▢

11. 597 ▢ 634
12. 389 ▢ 199
13. 927 ▢ 792
14. 362 ▢ 381
15. 617 ▢ 632
16. 392 ▢ 97

EXTENSIONS

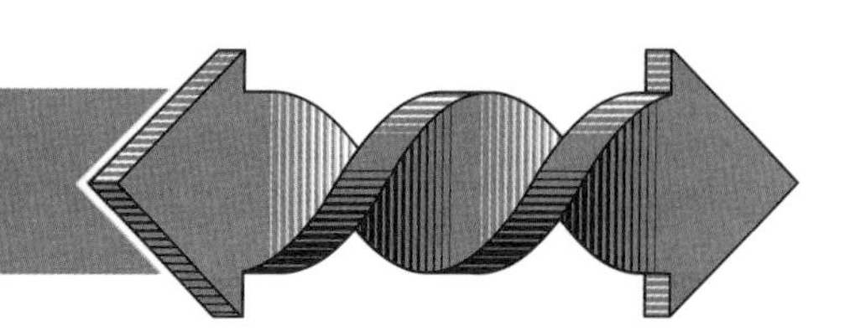

Ordering Numbers to 999

Which book has the most pages?

191 pages

189 pages

197 pages

Compare the first two numbers. → Compare the greater number with the third number. → Choose the greater number.

191
189
197 — 191 is greater

191
189
197 — 197 is greater

197 is the greatest number.

The book of bad jokes has the most pages.

Write the greatest number of the three.

1.	2.	3.
287	297	327
564	201	236
583	286	271

Write the least number of the three.

4.	5.	6.
892	326	922
907	407	647
865	385	256

Write the numbers from greatest to least.

7. 647, 286, 509

8. 287, 507, 582

PROBLEM SOLVING

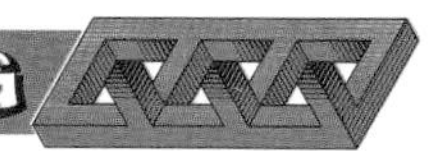

Who read the most pages?
Who read the least pages?

CHECK-UP

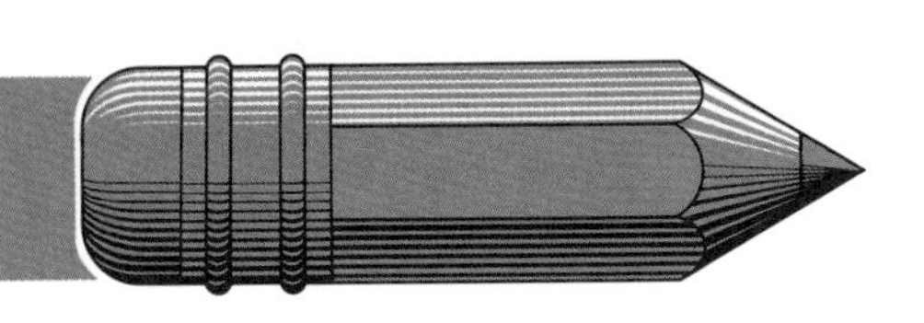

1. How much money is shown?

Write the numeral.

2. 5 tens, 4 ones
3. 3 tens, 7 ones
4. 50 + 6

Write the meaning of each coloured digit.

5. **2**9
6. 8**7**

Write the number that is greater.

7. 72
   68

Write the number that is less.

8. 29
   24

9. How many cubes are shown?

Hundreds	Tens	Ones

Write the numeral.

10. 3 hundreds, 2 tens, 4 ones
11. 6 hundreds, 5 ones
12. 200 + 30 + 7

Write the meaning of each coloured digit.

13. **3**61
14. 5**4**0

Write the number that is greater.

15. 834
    756

Write the number that is less.

16. 947
    952

ADDITION OF 2-DIGIT NUMBERS

INDIAN SUMMER

Along the line of smoky hills
  The crimson forest stands,
And all the day the blue-jay calls
  Throughout the autumn lands.

Now by the brook the maple leans
  With all his glory spread,
And all the sumachs on the hills
  Have turned their green to red.

Now by great marshes wrapt in mist,
  Or past some river's mouth,
Throughout the long, still autumn day
  Wild birds are flying south.

— Wilfred Campbell

Three Addends

How many monsters are there in all?

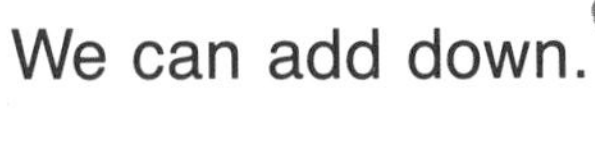

We can add down.

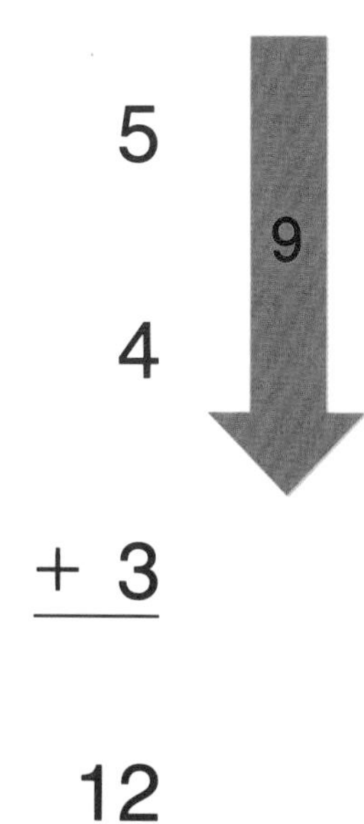

$$\begin{array}{r} 5 \\ 4 \\ +\ 3 \\ \hline 12 \end{array}$$

We can add up.

$$\begin{array}{r} 5 \\ 4 \\ +\ 3 \\ \hline 12 \end{array}$$

There are 12 monsters in all.

Add.

1. $\begin{array}{r} 2 \\ 3 \\ +6 \\ \hline \end{array}$
2. $\begin{array}{r} 5 \\ 4 \\ +2 \\ \hline \end{array}$
3. $\begin{array}{r} 7 \\ 2 \\ +6 \\ \hline \end{array}$
4. $\begin{array}{r} 5 \\ 3 \\ +6 \\ \hline \end{array}$
5. $\begin{array}{r} 6 \\ 3 \\ +5 \\ \hline \end{array}$
6. $\begin{array}{r} 8 \\ 1 \\ +7 \\ \hline \end{array}$
7. $\begin{array}{r} 6 \\ 3 \\ +4 \\ \hline \end{array}$
8. $\begin{array}{r} 7 \\ 1 \\ +8 \\ \hline \end{array}$
9. $\begin{array}{r} 8 \\ 0 \\ +9 \\ \hline \end{array}$
10. $\begin{array}{r} 4 \\ 2 \\ +7 \\ \hline \end{array}$
11. $\begin{array}{r} 8 \\ 2 \\ +6 \\ \hline \end{array}$
12. $\begin{array}{r} 6 \\ 4 \\ +5 \\ \hline \end{array}$
13. $\begin{array}{r} 7 \\ 3 \\ +6 \\ \hline \end{array}$
14. $\begin{array}{r} 9 \\ 1 \\ +8 \\ \hline \end{array}$
15. $\begin{array}{r} 5 \\ 5 \\ +7 \\ \hline \end{array}$

How many points are in each hand?

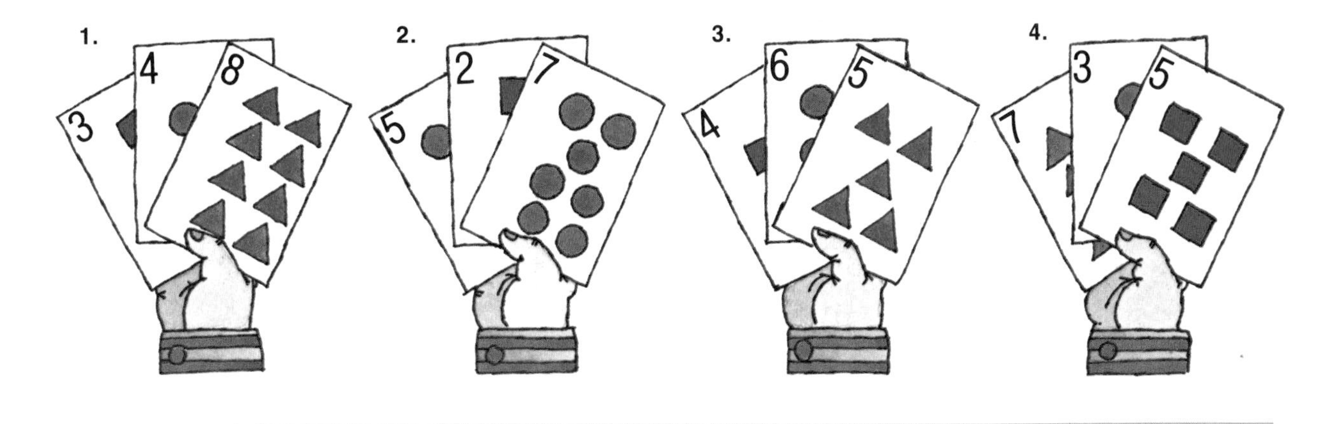

How many points were scored on each dart board?

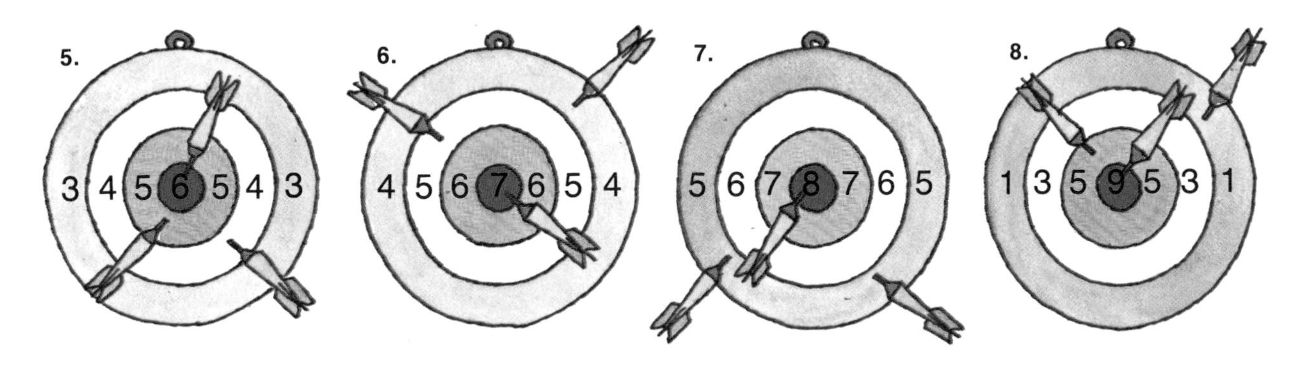

9. Copy and complete so the sum of the numbers on each side is 12.

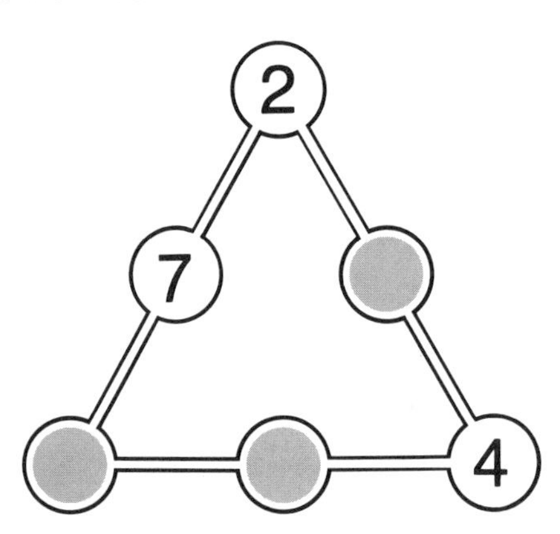

Copy and fill in the numbers from 3 to 6 so the sum on each side is 9.

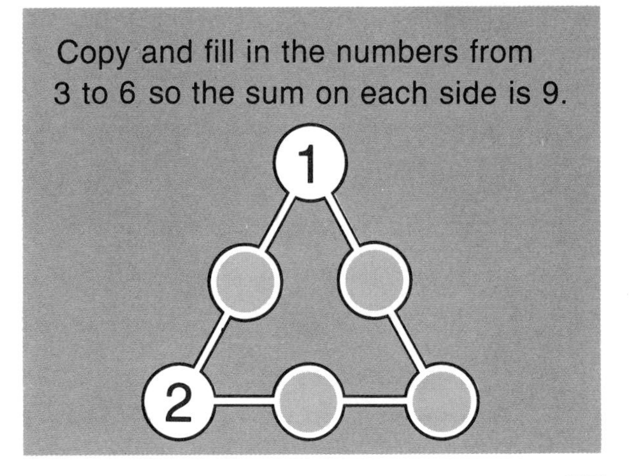

PROBLEM SOLVING

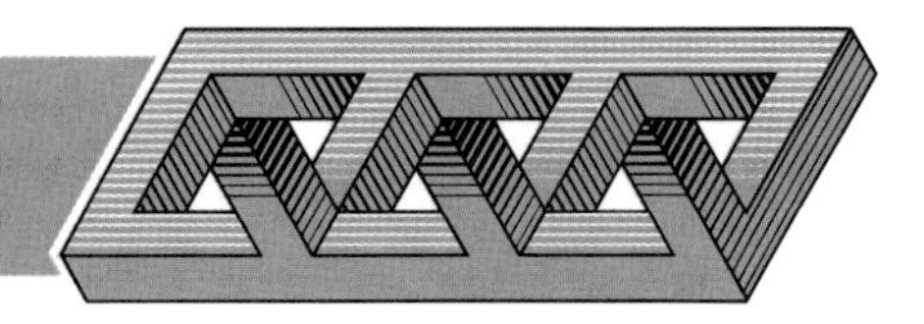

Looking for Possibilities

The paths out of the maze with a sum of 18 are safe.
How many safe paths can you find?
One safe path is shown.
It is 1 + 9 + 8.

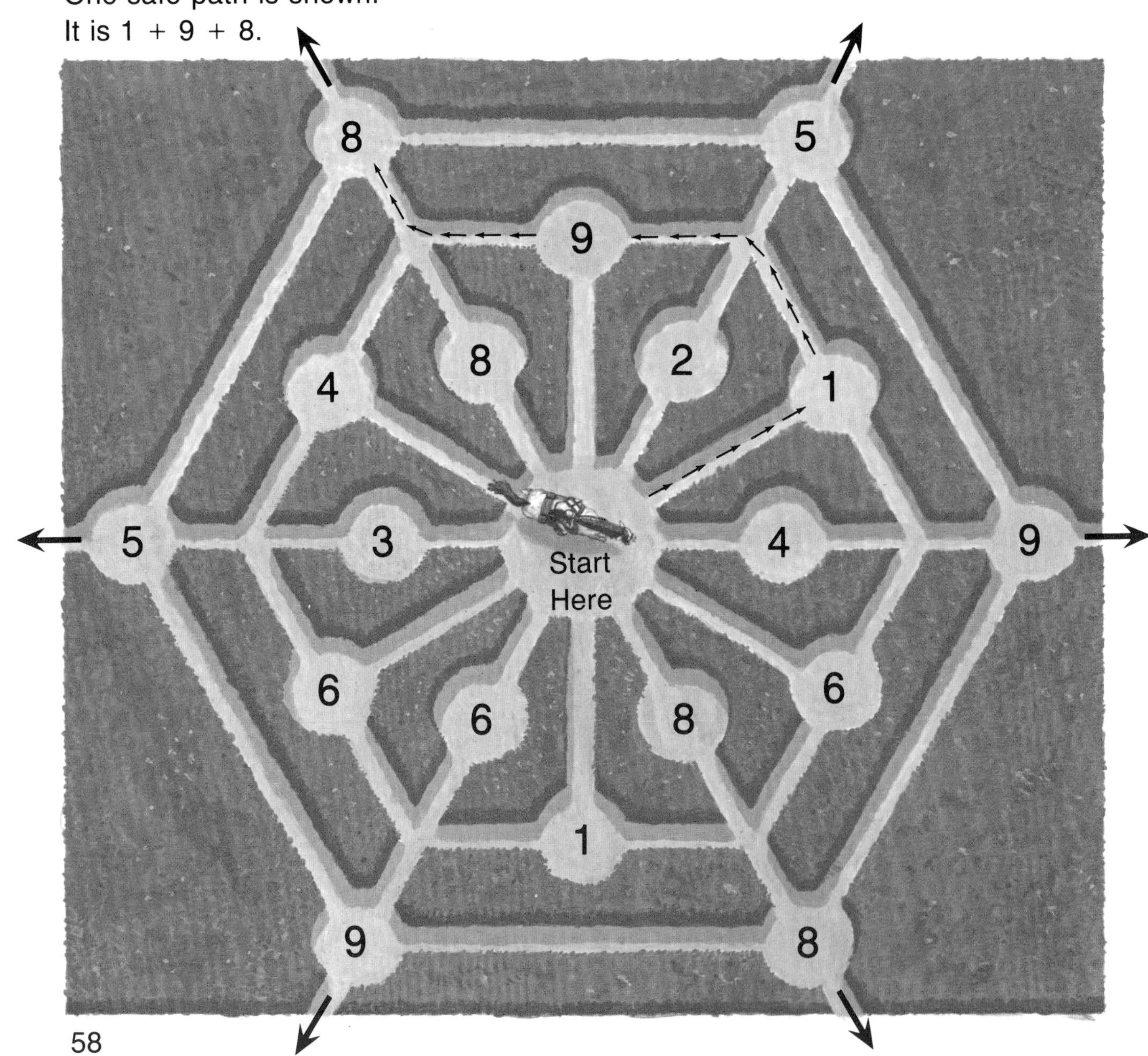

Four Addends

How many points are there in each hand?

1. 

2. 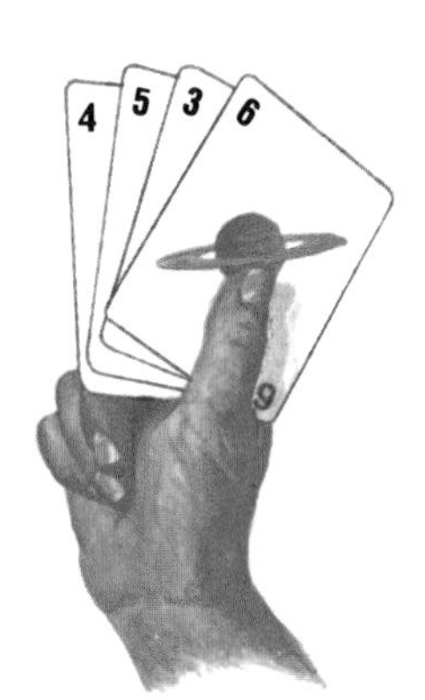

3. 

4. 

Add.

5.	6.	7.	8.	9.	10.
7	8	9	6	5	8
3	5	2	4	3	4
4	1	6	2	4	0
+2	+3	+1	+3	+2	+1

11.	12.	13.	14.	15.	16.
2	3	1	3	2	1
4	5	2	4	3	0
3	1	6	2	4	4
+7	+8	+9	+6	+5	+8

17. Fill in the missing numbers so the sum on each side is 17.

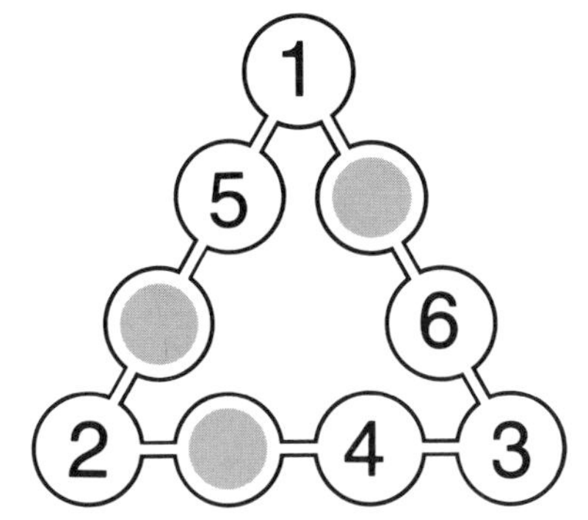

PROBLEM SOLVING

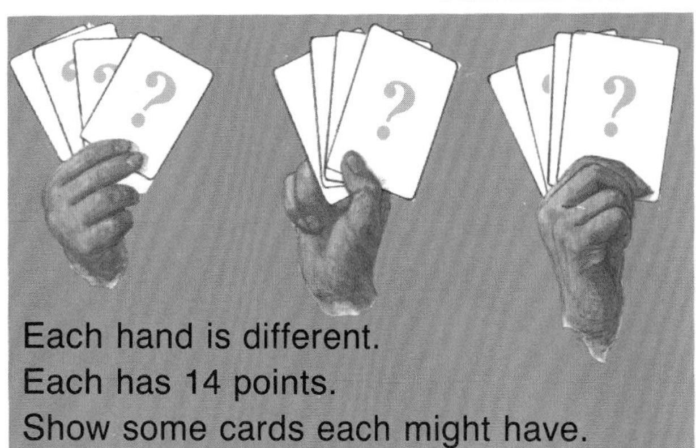

Adding Without Trading

How many stickers are there in all?

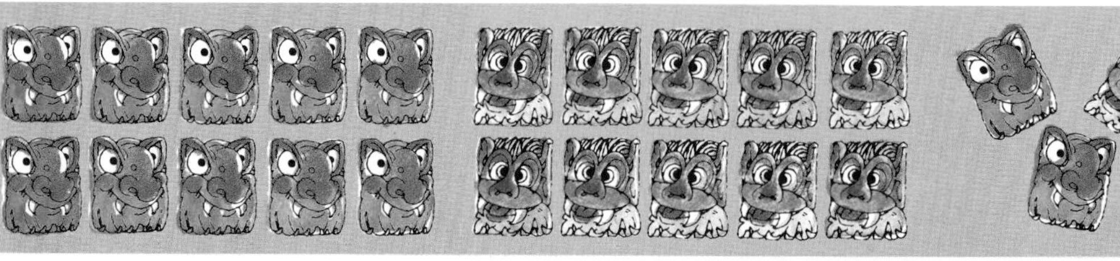

23 stickers

5 stickers

Add the ones.	Add the tens.
$\begin{array}{r} 23 \\ +\ 5 \\ \hline 8 \end{array}$	$\begin{array}{r} 23 \\ +\ 5 \\ \hline 28 \end{array}$

There are 28 stickers in all.

Add.

1. $\begin{array}{r} 12 \\ +\ 7 \\ \hline \end{array}$
2. $\begin{array}{r} 21 \\ +\ 6 \\ \hline \end{array}$
3. $\begin{array}{r} 53 \\ +\ 4 \\ \hline \end{array}$
4. $\begin{array}{r} 25 \\ +\ 2 \\ \hline \end{array}$
5. $\begin{array}{r} 14 \\ +\ 5 \\ \hline \end{array}$
6. $\begin{array}{r} 76 \\ +\ 3 \\ \hline \end{array}$
7. $\begin{array}{r} 41 \\ +\ 7 \\ \hline \end{array}$
8. $\begin{array}{r} 52 \\ +\ 6 \\ \hline \end{array}$
9. $\begin{array}{r} 27 \\ +\ 1 \\ \hline \end{array}$

Add.
Look for a pattern. Write the next 6 questions in each pattern.

1. $\begin{array}{r} 11 \\ +\ 3 \\ \hline \end{array}$ $\begin{array}{r} 21 \\ +\ 3 \\ \hline \end{array}$ $\begin{array}{r} 31 \\ +\ 3 \\ \hline \end{array}$ ...

2. $\begin{array}{r} 13 \\ +\ 6 \\ \hline \end{array}$ $\begin{array}{r} 23 \\ +\ 6 \\ \hline \end{array}$ $\begin{array}{r} 33 \\ +\ 6 \\ \hline \end{array}$ ...

Make your own pattern like these.

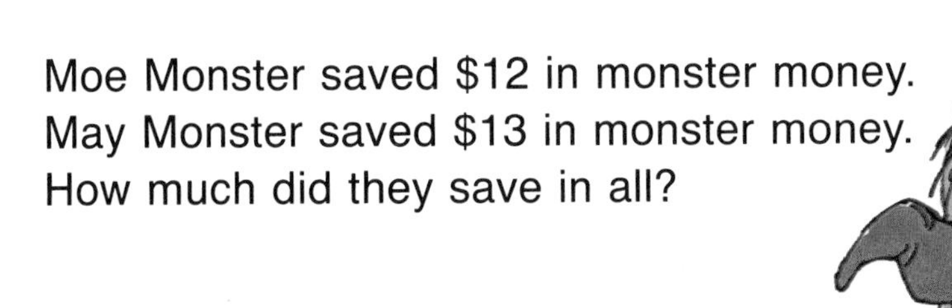

Moe Monster saved $12 in monster money.
May Monster saved $13 in monster money.
How much did they save in all?

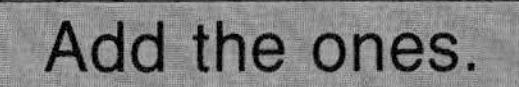

Add the ones.	Add the tens.
$\begin{array}{r} \$12 \\ +13 \\ \hline \$\ \ 5 \end{array}$	$\begin{array}{r} \$12 \\ +13 \\ \hline \$25 \end{array}$

They saved $25.

Add.

1. $\begin{array}{r} 26 \\ +31 \\ \hline \end{array}$	2. $\begin{array}{r} 42 \\ +35 \\ \hline \end{array}$	3. $\begin{array}{r} 27 \\ +42 \\ \hline \end{array}$	4. $\begin{array}{r} 81 \\ +17 \\ \hline \end{array}$	5. $\begin{array}{r} 63 \\ +22 \\ \hline \end{array}$	6. $\begin{array}{r} 37 \\ +21 \\ \hline \end{array}$
7. $\begin{array}{r} 52 \\ +34 \\ \hline \end{array}$	8. $\begin{array}{r} 27 \\ +32 \\ \hline \end{array}$	9. $\begin{array}{r} 55 \\ +32 \\ \hline \end{array}$	10. $\begin{array}{r} 63 \\ +24 \\ \hline \end{array}$	11. $\begin{array}{r} 37 \\ +22 \\ \hline \end{array}$	12. $\begin{array}{r} 51 \\ +18 \\ \hline \end{array}$
13. $\begin{array}{r} 62 \\ +25 \\ \hline \end{array}$	14. $\begin{array}{r} 73 \\ +16 \\ \hline \end{array}$	15. $\begin{array}{r} 27 \\ +31 \\ \hline \end{array}$	16. $\begin{array}{r} 93 \\ +\ \ 5 \\ \hline \end{array}$	17. $\begin{array}{r} 83 \\ +16 \\ \hline \end{array}$	18. $\begin{array}{r} 24 \\ +\ \ 3 \\ \hline \end{array}$

Trading 10 Ones for a Ten

How many cubes are there?

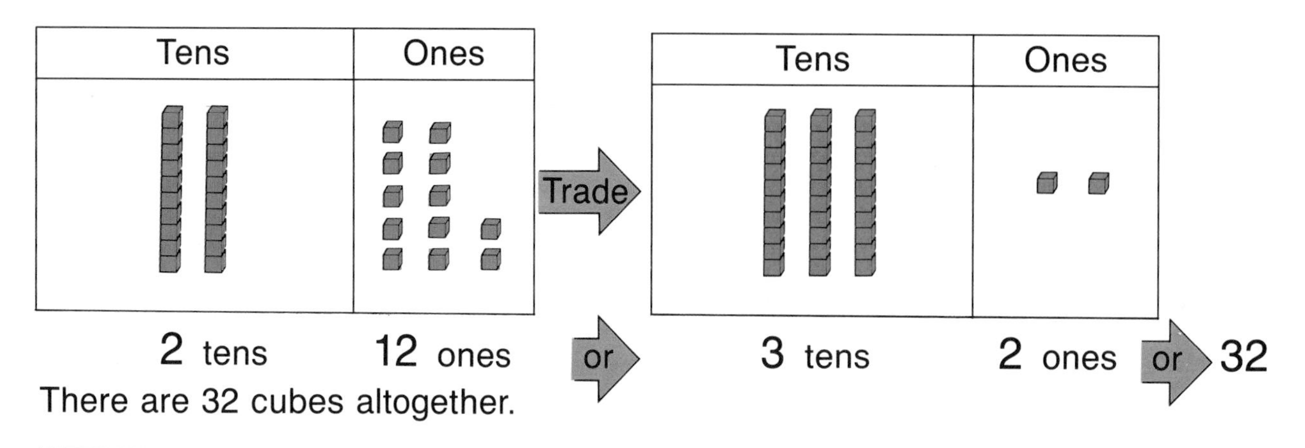

There are 32 cubes altogether.

How many cubes are there? Trade.

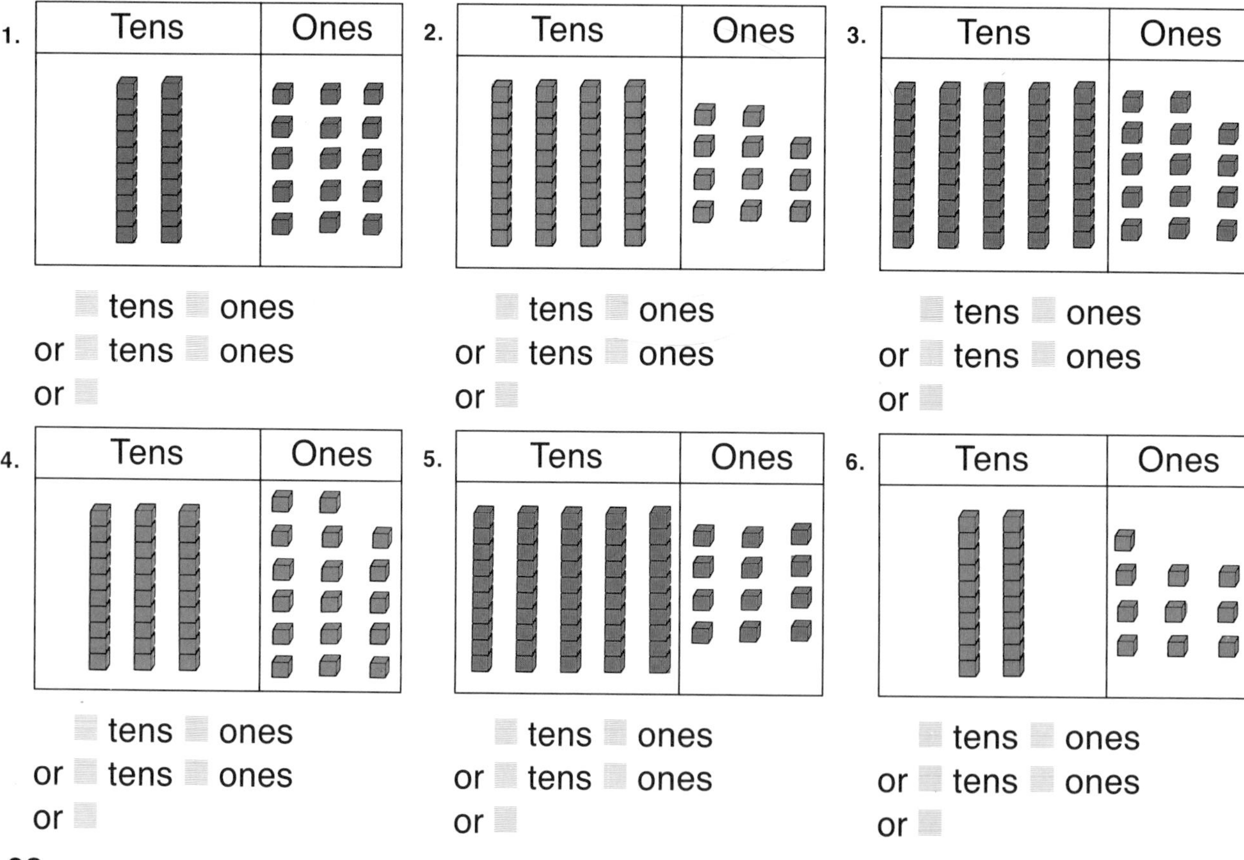

Adding with Trading

How many cubes are there in all?

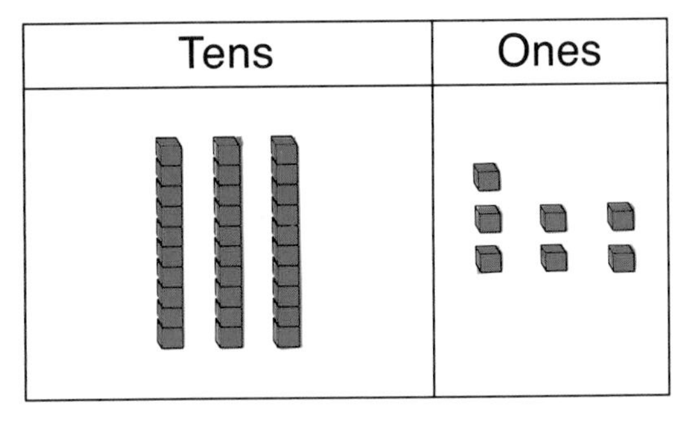

Tens	Ones

37 + 5

Add the ones. → Trade. → Add the tens.

$$\begin{array}{r} 37 \\ +\ 5 \\ \hline \end{array}$$

12 ones

$$\begin{array}{r} ^{1}\ \ \\ 37 \\ +\ 5 \\ \hline 2 \end{array}$$

$$\begin{array}{r} ^{1}\ \ \\ 37 \\ +\ 5 \\ \hline 42 \end{array}$$

There are 42 cubes in all.

Add. Use materials if you wish.

1. $\begin{array}{r} 12 \\ +\ 9 \\ \hline \end{array}$ 2. $\begin{array}{r} 47 \\ +\ 6 \\ \hline \end{array}$ 3. $\begin{array}{r} 84 \\ +\ 7 \\ \hline \end{array}$ 4. $\begin{array}{r} 28 \\ +\ 3 \\ \hline \end{array}$ 5. $\begin{array}{r} 55 \\ +\ 9 \\ \hline \end{array}$ 6. $\begin{array}{r} 39 \\ +\ 2 \\ \hline \end{array}$

7. $\begin{array}{r} 76 \\ +\ 6 \\ \hline \end{array}$ 8. $\begin{array}{r} 63 \\ +\ 8 \\ \hline \end{array}$ 9. $\begin{array}{r} 31 \\ +\ 9 \\ \hline \end{array}$ 10. $\begin{array}{r} 87 \\ +\ 5 \\ \hline \end{array}$ 11. $\begin{array}{r} 42 \\ +\ 3 \\ \hline \end{array}$ 12. $\begin{array}{r} 58 \\ +\ 4 \\ \hline \end{array}$

13. $\begin{array}{r} 33 \\ +\ 8 \\ \hline \end{array}$ 14. $\begin{array}{r} 22 \\ +\ 6 \\ \hline \end{array}$ 15. $\begin{array}{r} 19 \\ +\ 5 \\ \hline \end{array}$ 16. $\begin{array}{r} 56 \\ +\ 8 \\ \hline \end{array}$ 17. $\begin{array}{r} 81 \\ +\ 9 \\ \hline \end{array}$ 18. $\begin{array}{r} 64 \\ +\ 7 \\ \hline \end{array}$

Adding: Two 2-Digit Numbers

Monster Munchies cost \$26.
Power Punch costs \$15.
How much would it cost to buy both?

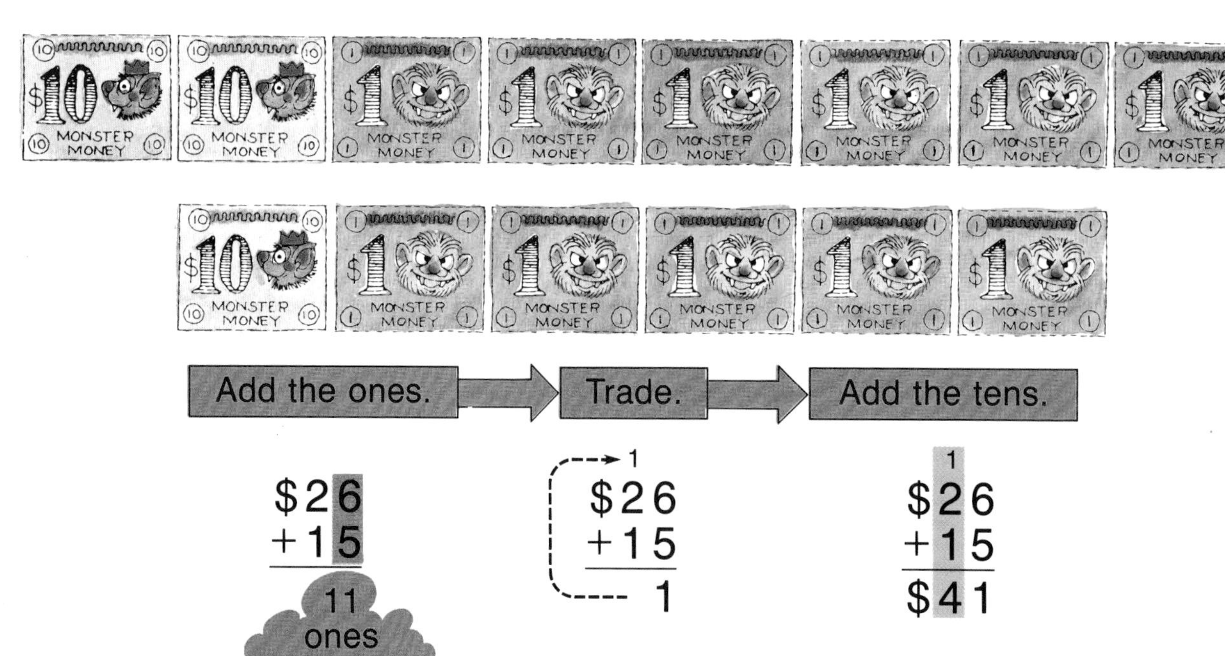

It would cost \$41 to buy both.

Add.

1. 26 + 18	2. 43 + 39	3. 37 + 42	4. 73 + 18	5. 35 + 37	6. 14 + 59
7. 19 + 65	8. 28 + 26	9. 57 + 36	10. 11 + 28	11. 46 + 35	12. 38 + 27

Add.

1. $\begin{array}{r} 36 \\ +39 \\ \hline \end{array}$ 2. $\begin{array}{r} 47 \\ +24 \\ \hline \end{array}$ 3. $\begin{array}{r} 14 \\ +68 \\ \hline \end{array}$ 4. $\begin{array}{r} 25 \\ +13 \\ \hline \end{array}$ 5. $\begin{array}{r} 63 \\ +29 \\ \hline \end{array}$ 6. $\begin{array}{r} 55 \\ +36 \\ \hline \end{array}$

7. $\begin{array}{r} 24 \\ +49 \\ \hline \end{array}$ 8. $\begin{array}{r} 78 \\ +11 \\ \hline \end{array}$ 9. $\begin{array}{r} 32 \\ +19 \\ \hline \end{array}$ 10. $\begin{array}{r} 69 \\ +12 \\ \hline \end{array}$ 11. $\begin{array}{r} 43 \\ +48 \\ \hline \end{array}$ 12. $\begin{array}{r} 14 \\ +38 \\ \hline \end{array}$

13. 26 monsters went shopping.
16 monsters stayed home.
How many monsters were there in all?

14. There are 13 more purple monsters than green monsters.
There are 19 green monsters.
How many purple monsters are there?

PROBLEM SOLVING

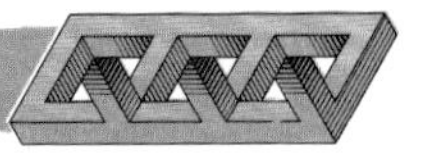

Merv Monster spent $73.
What did he buy?

Fangs $39

Vampire Shirt $34

Monster Wings $48

Monster Make-up $25

PROBLEM SOLVING

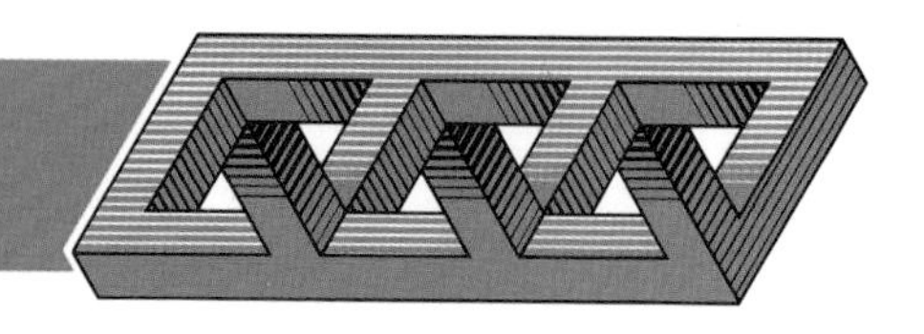

Interpreting Information on a Graph

The graph shows how many leaves were collected by Miss Willson's class.

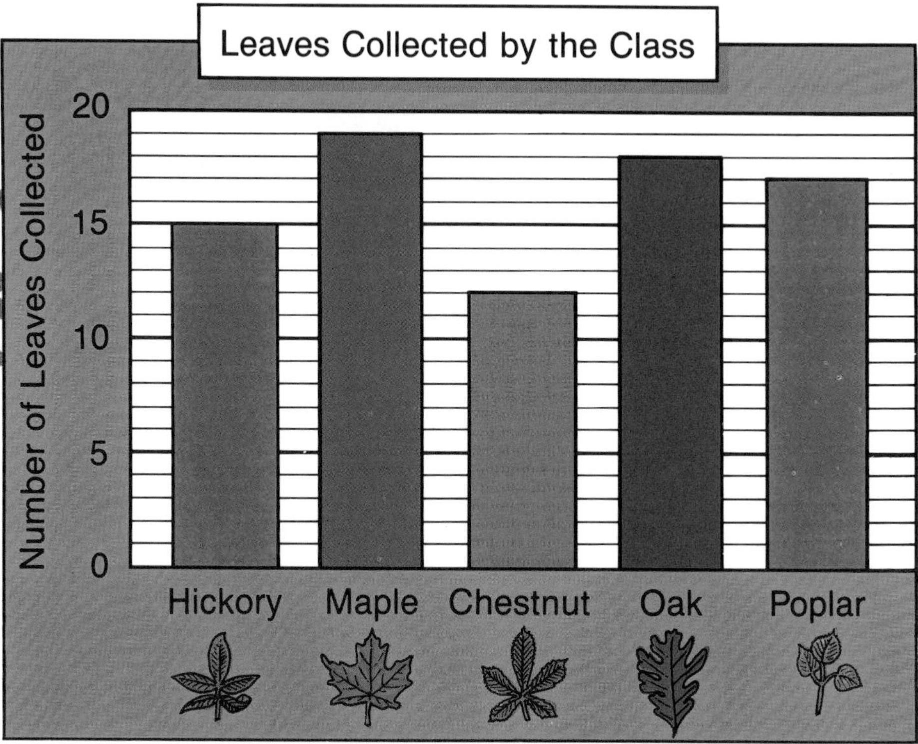

1. Which leaves did the class collect the most?

2. How many hickory leaves were collected?

3. How many poplar leaves were collected?

4. How many yellow leaves were collected in all?

5. How many hickory and maple leaves were collected in all?

6. How many leaves were not yellow or red?

7. How many yellow and green leaves were collected in all?

8. How many leaves were not yellow or green?

9. How many leaves were not yellow?

10. Write your own problem about this graph.

JUST FOR FUN

RIDDLE: What do you say to a two-headed monster?

Add.

A. $67 + 18$	D. $23 + 49$	O. $68 + 23$
N. $23 + 48$	L. $37 + 49$	W. $25 + 69$
M. $37 + 25$	H. $46 + 52$	T. $28 + 37$
R. $47 + 19$	G. $88 + 9$	L. $21 + 49$
I. $42 + 38$	Y. $48 + 28$	C. $29 + 38$
	E. $64 + 9$	S. $83 + 14$
	L. $67 + 26$	E. $19 + 49$
	P. $18 + 25$	O. $16 + 17$
Y. $81 + 7$	M. $28 + 39$	H. $13 + 40$

To answer the riddle, write the letters of the questions with these answers.

__ __ __ __ __   __ __ __ __ __

98 73 86 93 91   53 68 70 86 33

Adding 2-Digit Numbers: 3-Digit Sums

Grandpa monster is 87 years old.

Great-grandpa monster is 45 years older.

How old is great-grandpa?

Add the ones. → Trade 10 ones for a ten. → Add the tens. → Write 10 tens as a hundred.

$$\begin{array}{r} 87 \\ +45 \\ \hline \end{array}$$
12 ones

$$\begin{array}{r} {}^{1} \\ 87 \\ +45 \\ \hline 2 \end{array}$$

$$\begin{array}{r} {}^{1} \\ 87 \\ +45 \\ \hline 132 \end{array}$$

Great-grandpa monster is 132 years old!

Add. Use materials if you wish.

1. 32 + 86	2. 58 + 61	3. 59 + 73	4. 28 + 82	5. 69 + 37	6. 51 + 76
7. 82 + 86	8. 79 + 93	9. 62 + 46	10. 73 + 87	11. 68 + 27	12. 59 + 87
13. 97 + 8	14. 9 + 96	15. 88 + 93	16. 24 + 37	17. 67 + 97	18. 85 + 67

JUST FOR FUN

RIDDLE: Where does a mean monster sit when he goes to the movies?

Add.

G. $\begin{array}{r} 83 \\ +82 \\ \hline \end{array}$	H. $\begin{array}{r} 73 \\ +88 \\ \hline \end{array}$	A. $\begin{array}{r} 68 \\ +37 \\ \hline \end{array}$	D. $\begin{array}{r} 96 \\ +13 \\ \hline \end{array}$	I. $\begin{array}{r} 26 \\ +97 \\ \hline \end{array}$	J. $\begin{array}{r} 32 \\ +88 \\ \hline \end{array}$
K. $\begin{array}{r} 87 \\ +79 \\ \hline \end{array}$	L. $\begin{array}{r} 78 \\ +88 \\ \hline \end{array}$	B. $\begin{array}{r} 59 \\ +26 \\ \hline \end{array}$	C. $\begin{array}{r} 73 \\ +28 \\ \hline \end{array}$	M. $\begin{array}{r} 44 \\ +77 \\ \hline \end{array}$	E. $\begin{array}{r} 70 \\ +90 \\ \hline \end{array}$
N. $\begin{array}{r} 73 \\ +96 \\ \hline \end{array}$	O. $\begin{array}{r} 38 \\ +78 \\ \hline \end{array}$	S. $\begin{array}{r} 38 \\ +99 \\ \hline \end{array}$	Q. $\begin{array}{r} 93 \\ +87 \\ \hline \end{array}$	P. $\begin{array}{r} 53 \\ +19 \\ \hline \end{array}$	F. $\begin{array}{r} 67 \\ +58 \\ \hline \end{array}$
T. $\begin{array}{r} 73 \\ +89 \\ \hline \end{array}$	U. $\begin{array}{r} 76 \\ +56 \\ \hline \end{array}$	V. $\begin{array}{r} 83 \\ +96 \\ \hline \end{array}$	W. $\begin{array}{r} 86 \\ +88 \\ \hline \end{array}$	Y. $\begin{array}{r} 98 \\ +99 \\ \hline \end{array}$	R. $\begin{array}{r} 55 \\ +45 \\ \hline \end{array}$

To answer the riddle,
write the letters of the questions with these answers.

___ ___ ___ ___ ___ ___ ___ ___   ___ ___   ___ ___ ___ ___ ___

105 169 197 174 161 160 100 160   161 160   174 105 169 162 137

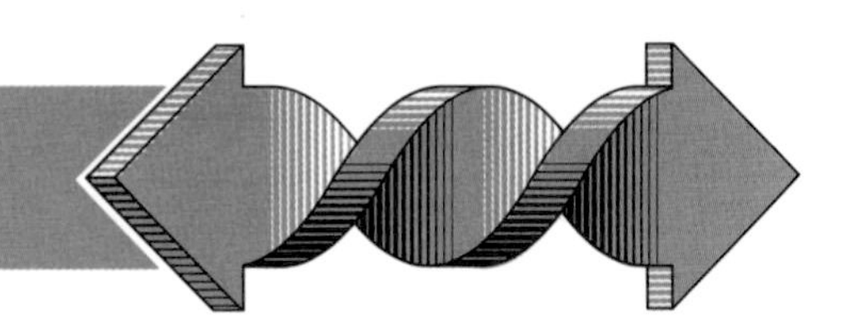

EXTENSIONS

Adding with Trading: 3 Addends

Kathy, David and Andrew sold apples. The table shows how many they sold each day.

	Thursday	Friday	Saturday
Kathy	24	35	69
David	36	27	56
Andrew	20	46	61

How many apples were sold on Saturday?

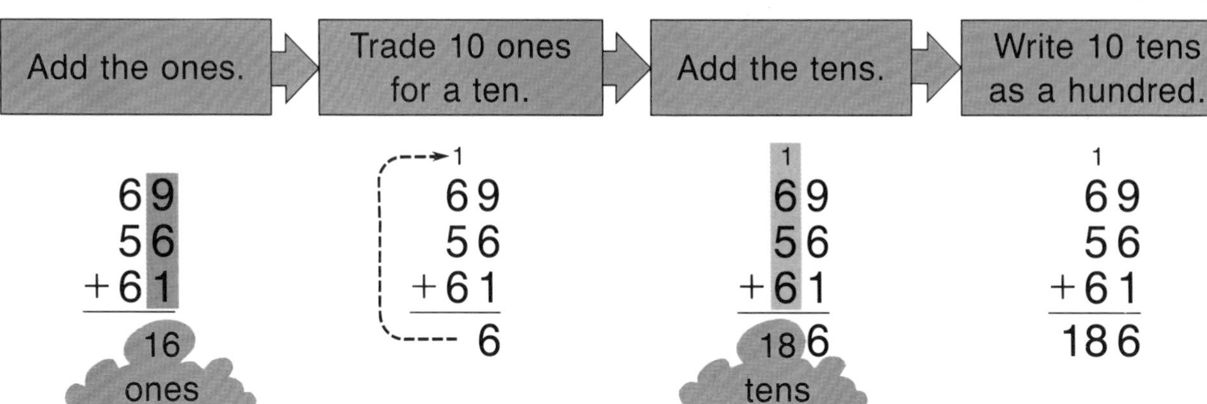

186 apples were sold on Saturday.

Add.

1. 26 + 32 + 17 =
2. 47 + 29 + 18 =
3. 63 + 75 + 41 =
4. 78 + 39 + 27 =
5. 53 + 42 + 70 =
6. 90 + 73 + 89 =

7. How many apples were sold on Friday?
8. Who sold the most apples?
9. On which day were the most apples sold?
10. Who sold 119 apples?

BITS AND BYTES

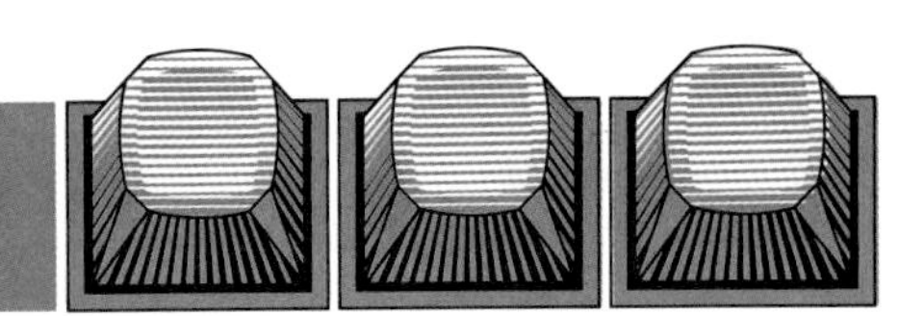

Look at this!

Find the sums for A to H by adding along the arrows.

Why is this called a magic square?

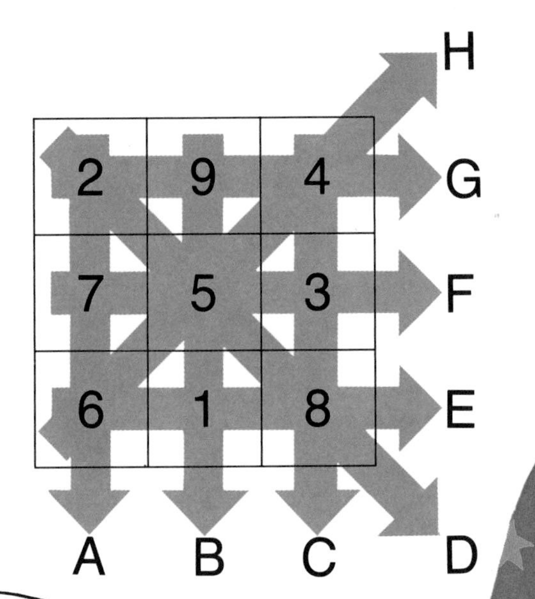

Use your calculator to see which of these are magic squares.

1.

42	49	14
7	35	63
56	21	28

2.

26	97	72
91	65	39
78	13	104

3.

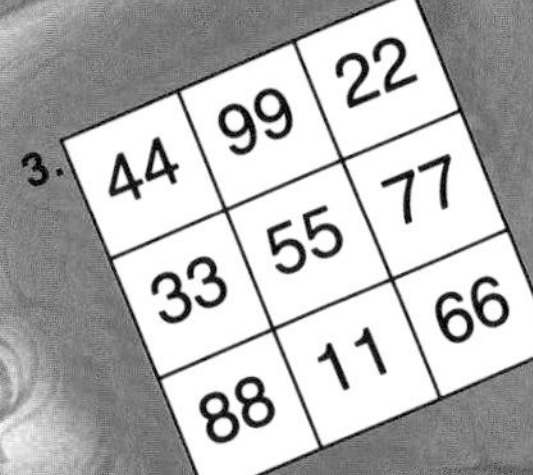

44	99	22
33	55	77
88	11	66

4.

34	69	62
83	55	27
48	41	76

The Human Skeleton

Did you know...
There are 206 bones in a human body.
To find out where some of them are, answer these questions.

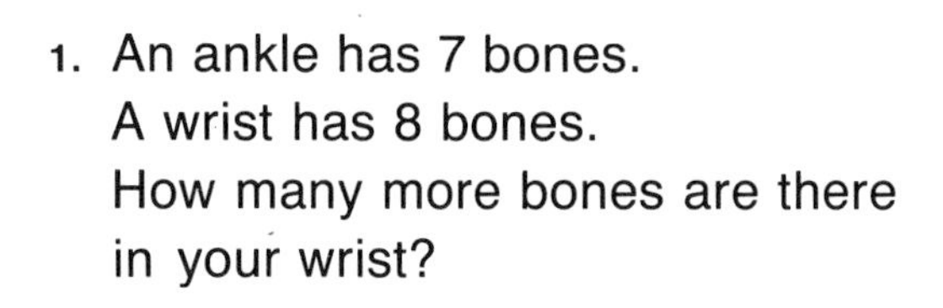

1. An ankle has 7 bones.
   A wrist has 8 bones.
   How many more bones are there in your wrist?

2. There are 14 bones in your face.
   It takes 15 more to make a complete skull.
   How many bones are there in your skull?

3. Your two ankles have 14 bones.
There are 38 bones in your feet.
How many bones are there in your ankles and your feet?

4. Your wrists have 16 bones.
Your hands have 22 more.
How many bones are there in your hands?

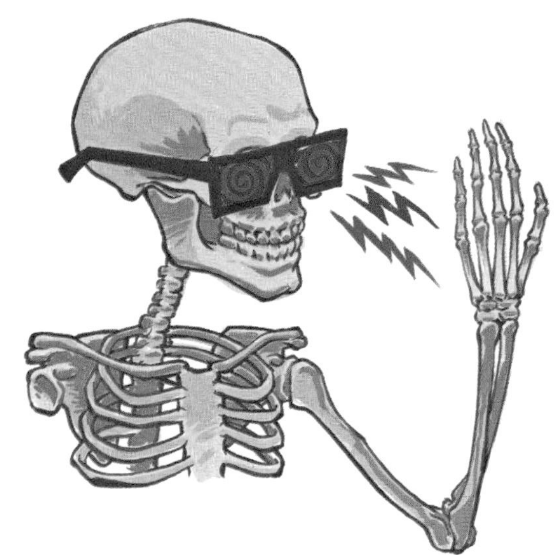

6. Your skull has 29 bones.
It is attached to the 26 bones in your spine.
How many bones in both your skull and your spine?

7. You can see the bones of your hand on an X-ray.
There are 19 bones.
How many are there in both hands?

8. Arms have 64 bones.
Legs have 62 bones.
How many are there in both arms and legs?

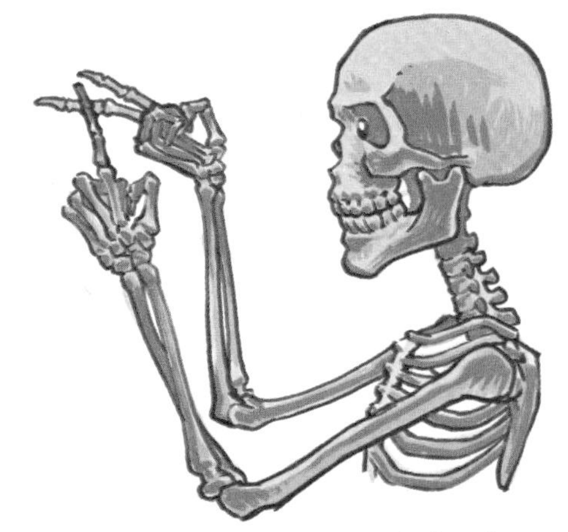

5. Inside your ears there are 6 tiny bones.
The rest of your face has 14 bones.
How many are there in your face and ears?

ESTIMATING

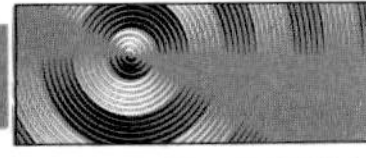

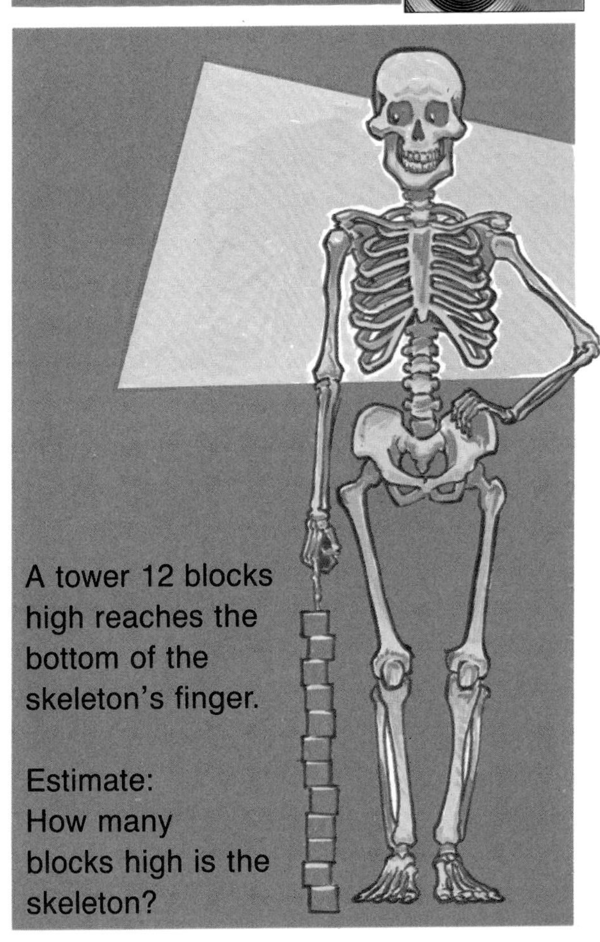

CHECK-UP

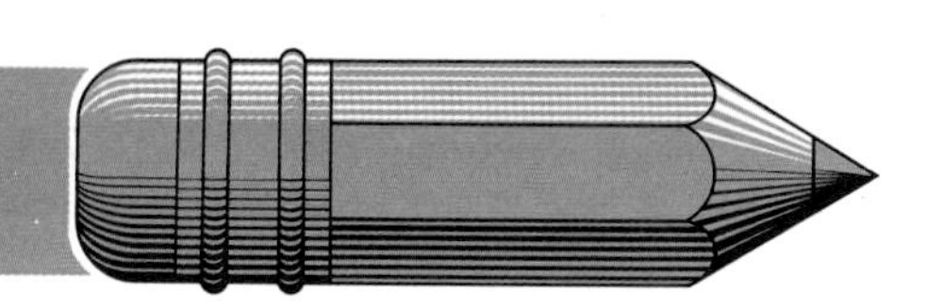

1. How many cubes are there?
Trade.

▢ tens ▢ ones
or ▢ tens ▢ ones
or ▢

Tens	Ones

Add.

2. $2 + 7 + 3$
3. $4 + 5 + 2$
4. $7 + 1 + 6$
5. $4 + 5 + 6 + 4$
6. $2 + 4 + 9 + 1$
7. $6 + 7 + 3 + 4$

8. $18 + 5$
9. $14 + 23$
10. $36 + 29$
11. $48 + 9$
12. $83 + 34$
13. $39 + 83$

14. $32 + 9$
15. $56 + 32$
16. $28 + 33$
17. $53 + 8$
18. $46 + 63$
19. $57 + 65$

20. $36 + 6$
21. $30 + 27$
22. $67 + 28$
23. $28 + 4$
24. $65 + 52$
25. $94 + 38$

26. Jill planted 23 yellow tulips.
Kim planted 47 yellow daffodils.
How many yellow flowers
were planted?

27. Mr. Lee planted 26 bushes one day.
He planted 18 the next day.
How many did he plant altogether?

CUMULATIVE REVIEW

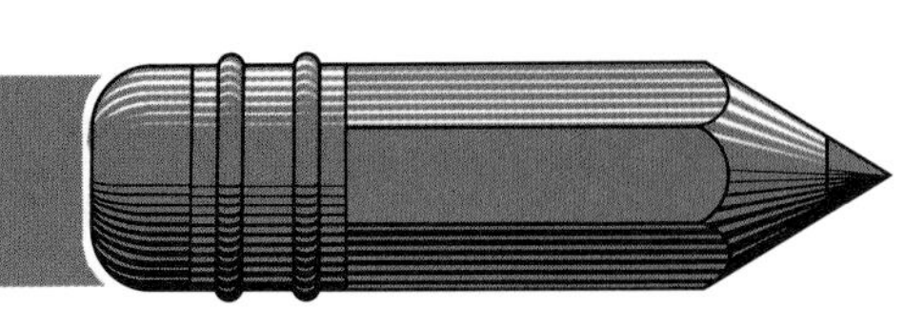

Add.

1. $\begin{array}{r} 4 \\ +3 \\ \hline \end{array}$ 2. $\begin{array}{r} 1 \\ +5 \\ \hline \end{array}$ 3. $\begin{array}{r} 6 \\ +6 \\ \hline \end{array}$ 4. $\begin{array}{r} 8 \\ +3 \\ \hline \end{array}$ 5. $\begin{array}{r} 8 \\ +5 \\ \hline \end{array}$ 6. $\begin{array}{r} 7 \\ +9 \\ \hline \end{array}$

Subtract.

7. $\begin{array}{r} 10 \\ -6 \\ \hline \end{array}$ 8. $\begin{array}{r} 5 \\ -4 \\ \hline \end{array}$ 9. $\begin{array}{r} 8 \\ -3 \\ \hline \end{array}$ 10. $\begin{array}{r} 17 \\ -8 \\ \hline \end{array}$ 11. $\begin{array}{r} 15 \\ -7 \\ \hline \end{array}$ 12. $\begin{array}{r} 14 \\ -9 \\ \hline \end{array}$

Complete. Watch the signs.

13. $\begin{array}{r} 7 \\ -3 \\ \hline \end{array}$ 14. $\begin{array}{r} 9 \\ +8 \\ \hline \end{array}$ 15. $\begin{array}{r} 14 \\ -6 \\ \hline \end{array}$ 16. $\begin{array}{r} 7 \\ +5 \\ \hline \end{array}$ 17. $\begin{array}{r} 6 \\ +8 \\ \hline \end{array}$ 18. $\begin{array}{r} 11 \\ -9 \\ \hline \end{array}$

Write 4 facts for each picture.

19. 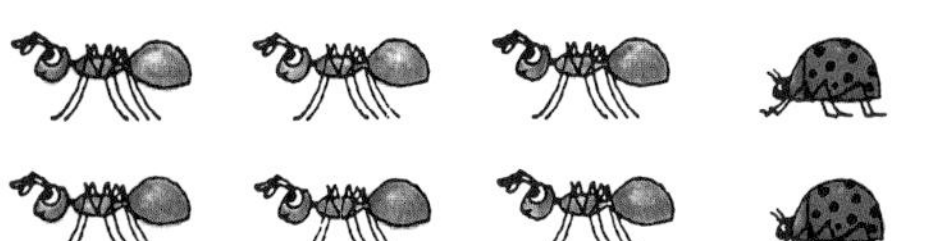

20. 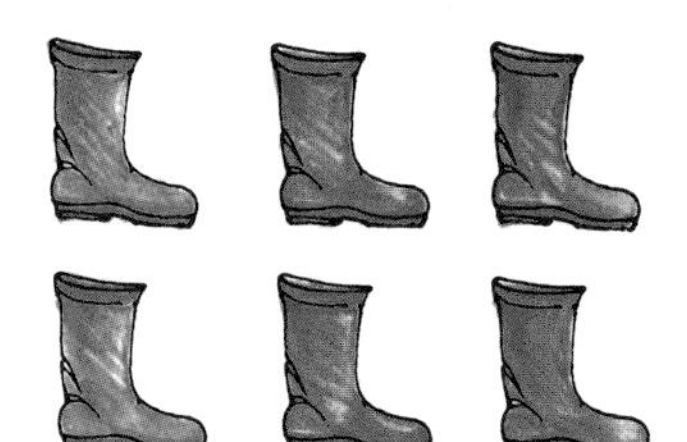

Copy and complete.

21. 30 + 8 = ▒ 22. 53 = 50 + ▒ 23. 862 = 800 + ▒ + ▒

Write the meaning of each coloured digit.

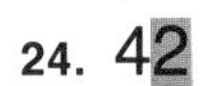

24. 4**2** 25. **6**0 26. **3**5 27. 5**1**8 28. 20**7** 29. **8**73

Write the missing numbers.

30. 62, 63, 64,...69
31. 15, 25, 35,...85
32. 795, 796, 797,...802
33. 436, 426, 416,...366

CUMULATIVE REVIEW

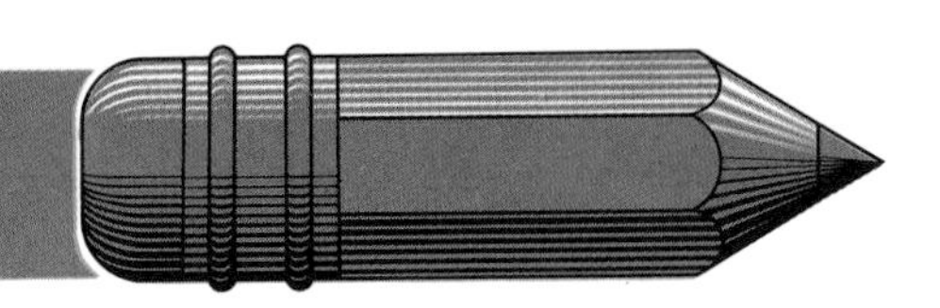

Add.

34. $\begin{array}{r} 63 \\ +24 \\ \hline \end{array}$

35. $\begin{array}{r} 25 \\ +13 \\ \hline \end{array}$

36. $\begin{array}{r} 59 \\ +32 \\ \hline \end{array}$

37. $\begin{array}{r} 72 \\ +19 \\ \hline \end{array}$

38. $\begin{array}{r} 86 \\ +35 \\ \hline \end{array}$

39. $\begin{array}{r} 24 \\ +93 \\ \hline \end{array}$

40. $\begin{array}{r} 4 \\ 7 \\ +2 \\ \hline \end{array}$

41. $\begin{array}{r} 4 \\ 3 \\ +6 \\ \hline \end{array}$

42. $\begin{array}{r} 9 \\ 5 \\ +7 \\ \hline \end{array}$

43. $\begin{array}{r} 2 \\ 5 \\ 3 \\ +1 \\ \hline \end{array}$

44. $\begin{array}{r} 1 \\ 4 \\ 2 \\ +6 \\ \hline \end{array}$

45. $\begin{array}{r} 5 \\ 2 \\ 2 \\ +3 \\ \hline \end{array}$

46. June planted 24 tulips.
Her mother planted 36 daffodils.
How many flowers did they plant together?

47. Garth had 69 marbles.
He won 32 more.
How many marbles did he have altogether?

Write the number that is greater.

48. 43
29

49. 82
88

50. 468
458

Look at the graph.

51. How many tigers are in the zoo?

52. How many more elephants than monkeys are there?

53. How many animals are there altogether?

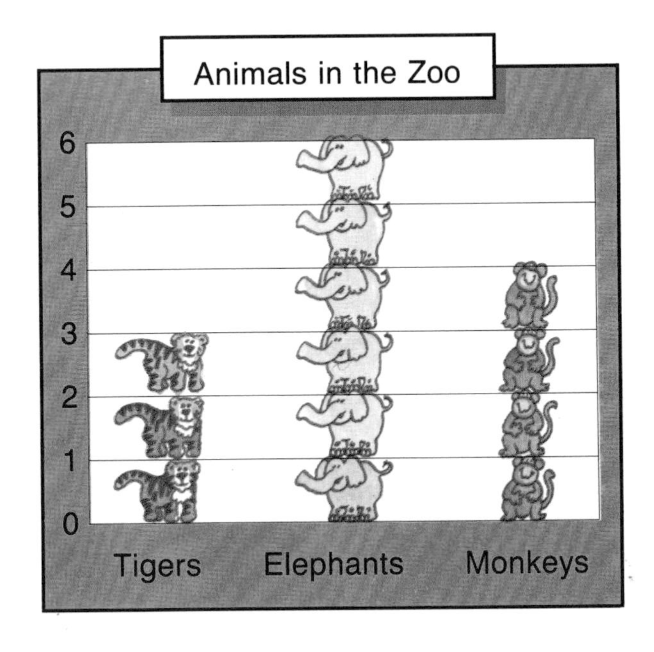

THE PUZZLE

Annie and Ernie
McGilligan Spock
Pedalled their tricycles
Round the block.

They pedalled and pedalled
And pedalled in pairs,
Till they came to a house
That was just like theirs.

In the same front yard
Stood the same small tree;
On the same brown table
The same pot of tea;

And the very same smells!
And the very same noise!
And the very same beds
With the very same toys!

They stood and they stared
And they stared and they stood;
The thing was too weird
To be understood:

How was it possible?
Think of the shock
Of Annie and Ernie
McGilligan Spock!

—Dennis Lee

Subtracting without Trading

Nancy had 34 stamps.

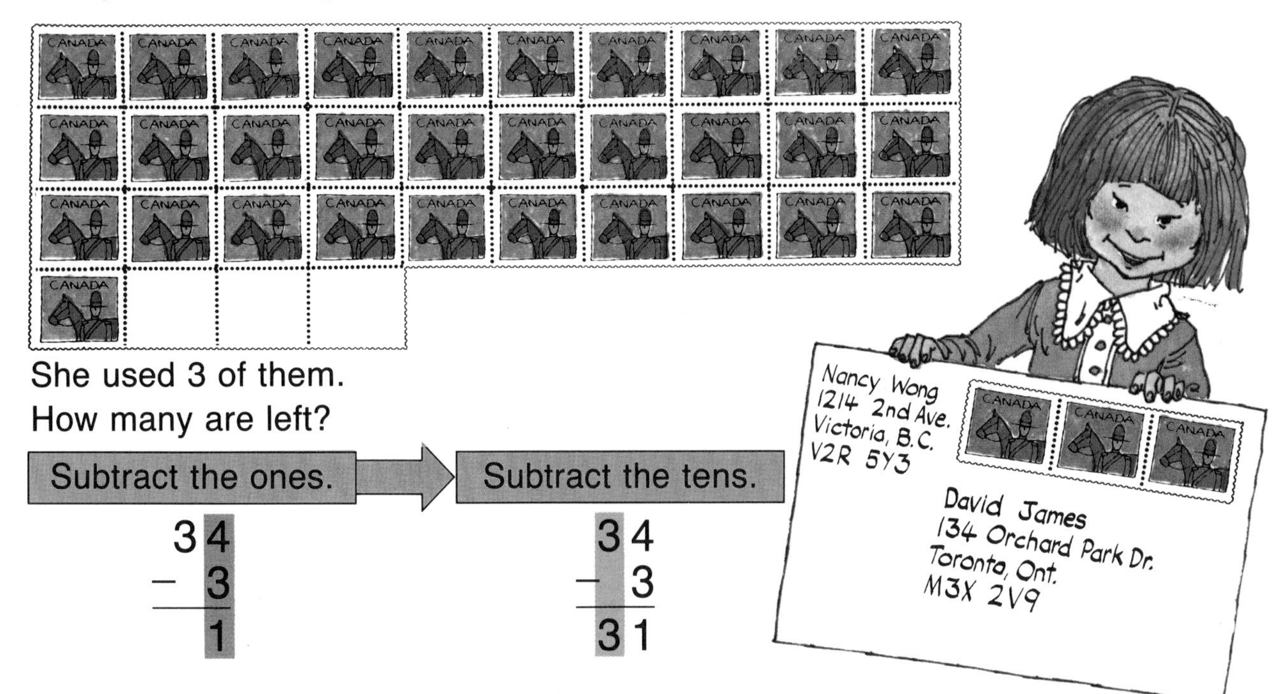

She used 3 of them.
How many are left?

Subtract the ones. → **Subtract the tens.**

$$\begin{array}{r} 34 \\ -\ \ 3 \\ \hline 1 \end{array} \qquad \begin{array}{r} 34 \\ -\ \ 3 \\ \hline 31 \end{array}$$

There are 31 stamps left.

Subtract.

1. $26 - 5$	2. $37 - 3$	3. $46 - 2$	4. $57 - 1$	5. $68 - 6$	6. $63 - 1$
7. $79 - 8$	8. $87 - 5$	9. $69 - 3$	10. $54 - 3$	11. $63 - 2$	12. $78 - 6$
13. $95 - 4$	14. $96 - 3$	15. $84 - 0$	16. $67 - 5$	17. $68 - 3$	18. $49 - 6$
19. $56 - 4$	20. $87 - 4$	21. $17 - 3$	22. $97 - 2$	23. $19 - 5$	24. $18 - 4$

Subtracting without Trading

28 books were returned to the library.
16 of them were put back on the shelves.
How many books are still to be put back?

Subtract the ones.	Subtract the tens.
$\begin{array}{r} 28 \\ -16 \\ \hline 2 \end{array}$	$\begin{array}{r} 28 \\ -16 \\ \hline 12 \end{array}$

There are 12 books still to be put back.

Subtract.

1. $\begin{array}{r} 86 \\ -34 \\ \hline \end{array}$	2. $\begin{array}{r} 93 \\ -62 \\ \hline \end{array}$	3. $\begin{array}{r} 57 \\ -35 \\ \hline \end{array}$	4. $\begin{array}{r} 73 \\ -41 \\ \hline \end{array}$	5. $\begin{array}{r} 68 \\ -40 \\ \hline \end{array}$	6. $\begin{array}{r} 90 \\ -20 \\ \hline \end{array}$
7. $\begin{array}{r} 87 \\ -56 \\ \hline \end{array}$	8. $\begin{array}{r} 73 \\ -20 \\ \hline \end{array}$	9. $\begin{array}{r} 58 \\ -28 \\ \hline \end{array}$	10. $\begin{array}{r} 67 \\ -15 \\ \hline \end{array}$	11. $\begin{array}{r} 49 \\ -13 \\ \hline \end{array}$	12. $\begin{array}{r} 69 \\ -7 \\ \hline \end{array}$
13. $\begin{array}{r} 96 \\ -5 \\ \hline \end{array}$	14. $\begin{array}{r} 88 \\ -77 \\ \hline \end{array}$	15. $\begin{array}{r} 64 \\ -14 \\ \hline \end{array}$	16. $\begin{array}{r} 58 \\ -13 \\ \hline \end{array}$	17. $\begin{array}{r} 76 \\ -33 \\ \hline \end{array}$	18. $\begin{array}{r} 84 \\ -31 \\ \hline \end{array}$
19. $\begin{array}{r} 65 \\ -22 \\ \hline \end{array}$	20. $\begin{array}{r} 86 \\ -44 \\ \hline \end{array}$	21. $\begin{array}{r} 69 \\ -33 \\ \hline \end{array}$	22. $\begin{array}{r} 51 \\ -30 \\ \hline \end{array}$	23. $\begin{array}{r} 76 \\ -36 \\ \hline \end{array}$	24. $\begin{array}{r} 83 \\ -23 \\ \hline \end{array}$

Trading a Ten for 10 Ones

Justin must pay $4 to Lana.

Trading a Ten for 10 Ones

We have:		We trade a ten for 10 ones.		We write:	
2 tens	3 ones	1 ten	13 ones	Tens	Ones
				1 ~~2~~	13 ~~3~~

Trade a ten for 10 ones.

1.

Tens	Ones
7	4

1. Tens	Ones
6	14
~~7~~	~~4~~

2.

Tens	Ones
5	3

3.

Tens	Ones
6	2

4.

Tens	Ones
8	6

5.

Tens	Ones
5	7

6.

Tens	Ones
1	9

7.

Tens	Ones
8	7

8.

Tens	Ones
3	0

9.

Tens	Ones
8	5

10.

Tens	Ones
7	7

11.

Tens	Ones
9	8

12.

Tens	Ones
7	0

13.

Tens	Ones
1	0

14.

Tens	Ones
9	6

Complete.

15. 5 tens 7 ones = 4 tens ▒ ones
16. 8 tens 6 ones = 7 tens ▒ ones
17. 9 tens 0 ones = 8 tens ▒ ones
18. 6 tens 7 ones = 5 tens ▒ ones
19. 4 tens 3 ones = ▒ tens 13 ones
20. 6 tens 8 ones = ▒ tens 18 ones

Subtracting with Trading

Justin had $23.

He traded a ten for 10 ones.

He paid $4 to Lana.

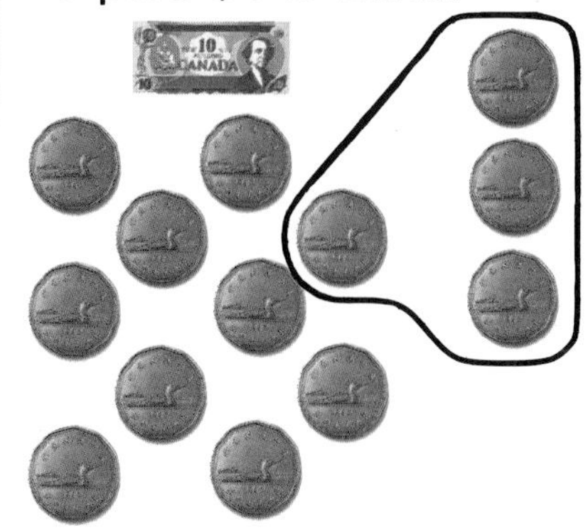

How much did he have left?

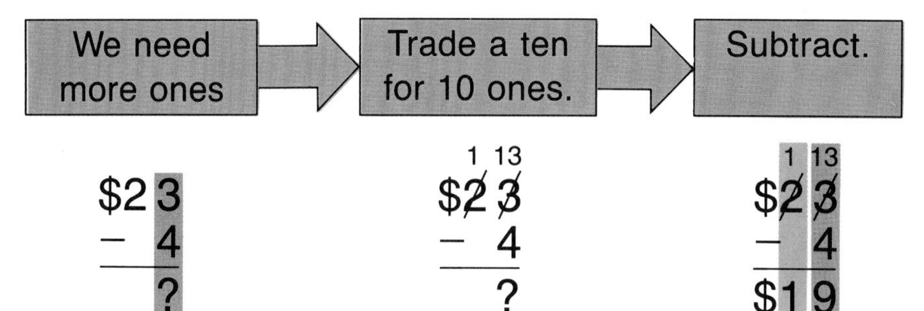

Justin had $19 left.

Subtract. Do only the questions that need trading.

1. 26 − 7	2. 45 − 6	3. 32 − 5	4. 21 − 8	5. 26 − 3	6. 70 − 9
7. 76 − 4	8. 84 − 5	9. 55 − 7	10. 49 − 4	11. 60 − 8	12. 47 − 4
13. 30 − 6	14. 34 − 7	15. 89 − 8	16. 16 − 9	17. 33 − 9	18. 17 − 8

There are 42 cards for the game.
17 cards have been used.

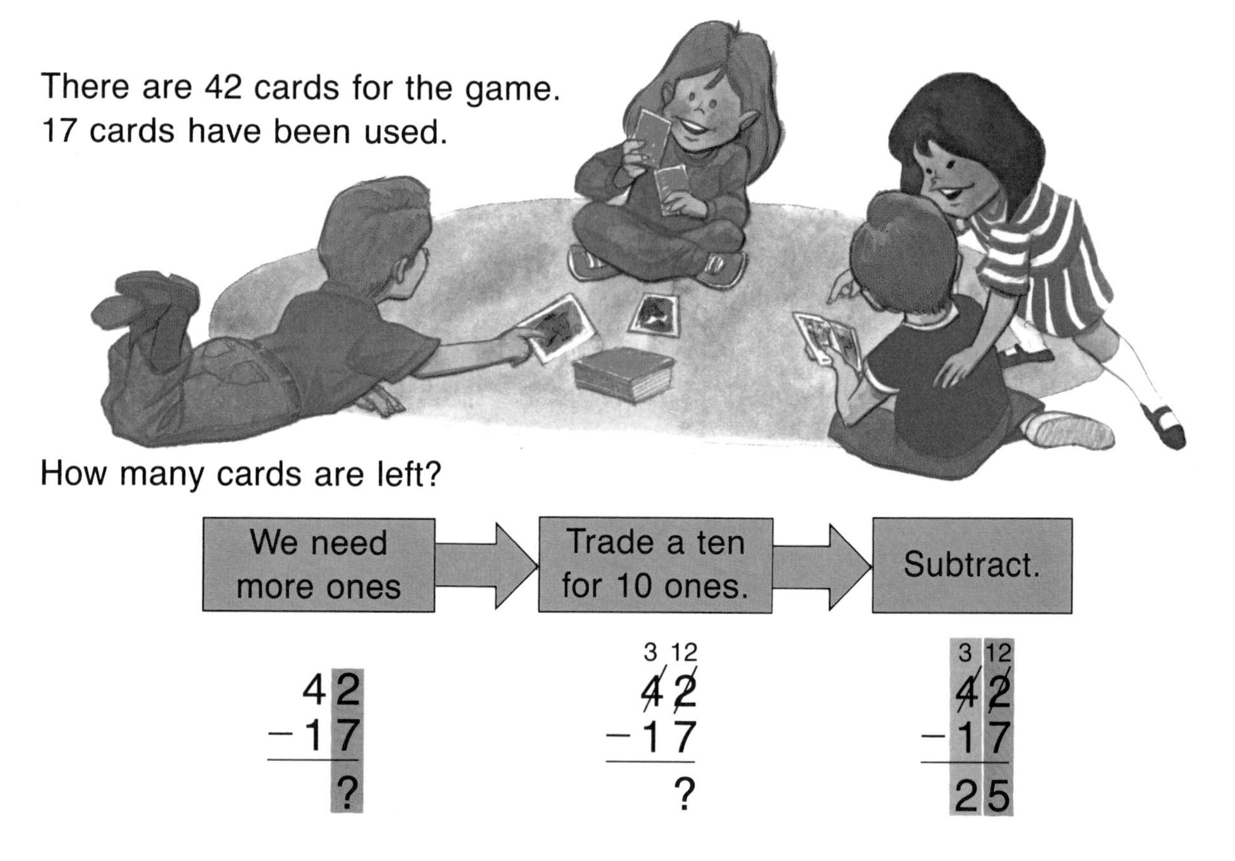

How many cards are left?

We need more ones → Trade a ten for 10 ones. → Subtract.

$$\begin{array}{r} 42 \\ -17 \\ \hline ? \end{array} \qquad \begin{array}{r} \scriptstyle 3\ 12 \\ \not{4}\not{2} \\ -17 \\ \hline ? \end{array} \qquad \begin{array}{r} \scriptstyle 3\ 12 \\ \not{4}\not{2} \\ -17 \\ \hline 25 \end{array}$$

There are 25 cards left.

Subtract. Trade only when you need to.

1. $\begin{array}{r} 36 \\ -27 \\ \hline \end{array}$	2. $\begin{array}{r} 42 \\ -38 \\ \hline \end{array}$	3. $\begin{array}{r} 56 \\ -28 \\ \hline \end{array}$	4. $\begin{array}{r} 69 \\ -32 \\ \hline \end{array}$	5. $\begin{array}{r} 70 \\ -14 \\ \hline \end{array}$	6. $\begin{array}{r} 86 \\ -27 \\ \hline \end{array}$
7. $\begin{array}{r} 45 \\ -31 \\ \hline \end{array}$	8. $\begin{array}{r} 55 \\ -38 \\ \hline \end{array}$	9. $\begin{array}{r} 80 \\ -23 \\ \hline \end{array}$	10. $\begin{array}{r} 57 \\ -42 \\ \hline \end{array}$	11. $\begin{array}{r} 69 \\ -39 \\ \hline \end{array}$	12. $\begin{array}{r} 81 \\ -75 \\ \hline \end{array}$
13. $\begin{array}{r} 67 \\ -9 \\ \hline \end{array}$	14. $\begin{array}{r} 50 \\ -23 \\ \hline \end{array}$	15. $\begin{array}{r} 37 \\ -19 \\ \hline \end{array}$	16. $\begin{array}{r} 86 \\ -79 \\ \hline \end{array}$	17. $\begin{array}{r} 93 \\ -46 \\ \hline \end{array}$	18. $\begin{array}{r} 87 \\ -9 \\ \hline \end{array}$
19. $\begin{array}{r} 54 \\ -45 \\ \hline \end{array}$	20. $\begin{array}{r} 99 \\ -32 \\ \hline \end{array}$	21. $\begin{array}{r} 86 \\ -18 \\ \hline \end{array}$	22. $\begin{array}{r} 47 \\ -31 \\ \hline \end{array}$	23. $\begin{array}{r} 78 \\ -69 \\ \hline \end{array}$	24. $\begin{array}{r} 57 \\ -29 \\ \hline \end{array}$

JUST FOR FUN

RIDDLE: How do you keep a rhinoceros from charging?

A. $66 - 37$	B. $53 - 28$	C. $72 - 19$
D. $90 - 36$	E. $82 - 9$	F. $73 - 17$
G. $52 - 41$	H. $80 - 53$	I. $67 - 8$
J. $35 - 15$	K. $60 - 29$	L. $87 - 16$
M. $60 - 12$	N. $32 - 28$	O. $31 - 24$
P. $92 - 16$	Q. $69 - 58$	R. $82 - 73$
S. $56 - 28$	T. $37 - 19$	U. $86 - 18$
V. $42 - 27$	W. $54 - 16$	X. $65 - 38$
Y. $78 - 39$	Z. $85 - 68$	

To answer the riddle, write the letters to the questions with these answers.

__ __ __ __   __ __ __ __   __ __ __   __ __ __ __ __ __   __ __ __ __

18 29 31 73   29 38 29 39   27 59 28   53 9 73 54 59 18   53 29 9 54

Story Problems

There are 5 types of rhinoceros in the world.
A full-grown Indian rhino is over 2 m (two metres) high.
It is about 5 m (five metres) long.
It lives to be about 47 years old.

1. The white rhino lives about 29 years. How much longer does the Indian rhino live?

2. A zoo had 27 rhinos. 4 more were born. How many were there in all?

3. One rhino is 51 years old. Its mate is 39 years old. How much younger is the mate?

4. There are 33 rhinos in a zoo. 15 of them are females. How many are male rhinos?

5. Look at the table. How many more white rhinos are there than Indian rhinos in Hornby Zoo?

Rhinos	Number in Hornby Zoo
Indian	26
White	41

PROBLEM SOLVING

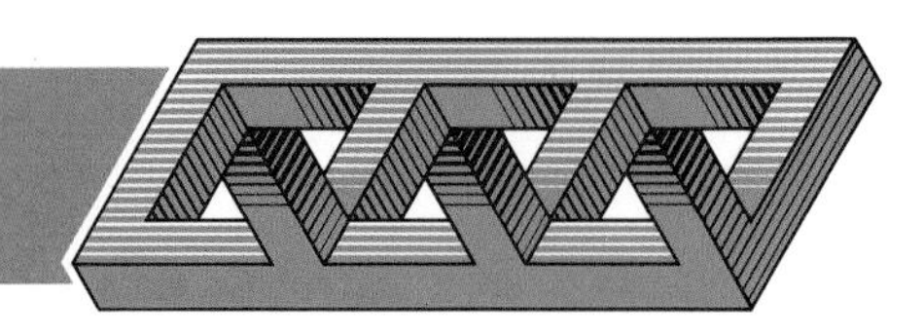

Obtaining Information from a Bar Graph

The life span of an animal is the number of years it lives. This bar graph shows the life span of some animals.

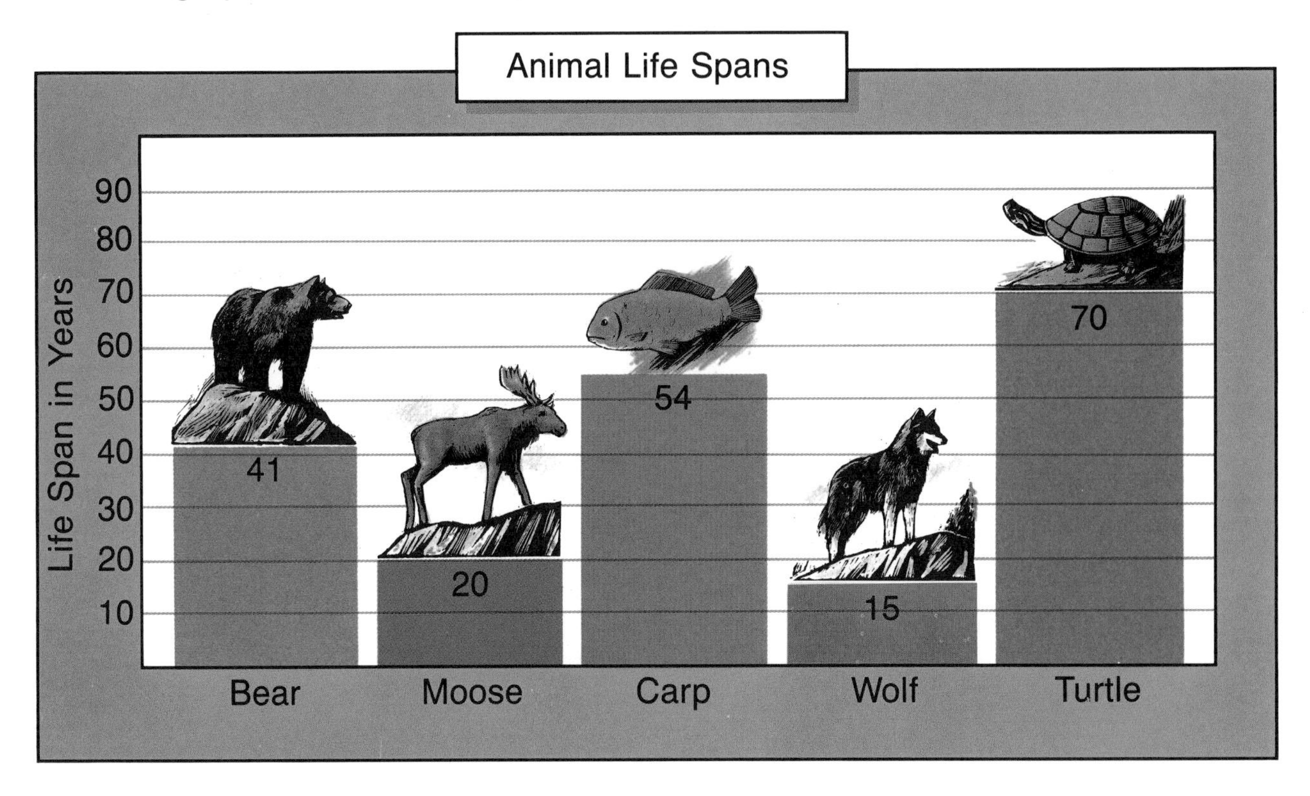

1. Which of these animals has the shortest life span?
2. What is the life span of the carp?
3. Which animal has a life span of 70 years?
4. Which two animals have life spans that differ by 13 years?
5. How much longer does a turtle live than a carp ?
6. How much shorter is the life of a bear than the life of a turtle?
7. Which animal lives 50 years longer than a moose?
8. Which animal lives 26 years longer than a wolf?

ESTIMATING

Estimating Number

Look twice.
Cover your eyes.
Estimate.

Here is a full set
of 100 animal cards.

Ruth

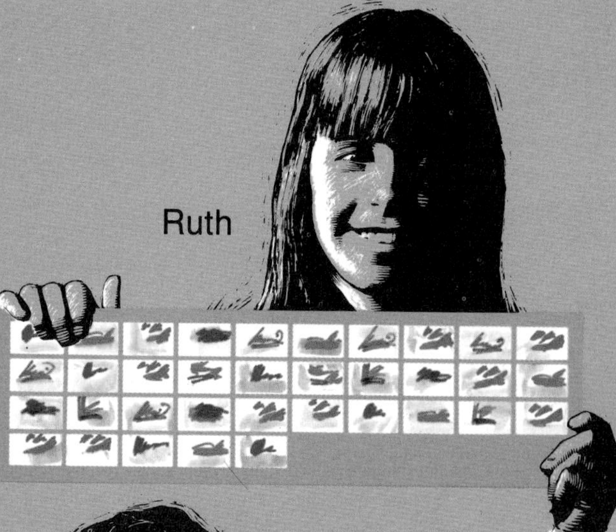

Estimate: How many cards does each person have?

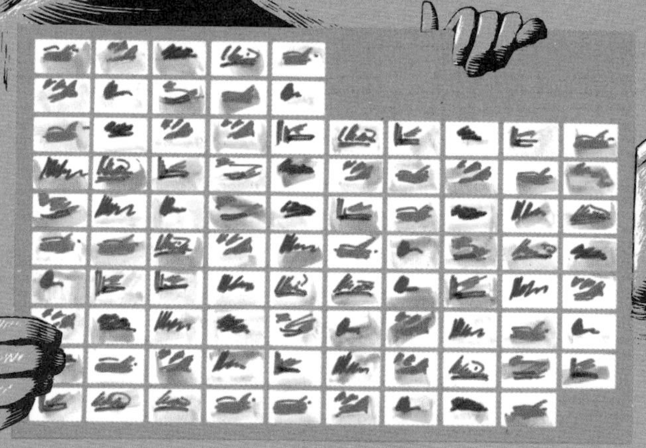

Liz

Dan

Check your estimates by calculating.

EXTENSIONS

Checking Subtraction by Addition

We can write a subtraction sentence to match the picture.

42	checkers in all
− 19	red checkers
23	black checkers

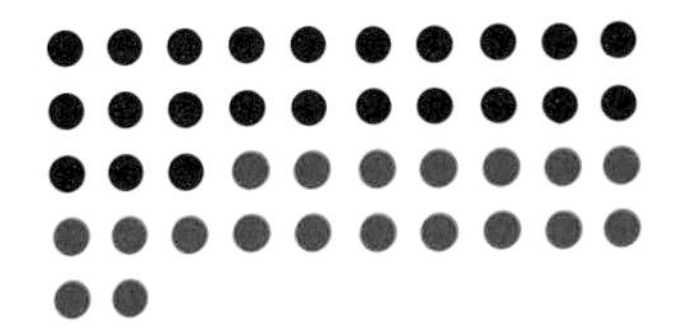

We can write an addition sentence to match the picture.

23	black checkers
+ 19	red checkers
42	checkers in all

Write a subtraction and an addition sentence to match each picture.

1.

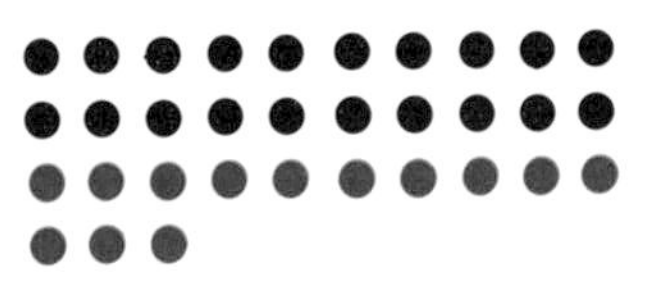

2.

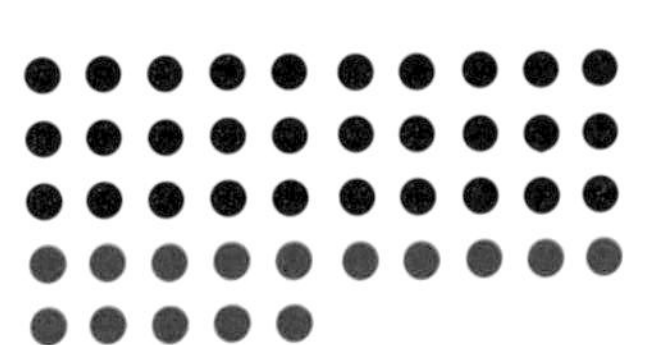

3.

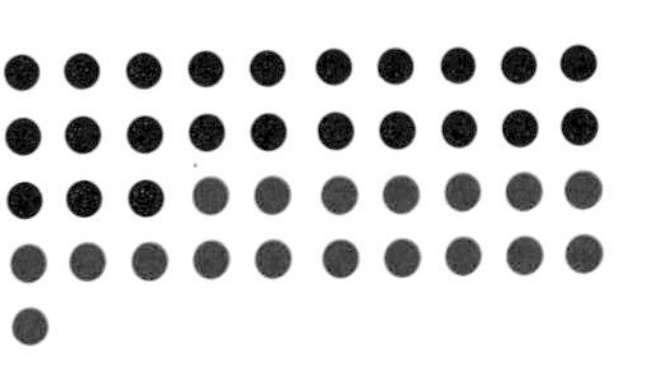

4.

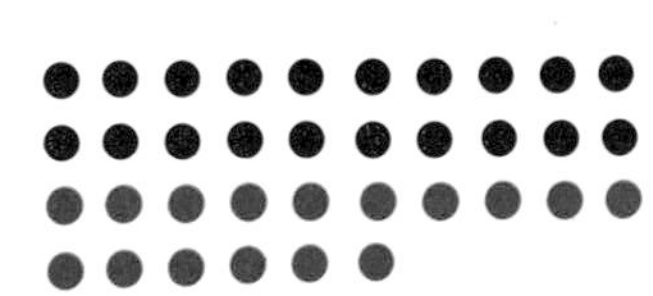

5.

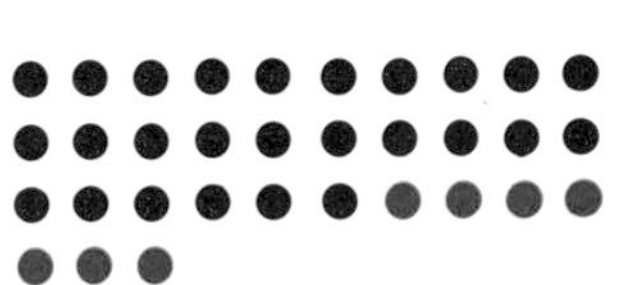

6.

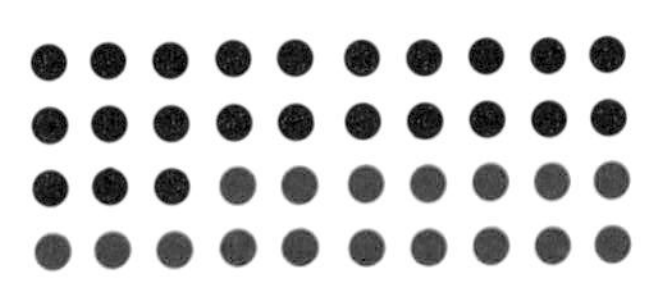

7.

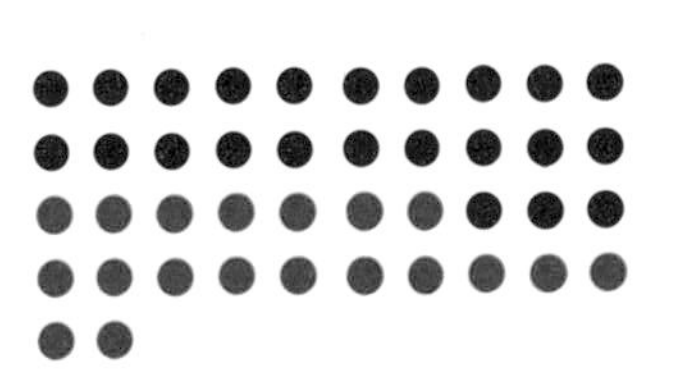

8.

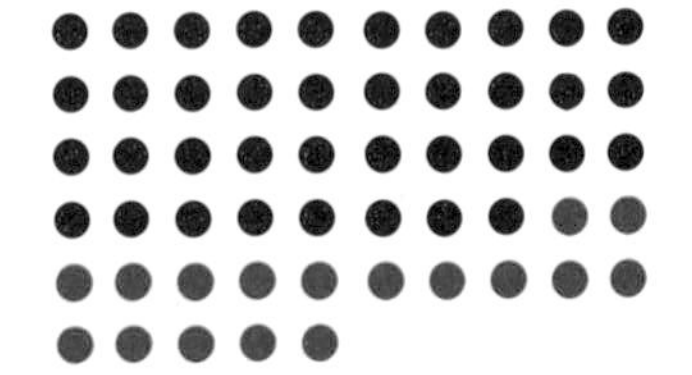

We can use addition to check subtraction.

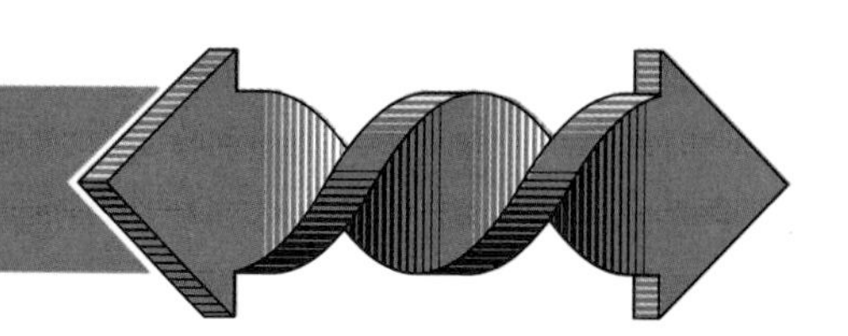

Subtract.

1. $\begin{array}{r} 82 \\ -16 \\ \hline \end{array}$
2. $\begin{array}{r} 47 \\ -28 \\ \hline \end{array}$
3. $\begin{array}{r} 90 \\ -17 \\ \hline \end{array}$
4. $\begin{array}{r} 68 \\ -39 \\ \hline \end{array}$

Write an addition sentence to match each subtraction.
Then add to check your subtraction.

Add to check these subtractions. Correct any wrong answers in your notebook.

5. $\begin{array}{r} 72 \\ -18 \\ \hline 54 \end{array}$
6. $\begin{array}{r} 76 \\ -15 \\ \hline 61 \end{array}$
7. $\begin{array}{r} 47 \\ -28 \\ \hline 29 \end{array}$
8. $\begin{array}{r} 52 \\ -33 \\ \hline 19 \end{array}$
9. $\begin{array}{r} 50 \\ -24 \\ \hline 36 \end{array}$
10. $\begin{array}{r} 71 \\ -28 \\ \hline 53 \end{array}$
11. $\begin{array}{r} 46 \\ -13 \\ \hline 23 \end{array}$
12. $\begin{array}{r} 27 \\ -19 \\ \hline 18 \end{array}$
13. $\begin{array}{r} 81 \\ -64 \\ \hline 17 \end{array}$
14. $\begin{array}{r} 93 \\ -57 \\ \hline 31 \end{array}$
15. $\begin{array}{r} 64 \\ -28 \\ \hline 36 \end{array}$
16. $\begin{array}{r} 71 \\ -25 \\ \hline 36 \end{array}$
17. $\begin{array}{r} 80 \\ -17 \\ \hline 63 \end{array}$
18. $\begin{array}{r} 92 \\ -77 \\ \hline 15 \end{array}$
19. $\begin{array}{r} 66 \\ -39 \\ \hline 37 \end{array}$
20. $\begin{array}{r} 82 \\ -9 \\ \hline 63 \end{array}$

CHECK-UP

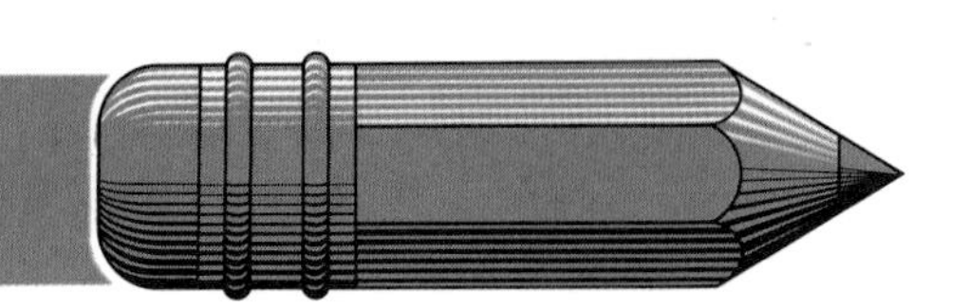

Trade a ten for 10 ones. Complete.

1. 4 tens 3 ones = 3 tens ▒ ones
2. 8 tens 2 ones = ▒ tens 12 ones

Subtract.

3.	4.	5.	6.	7.	8.
$\begin{array}{r} 37 \\ -\ 3 \\ \hline \end{array}$	$\begin{array}{r} 57 \\ -19 \\ \hline \end{array}$	$\begin{array}{r} 56 \\ -21 \\ \hline \end{array}$	$\begin{array}{r} 25 \\ -\ 7 \\ \hline \end{array}$	$\begin{array}{r} 64 \\ -23 \\ \hline \end{array}$	$\begin{array}{r} 84 \\ -27 \\ \hline \end{array}$
9. $\begin{array}{r} 59 \\ -\ 8 \\ \hline \end{array}$	**10.** $\begin{array}{r} 43 \\ -27 \\ \hline \end{array}$	**11.** $\begin{array}{r} 38 \\ -25 \\ \hline \end{array}$	**12.** $\begin{array}{r} 33 \\ -\ 8 \\ \hline \end{array}$	**13.** $\begin{array}{r} 85 \\ -42 \\ \hline \end{array}$	**14.** $\begin{array}{r} 41 \\ -28 \\ \hline \end{array}$
15. $\begin{array}{r} 36 \\ -\ 4 \\ \hline \end{array}$	**16.** $\begin{array}{r} 50 \\ -13 \\ \hline \end{array}$	**17.** $\begin{array}{r} 42 \\ -31 \\ \hline \end{array}$	**18.** $\begin{array}{r} 44 \\ -\ 5 \\ \hline \end{array}$	**19.** $\begin{array}{r} 96 \\ -15 \\ \hline \end{array}$	**20.** $\begin{array}{r} 60 \\ -28 \\ \hline \end{array}$

21. Last month the Cycle Shop sold 69 bicycles.
    The month before, they sold 58.
    How many more did they sell last month?

22. 24 children went on a field trip.
    19 brought their own lunch.
    How many did not bring lunch?

23. There are 73 seats on the bus.
    17 are empty.
    How many children are on the bus?

24. There are 31 days in May.
    Denise went to school on 23 days.
    How many days did she not go to school?

Two Boxes

Two boxes met upon the road.
Said one unto the other,
"If you're a box,
And I'm a box,
Then you must be my brother.
Our sides are thin,
We're cavin' in,
And we must get no thinner."
And so two boxes, hand in hand,
Went home to have their dinner.

—*Shel Silverstein*

from *Where the Sidewalk Ends*

Solids

How are these solids alike? How are they different?

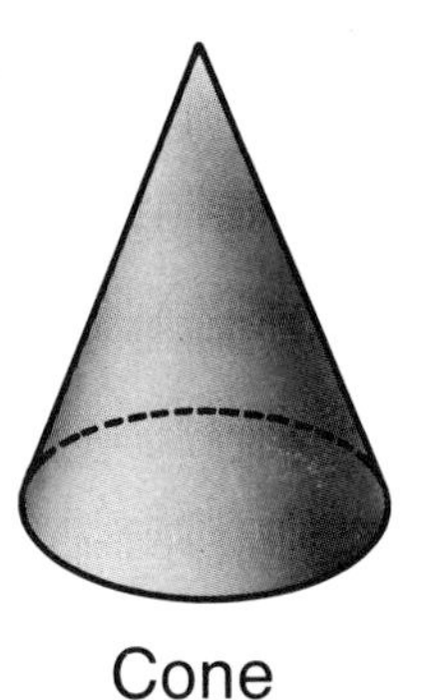

Cone

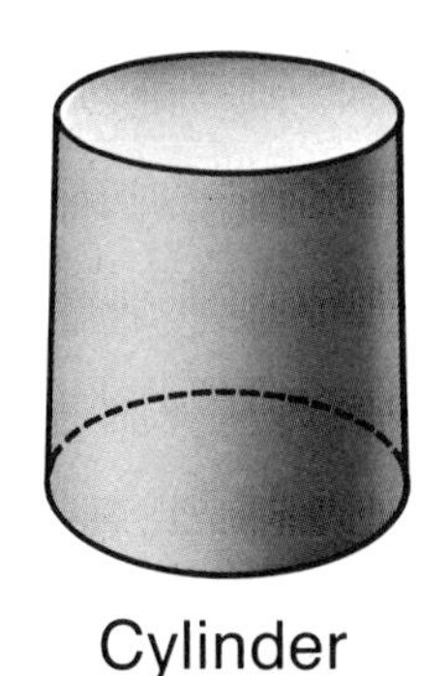

Cylinder

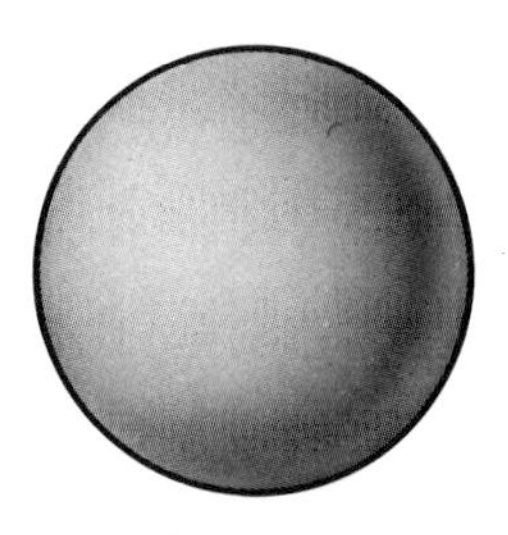

Sphere

1. Look at each picture.
   Find something shaped like
   one of the solids.
   Write the name of the solid.

2. Think of a different thing that is
   shaped like each solid.
   Name or draw a picture of each.

How are these solids alike? How are they different?

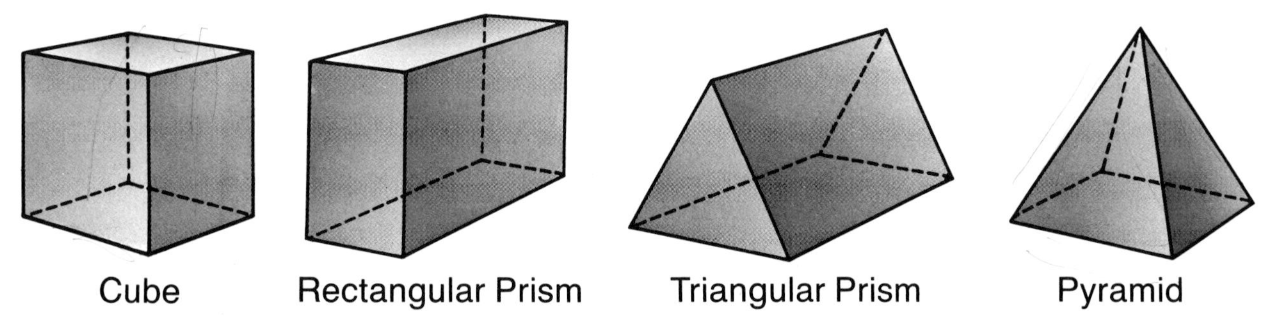

1. Look at each picture.
   Find something shaped like one of the solids.
   Write the name of the solid.

2. This is one face of the cube.

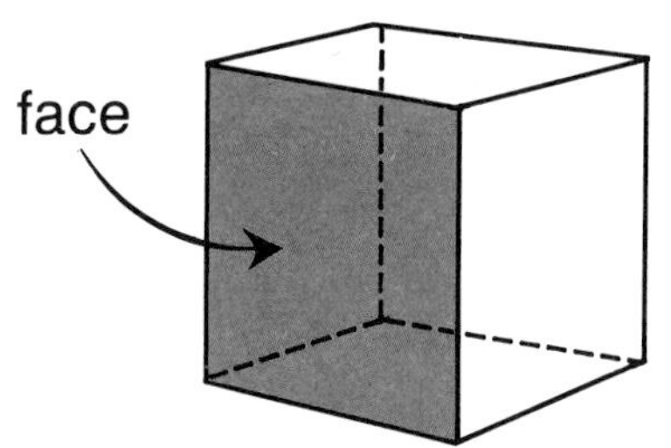

How many faces does a cube have?

3. Copy and complete the table.

Solid	Number of Faces
Cube	
Rectangular Prism	
Triangular Prism	
Pyramid	

Counting Solids

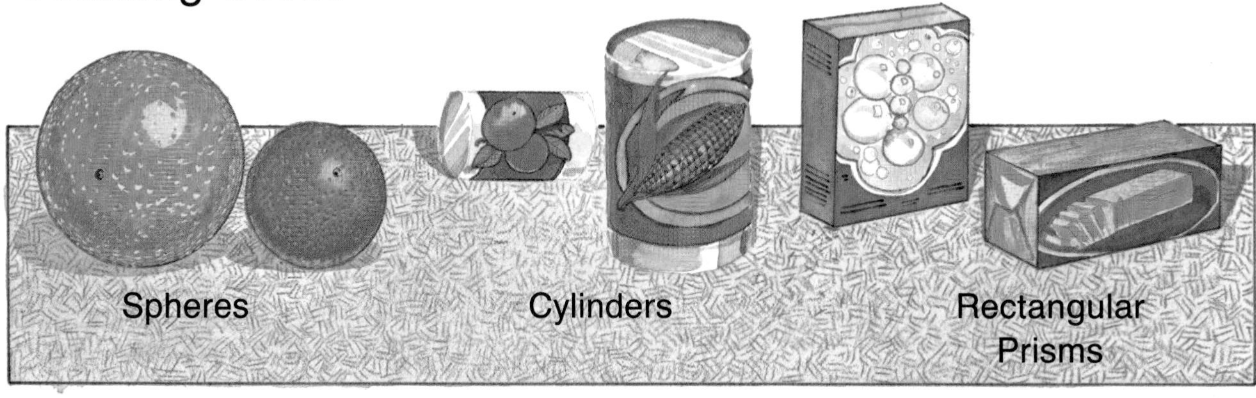

How many of each solid are in the picture?
Record your answer in a table like this.

	How many?
Spheres	▒
Cylinders	▒
Rectangular Prisms	▒

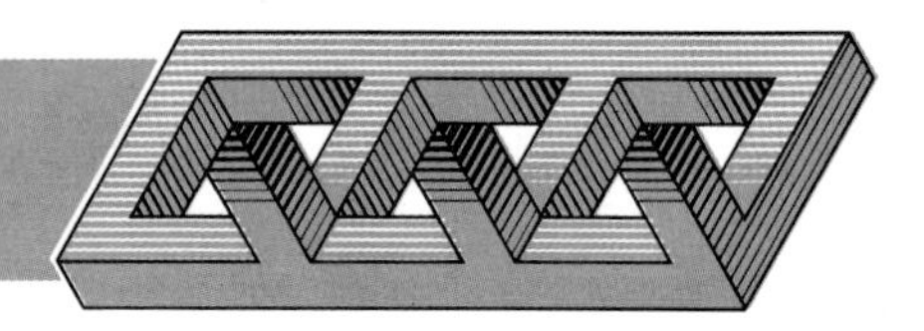

PROBLEM SOLVING

Interpreting Information from a Pictograph

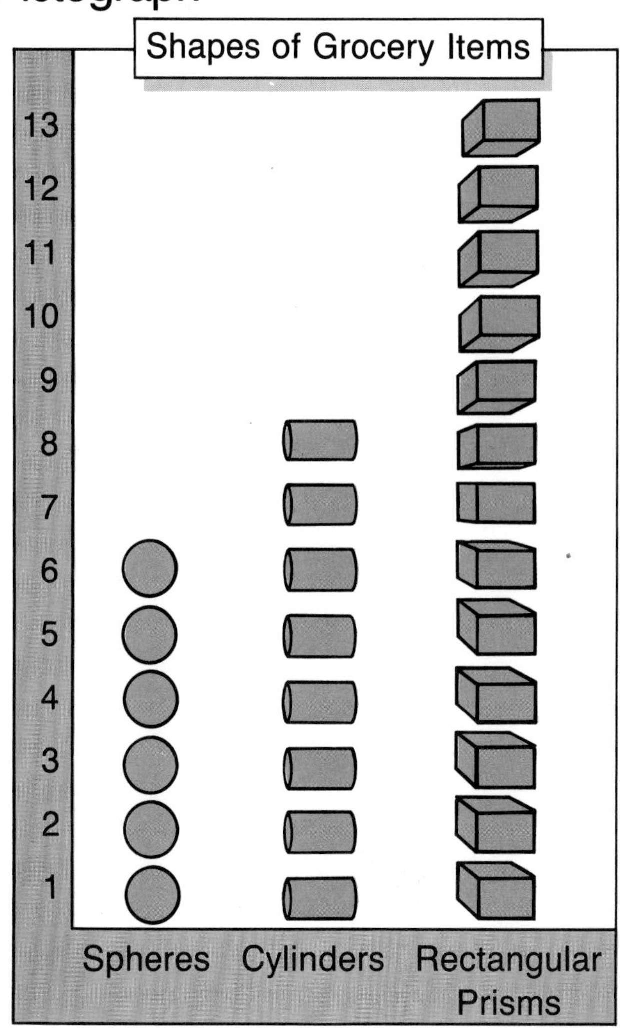

This pictograph shows how many items of each shape are in the basket.

1. Use the pictograph to complete this table.

	How many?
Spheres	▒
Cylinders	▒
Rectangular Prisms	▒

2. How many more prisms are there than spheres?
3. How many more prisms are there than cylinders?
4. How many items are there in all?
5. How many items are not prisms?

JUST FOR FUN

Make models like these.

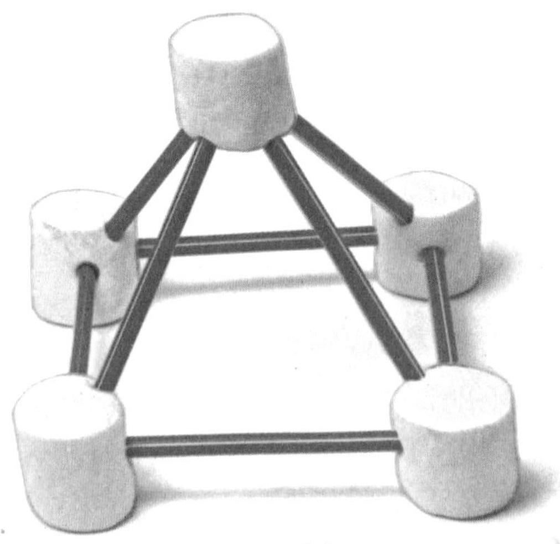

Pyramid

Tetrahedron

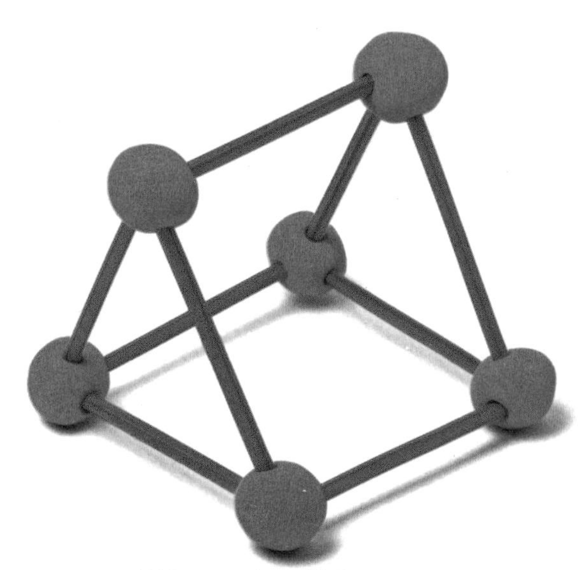

Triangular Prism

Cube

Can you make a model using
9 marshmallows and 16 toothpicks?

Corners and Edges

Each ——— is an edge.

Each ▯ is a corner.

Count the number of edges and corners in each model. Write your answers in a table.

	Number of Edges	Number of Corners
Tetrahedron		
Cube		
Pyramid		
Triangular Prism		

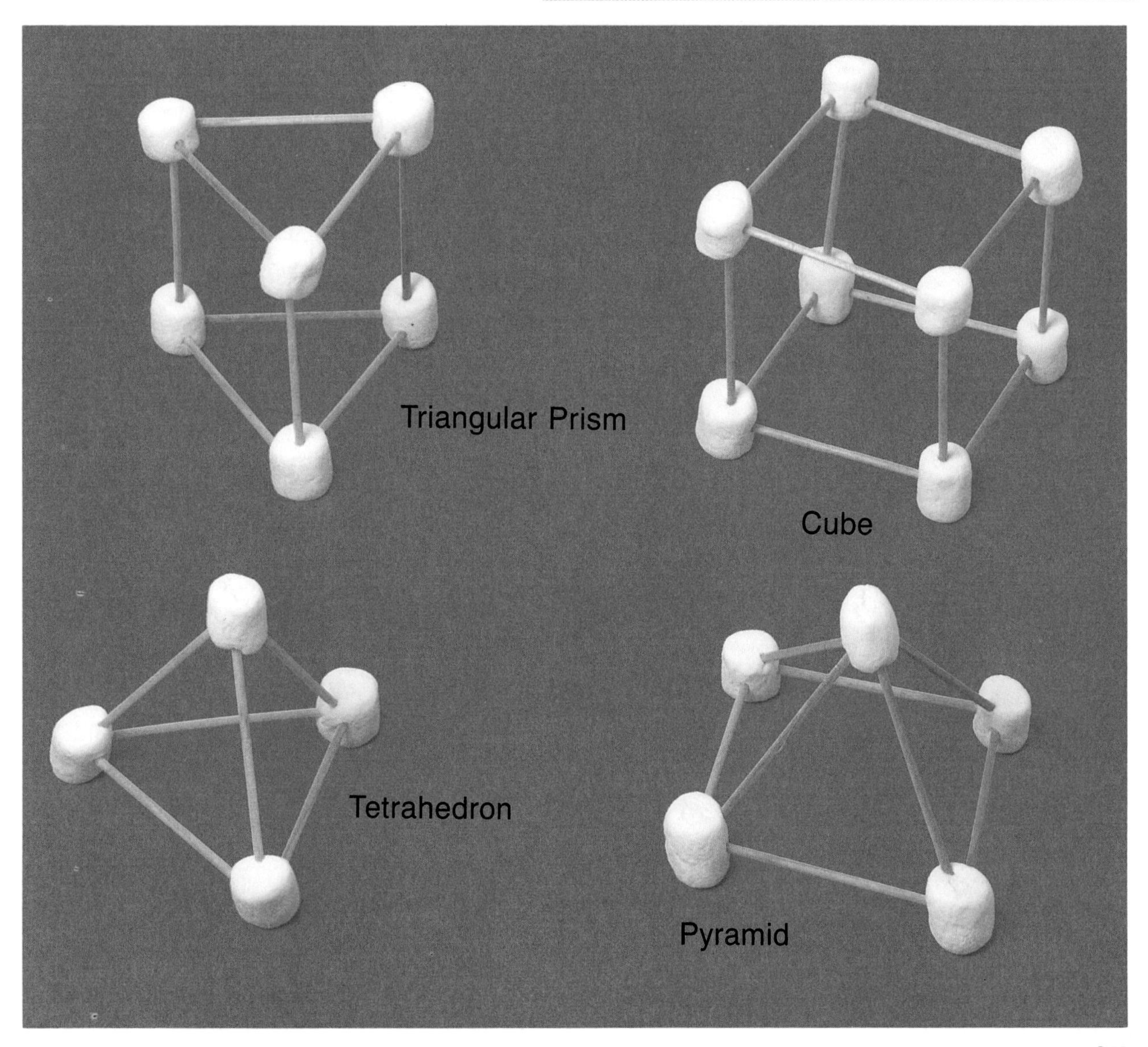

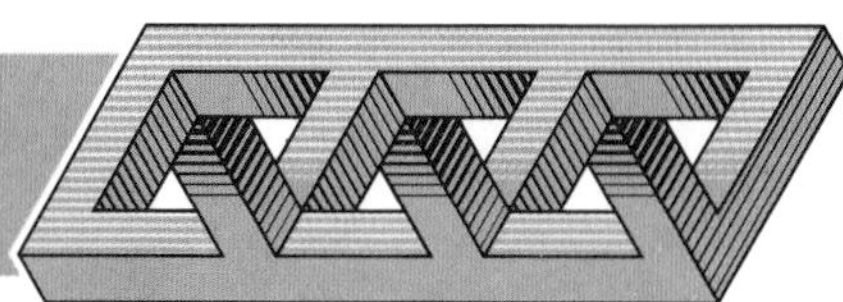

Making a Model

Emily traced this pattern.
She cut along the solid lines.
Then she folded along
the dotted lines.
She taped or glued the edges.

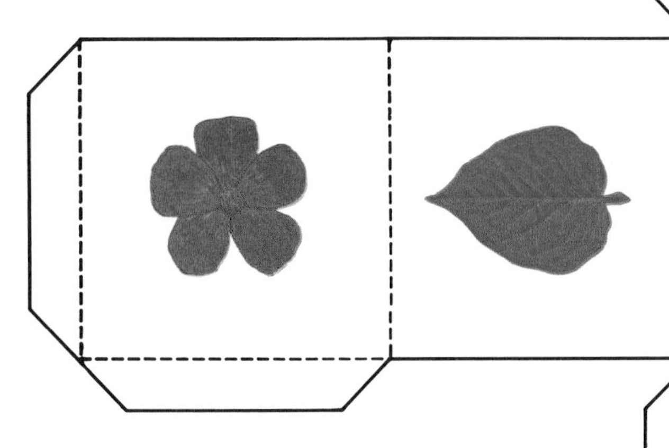

This made a cube.

1. This is one view of the cube.
   Which picture is opposite the flower?

2. This is another view.
   Which picture is opposite the ball?

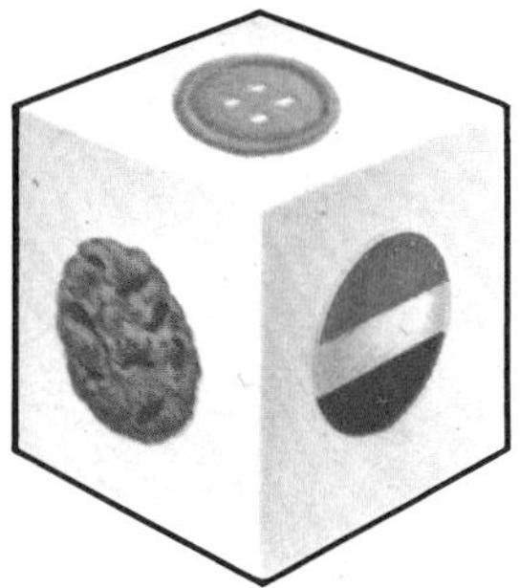

PROBLEM SOLVING

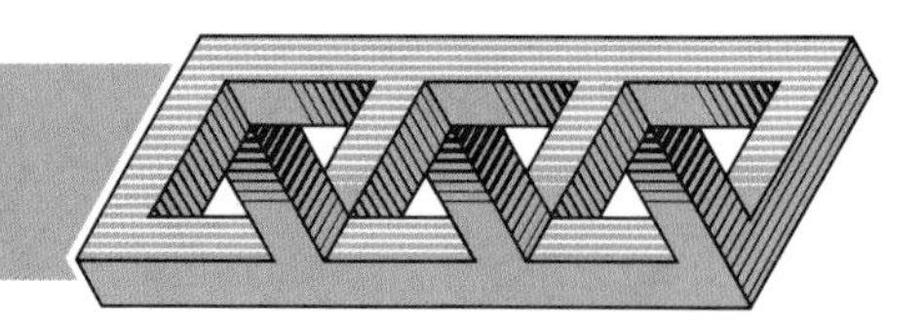

Interpreting Information

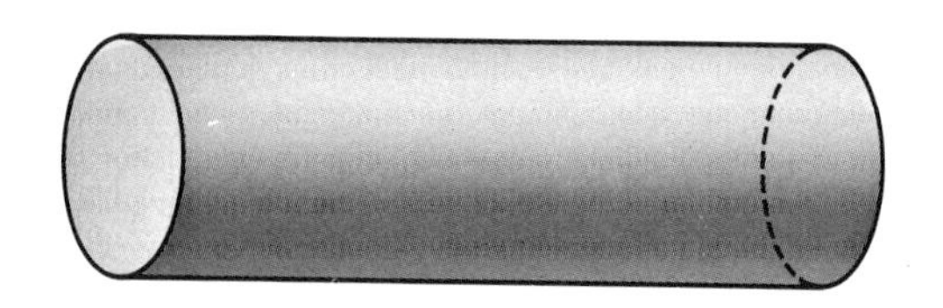

Cylinder

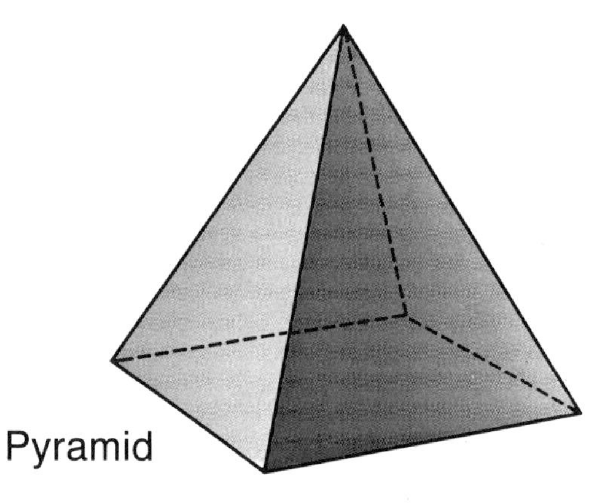

Pyramid

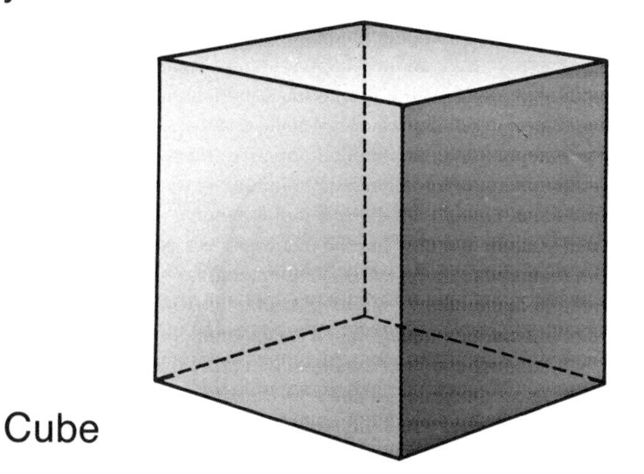

Cube

Sphere

Solve each riddle.

1. Place me on my side,
   And I can surely roll.
   Stand me on one end,
   I look just like a pole.
   Who am I?

2. I play baseball in the summer.
   I play soccer in the fall.
   And often when you see me,
   Your eyes are on the ball.
   Who am I?

3. If you peer amid my edges
   Five faces you will see.
   I can slide but I can't roll
   And you cannot stack on me!
   Who am I?

4. All six faces are flat
   And so I cannot roll.
   And if I'm in a stack
   My corners do not show.
   Who am I?

ESTIMATING

Estimating which Container Holds More

1. Make containers shaped like a cube and a cone. Follow these steps.
   - Copy figures A and B onto construction paper.
   - Cut along the solid lines.
   - Fold along the dotted lines.
   - Tape the tabs.

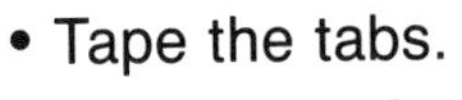

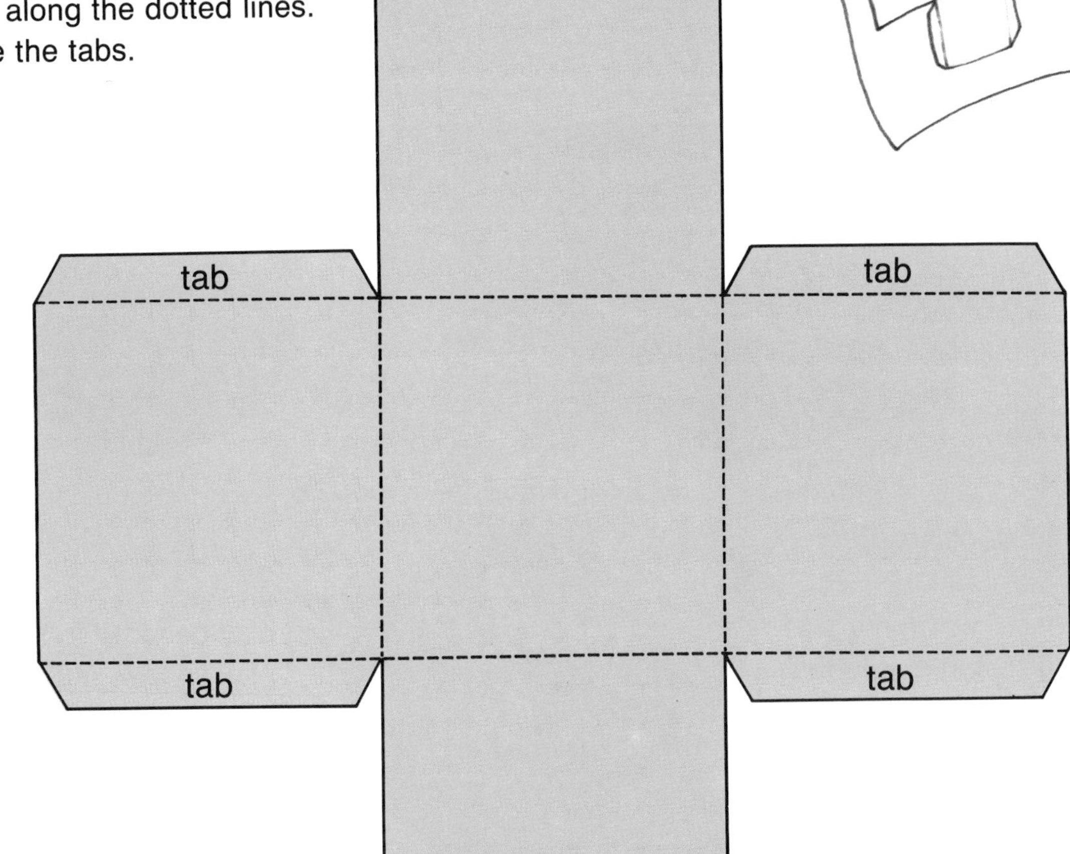

A.

2. Estimate: Which container holds more?

3. Check by filling each container.
   Use sand, rice, or dried beans.
   Hold in the sides of the cube
   so they won't bulge.

4. Compare.
   Which held more?

CHECK-UP

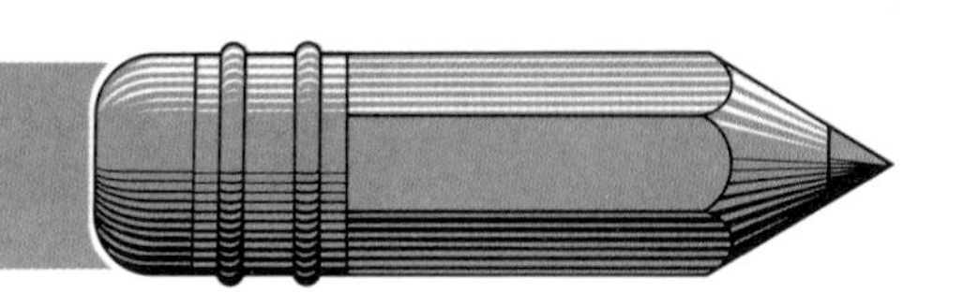

Is the solid a cube, sphere, rectangular prism, or cylinder?

1.

2.

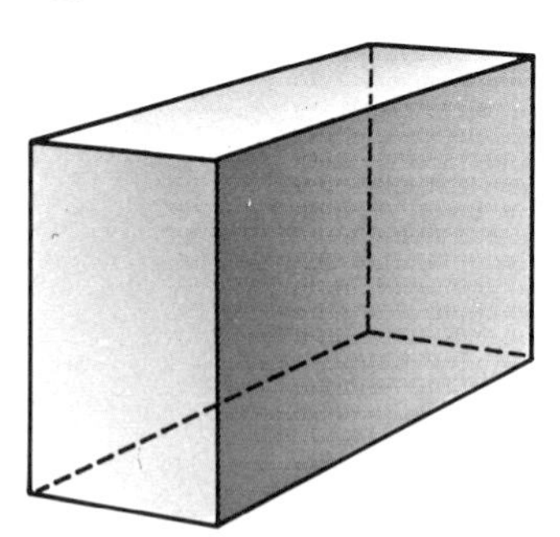

3.

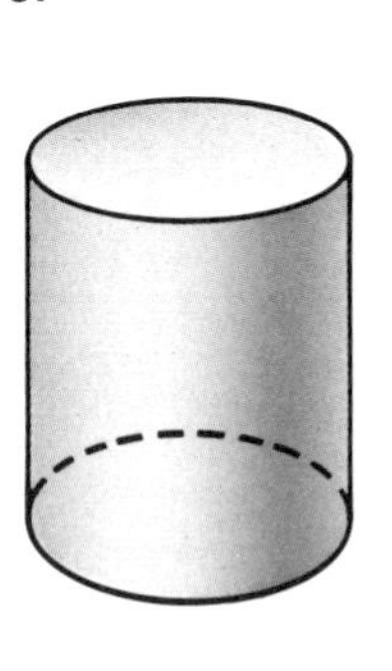

4.

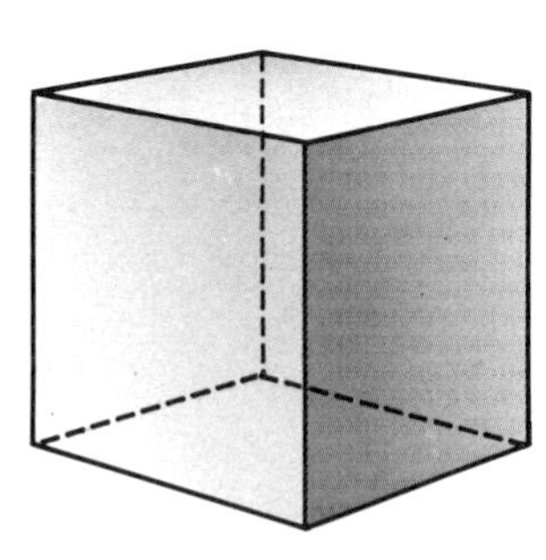

Which solid do you see in each picture?

5.

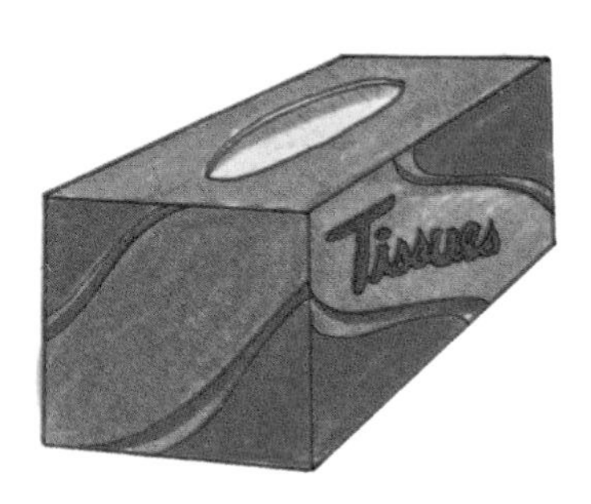

6.

7.

8.

9. How many faces does a pyramid have?

10. How many edges does a pyramid have?

11. How many corners does a pyramid have?

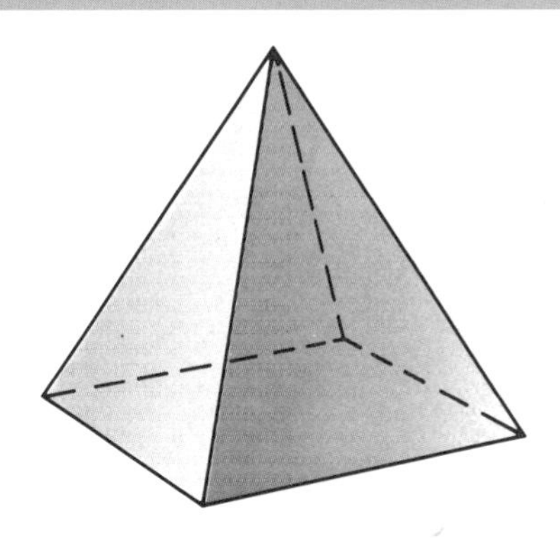

Me and My Giant

I have a friend who is a giant,
And he lives where the tall weeds grow.
He's high as a mountain and wide as a barn,
And I only come up to his toe, you know,
I only come up to his toe.

When the daylight grows dim, I talk with him
Way down in the marshy sands,
And his ear is too far away to hear,
But still he understands, he 'stands,
I know he understands.

—*Shel Silverstein*

from *Where the Sidewalk Ends*

Lengths in Centimetres

This line —— is one centimetre long.

We say: one centimetre
We write: 1 cm

This neon fish is 1 cm long.

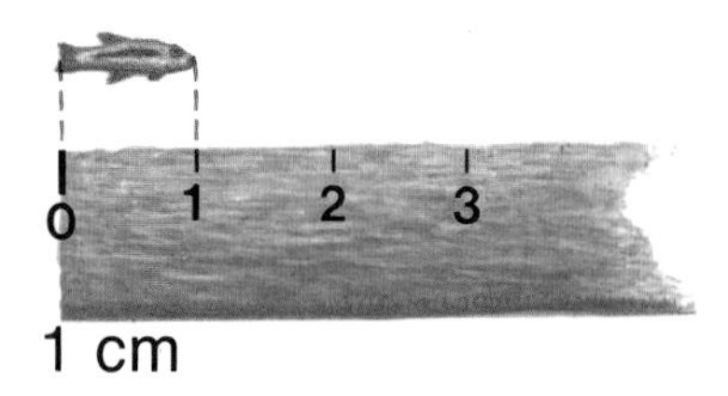

1 cm

This bumblebee fish is 3 cm long.

3 cm

Measure the length of each fish in the tank. Record your answers in a table like this one.

Fish	Length
Goldfish	cm
Moonfish	cm

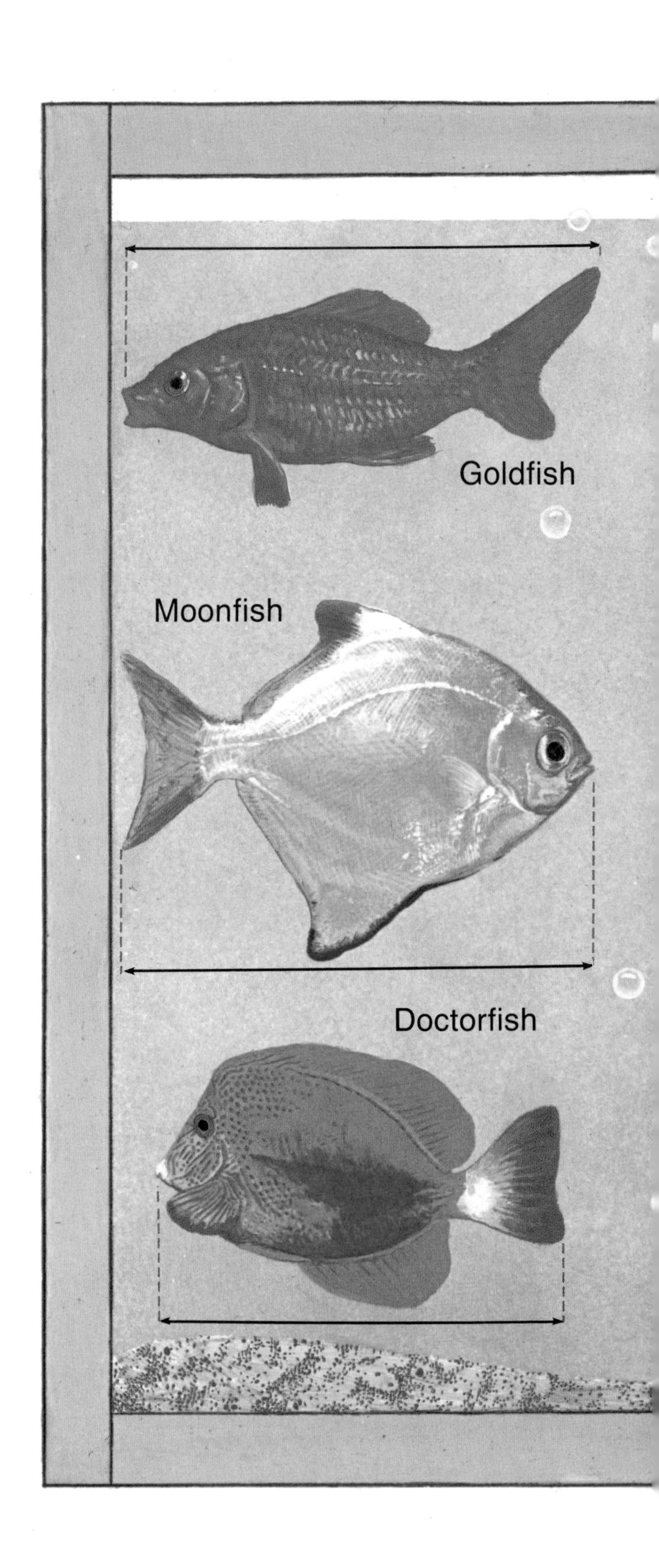

Tooth Carp
Longfin
Angelfish
Tigerfish
Platy
Scat

Heights in Centimetres

1. Read the height of each child in centimetres.
   Record in a table like this. →

	Height
Lynn	130 cm
Greg	▒ cm
Andy	▒ cm
Kim	▒ cm
Ted	▒ cm

2. Work with a partner.
   Measure your height in centimetres.
   Record it in your notebook.

3. Which children in the picture are not as tall as you?

PROBLEM SOLVING

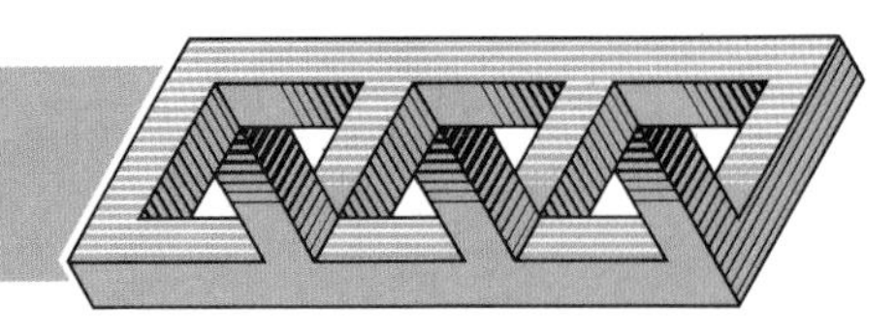

Selecting Relevant Information from a Picture

Answer these questions.
Use the pictures.

1. Which child is tallest?
2. Who is shortest?
3. How much taller is Kim than Andy?
4. Which children are tall enough to ride the rocket?
5. How many more centimetres must Ted grow to ride the rocket?
6. Kim rode the rocket 3 times. How much did it cost?
7. Make up a question about this picture.

Measuring to the Nearest Centimetre

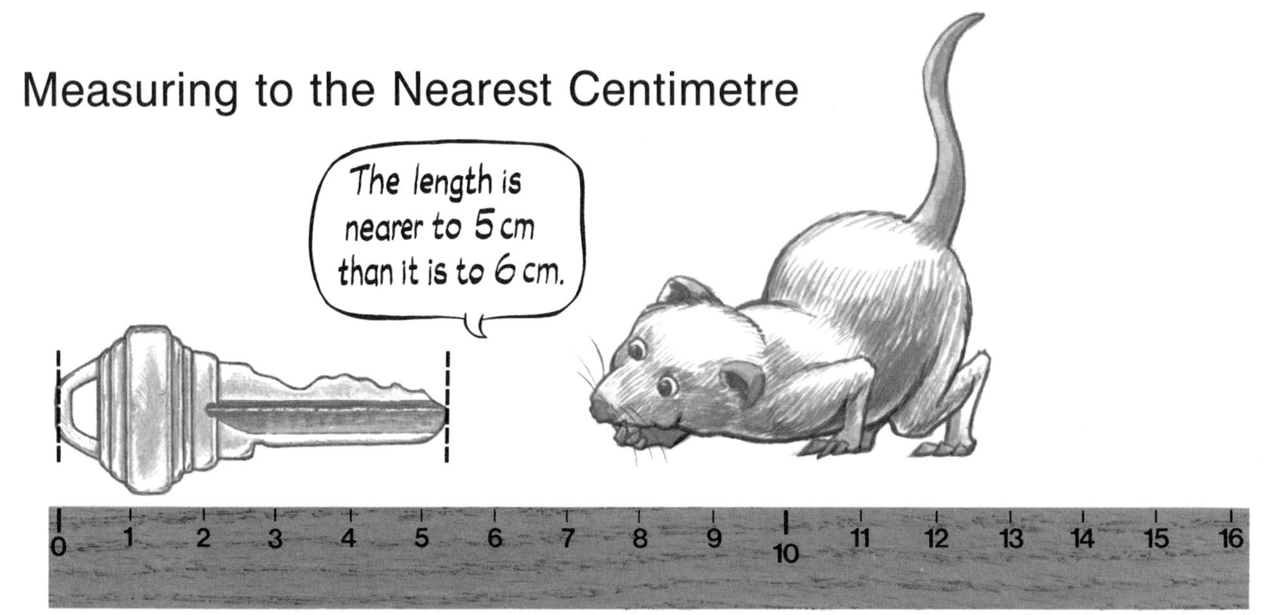

This key is 5 cm long,
to the nearest centimetre.

Measure to the nearest centimetre.
Record each length in a table.

	Length to Nearest Centimetre
1. Nail	▒ cm
2. Clip	▒ cm
3. Match	▒ cm

1.

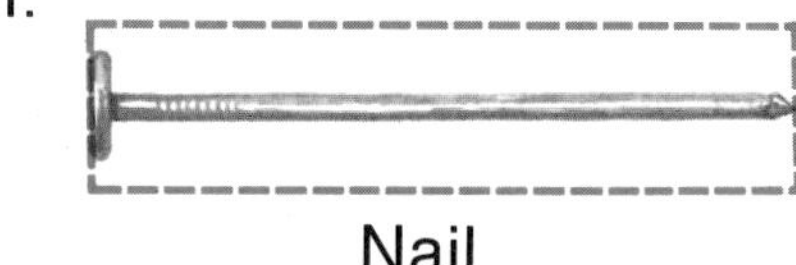
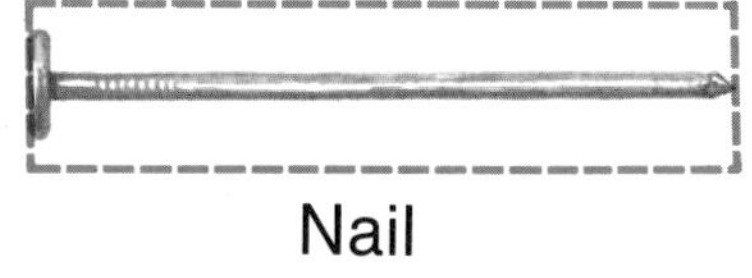

Nail

2.

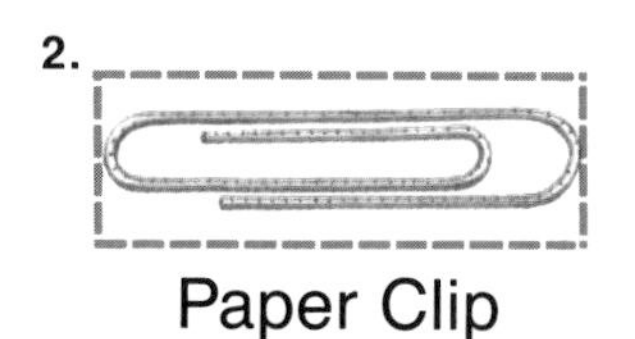

Paper Clip

3.

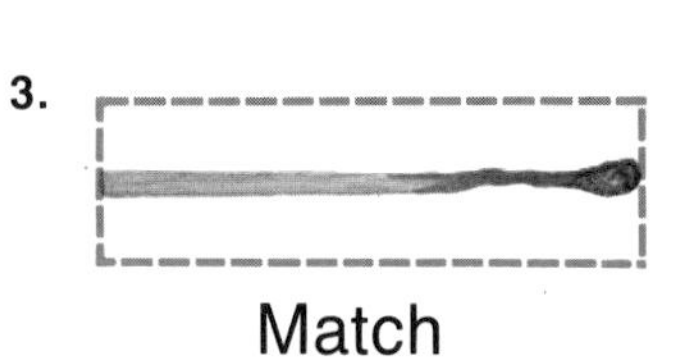

Match

4.

Safety Pin

5.

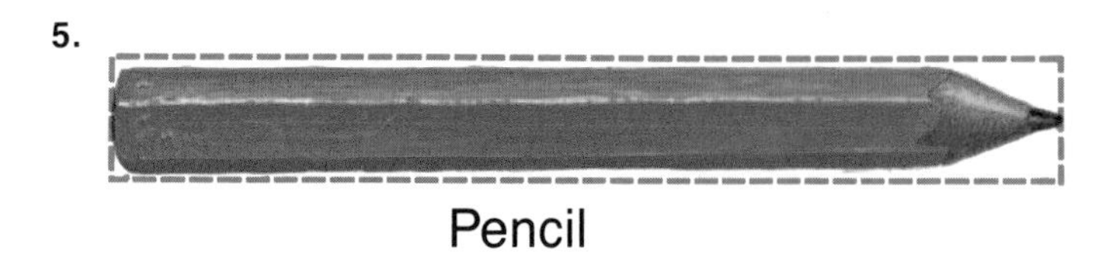

Pencil

6.

Push Pin

Making a Bar Graph

1. • Cut a strip of paper the same length as each rectangle on page 108.

   • Label each rectangle. Then colour it.

   • Glue each rectangle to centimetre graph paper to make a bar graph.

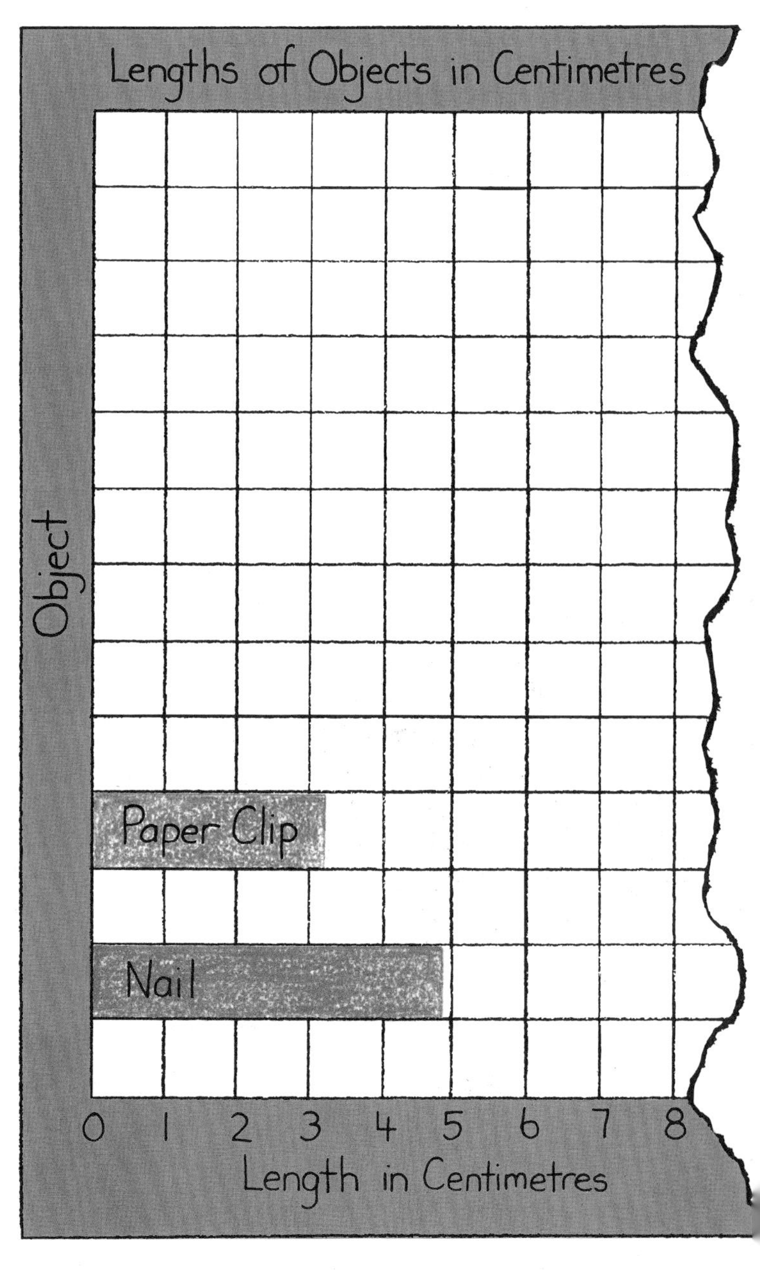

Use the graph to answer these questions.

2. Which object is longest?

3. Which object is shortest?

4. List the objects in order from longest to shortest.

5. About how much longer than the push pin is the safety pin?

ESTIMATING

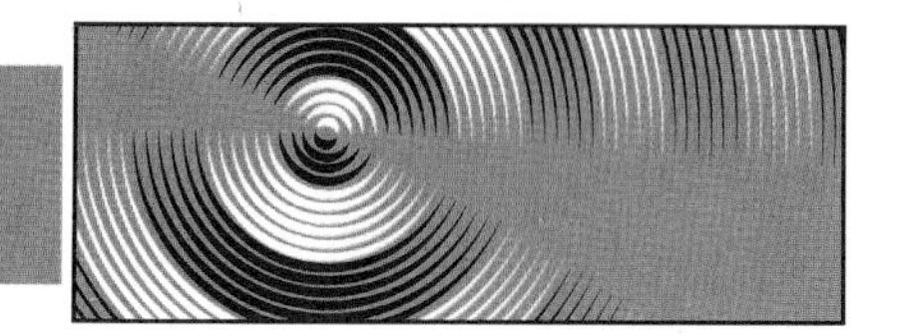

Estimating Length in Centimetres

This bar is 10 cm long.

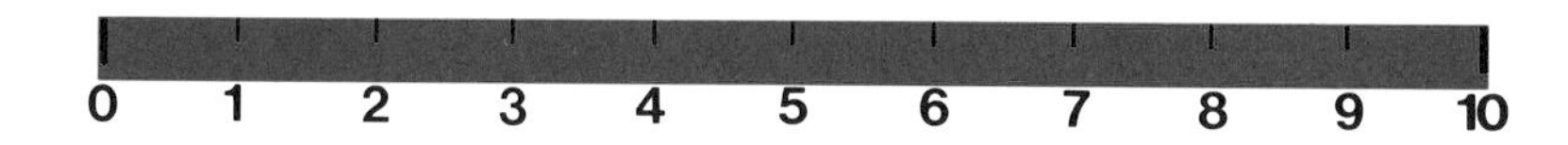

Estimate the length of each toy to the nearest centimetre.
Record your estimate in a table.
Measure to check.

Toy	Estimate	Measurement
A.	▒ cm	▒ cm

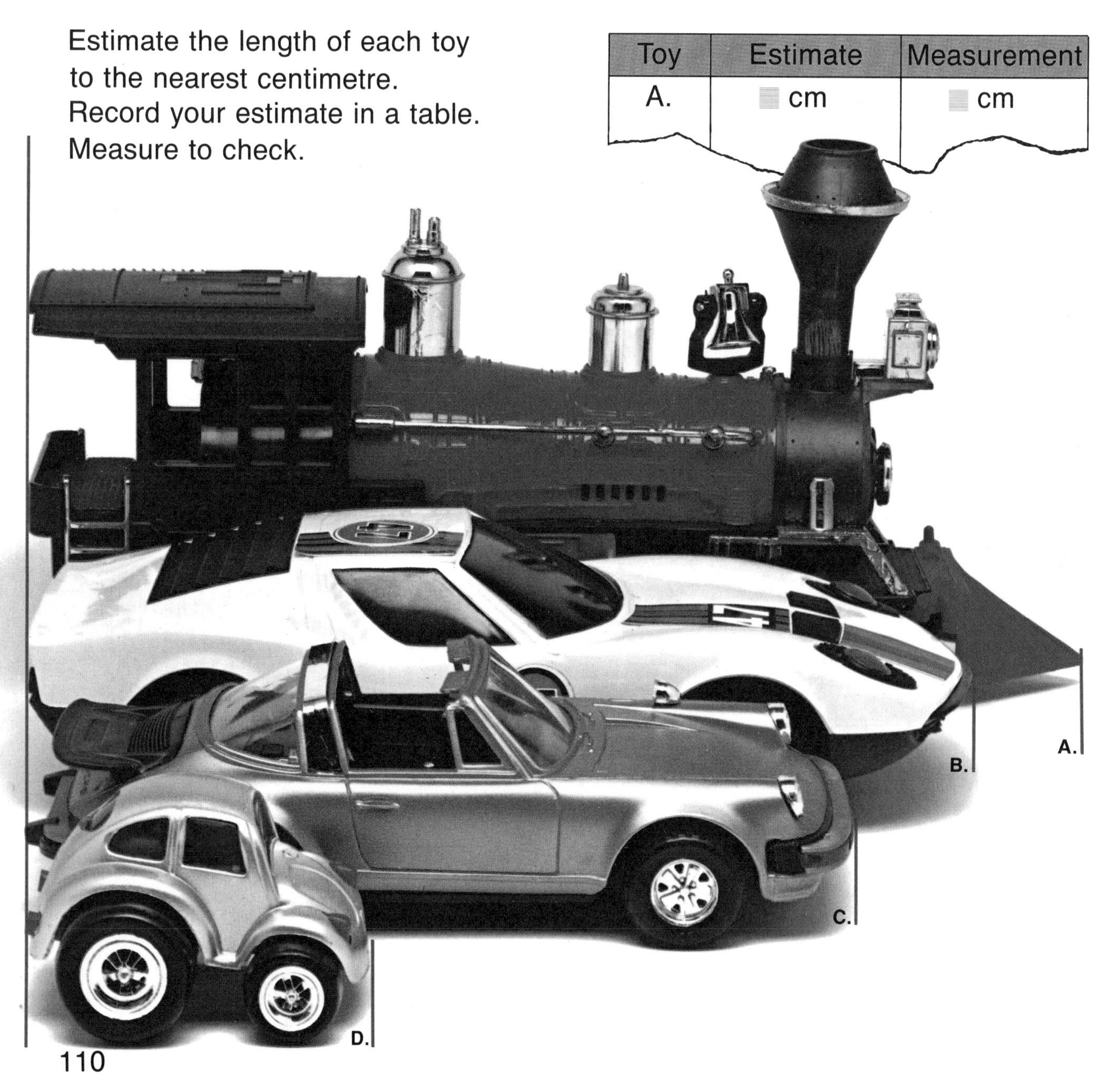

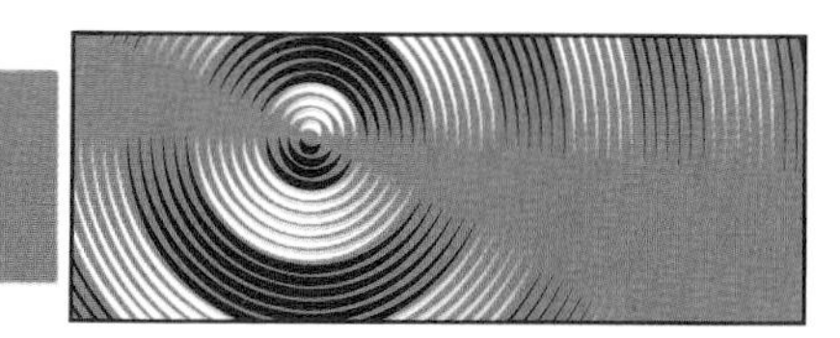

This bar is 10 cm long.

Estimate the distance from the floor to the top of each bear. Record your estimate to the nearest centimetre in a table.

Toy	Estimate	Measurement
A.	cm	cm

ESTIMATING

Estimating Length in Centimetres

Read the pirate's clue.
Which tree do you think is nearest
to the buried treasure?

Use a centimetre ruler to check your answer.

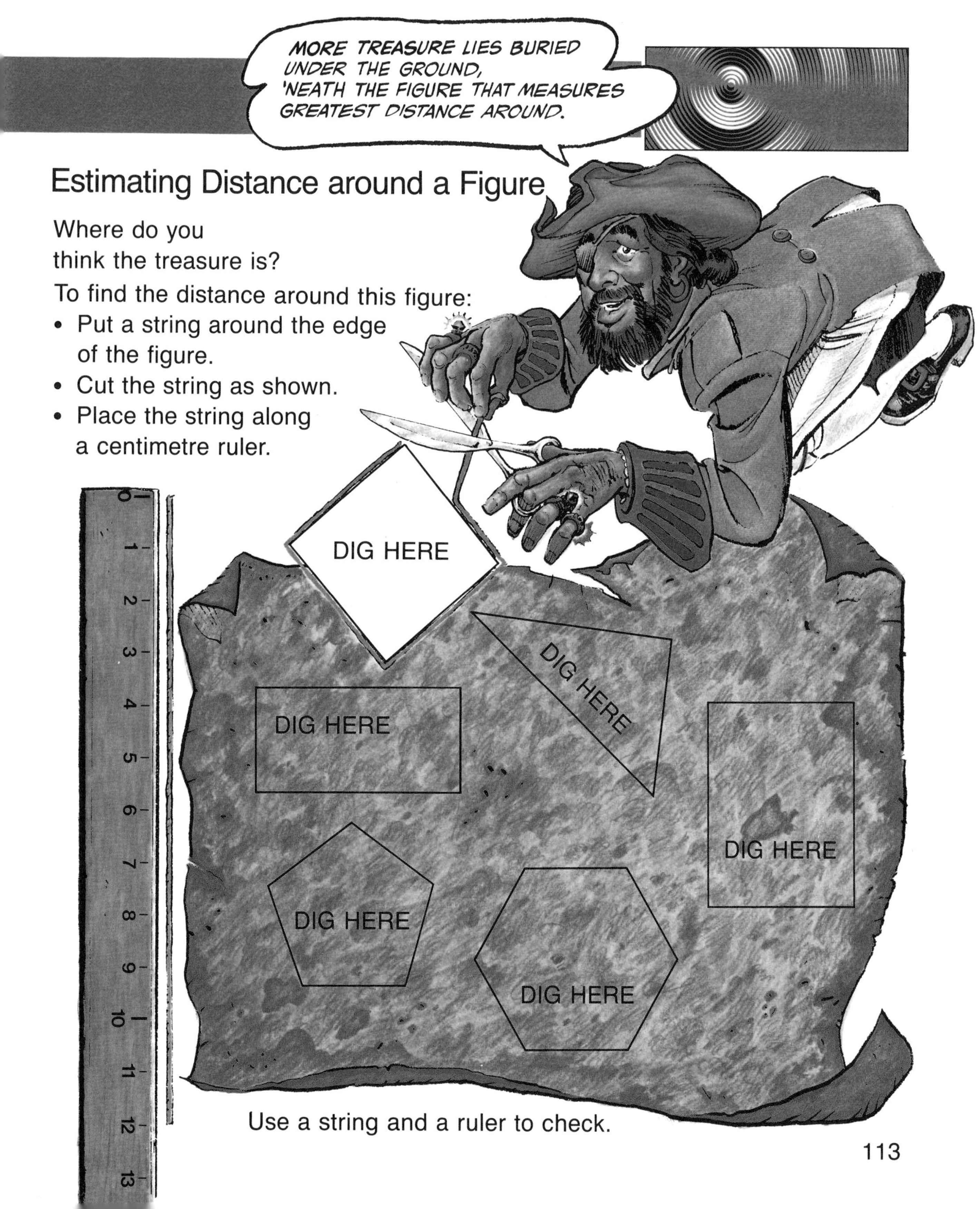

Estimating Distance around a Figure

Where do you think the treasure is?

To find the distance around this figure:

- Put a string around the edge of the figure.
- Cut the string as shown.
- Place the string along a centimetre ruler.

Use a string and a ruler to check.

Perimeter

An ant walked all the way around the outside of this cracker.
How far did it walk?

Side 3

Side 4

Side 2

Side 1

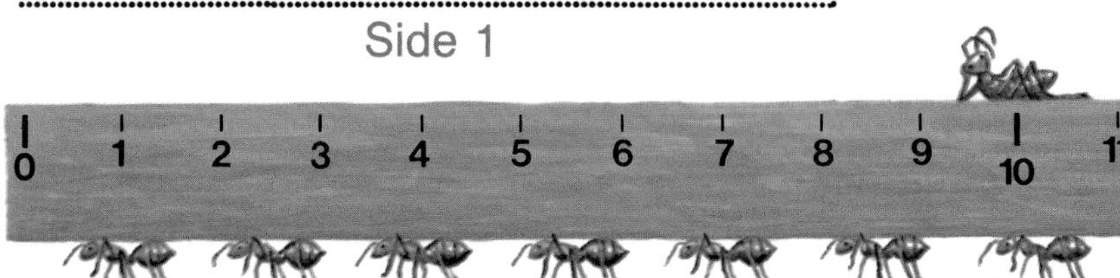

By measuring, we find:

Side 1		Side 2		Side 3		Side 4		
8 cm	+	4 cm	+	8 cm	+	4 cm	=	24 cm

The ant walked 24 cm.

The distance around the cracker is 24 cm.

We say: The perimeter of the cracker is 24 cm.

Find the perimeter of each cracker.
Use your ruler.

1.

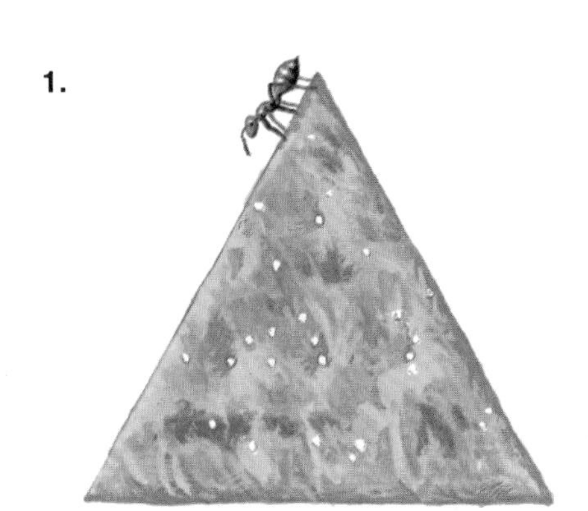

2.

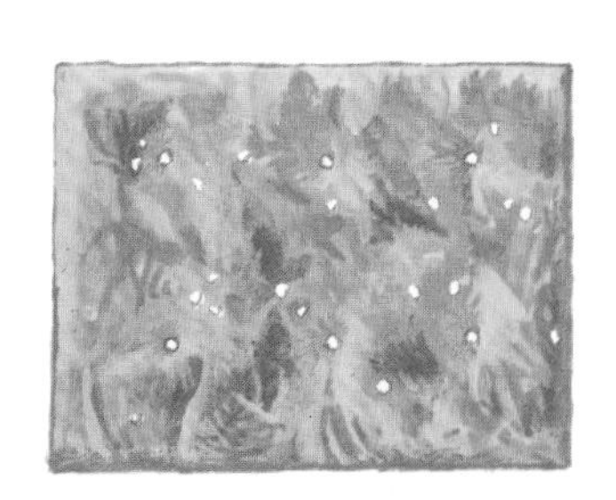

3.

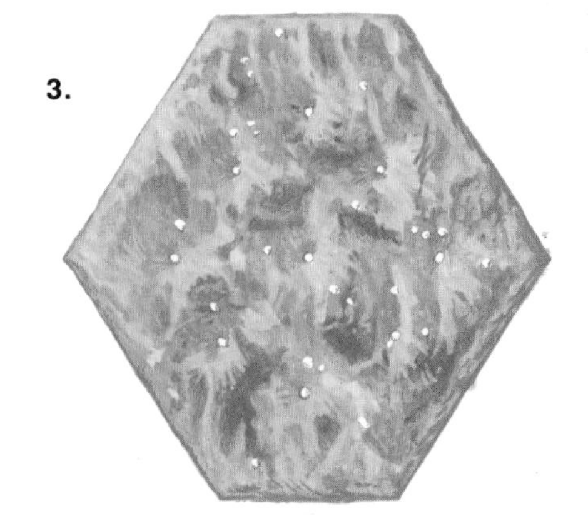

Find the perimeter of each figure.
Use your ruler.

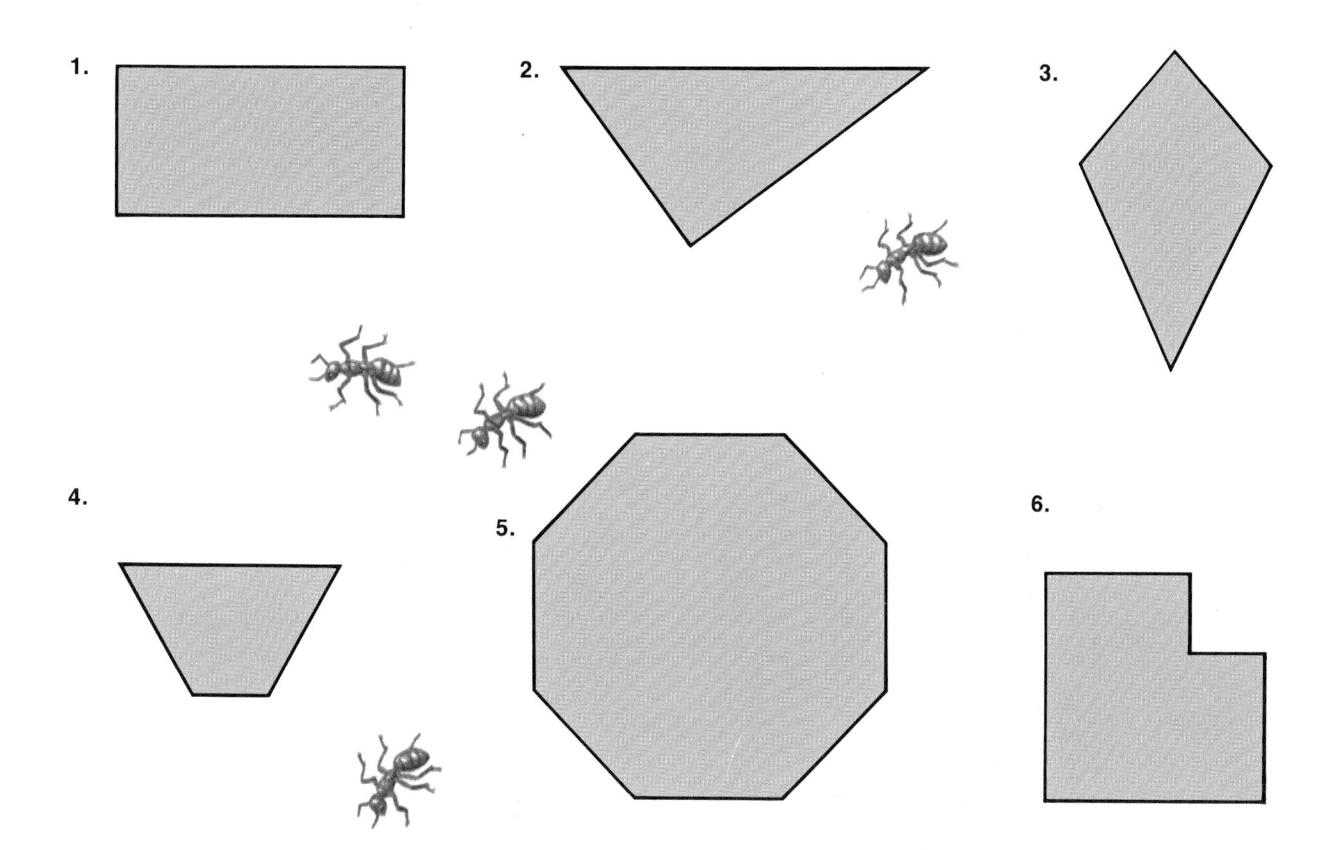

Draw rectangles with
the following perimeters.

7. 14 cm

8. 8 cm

9. 24 cm

PROBLEM SOLVING

Draw as many rectangles as you can with a perimeter of 16 cm.

Measuring in Metres

A length of 100 cm is called one metre.

We say: one metre
We write: 1 m

A shovel is about 1 m long.

A hockey net is about 1 m high.

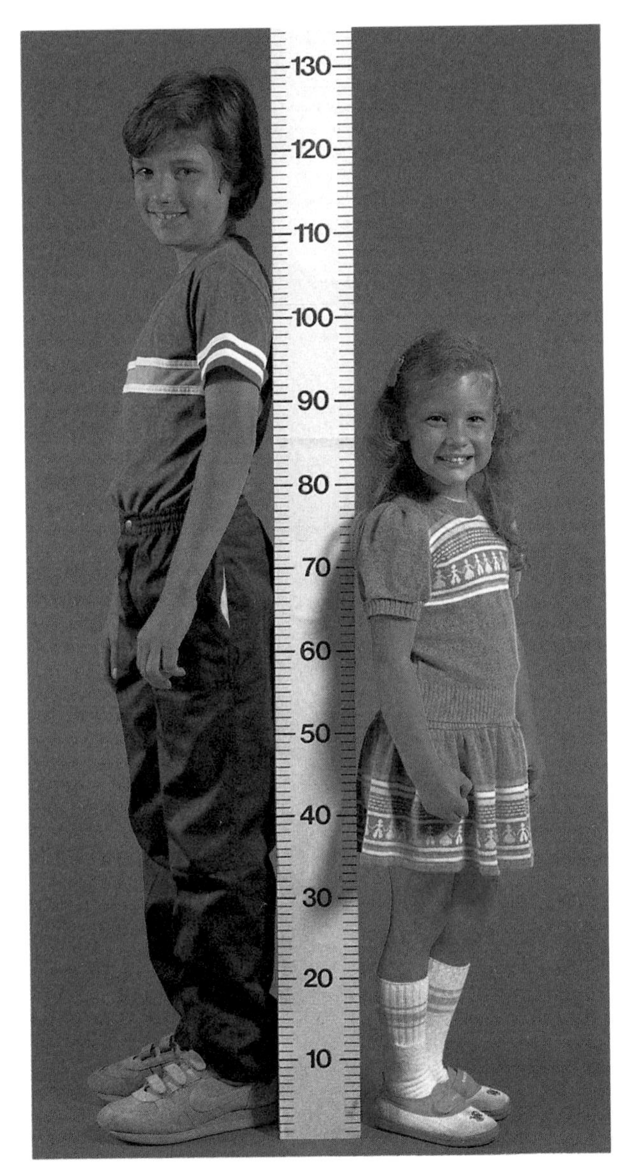

Aaron is more than 1 m tall.

Shannon is about 1 m tall.

Which of these things are longer than 1 m?

A. Garden Hose	B. Hammer	C. Teacher's Desk
D. Bicycle	E. Tennis Racket	F. Baseball Bat

Lengths to the Nearest Metre

The window is about 1 m wide.

Measure these lengths to
the nearest metre.
Then complete each sentence.

1\.

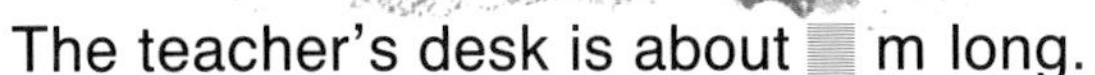

The teacher's desk is about ▒ m long.

2\.

The teacher's desk is about ▒ m wide.

3\.

The chalkboard is about ▒ m long.

4\.

The classroom is about ▒ m wide.

Heights to the Nearest Metre

Read the height of each animal to the nearest metre. Record in a table like this.

	Height to Nearest Metre
Child	1 m
Giraffe	▒ m

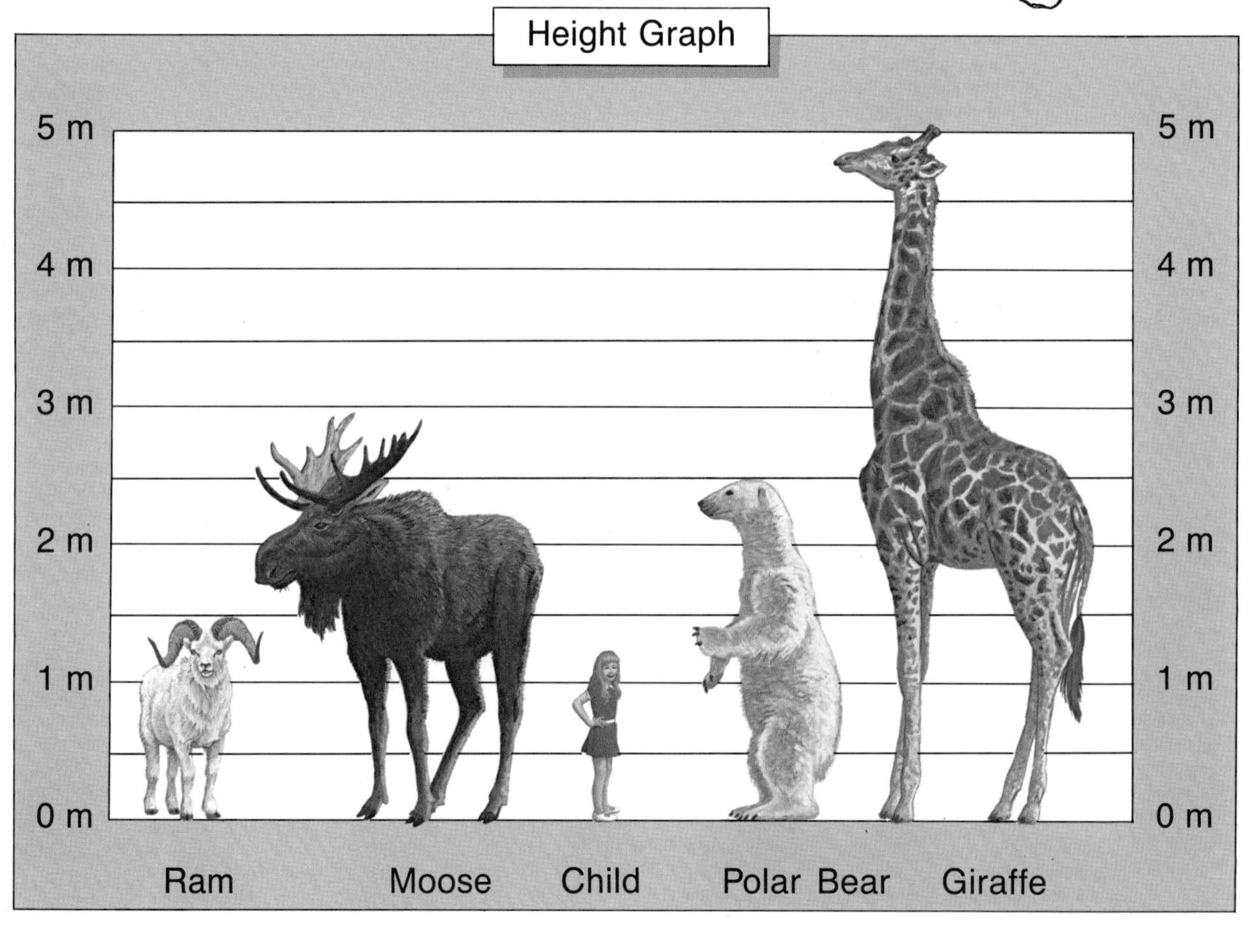

1. Which of the animals is the tallest?
2. Which of the animals is about the same height as the child?
3. Would a tower of 2 children be as tall as a moose?
4. Would a tower of 3 children be as tall as a giraffe?
5. Write your own question about this graph.

EXTENSIONS

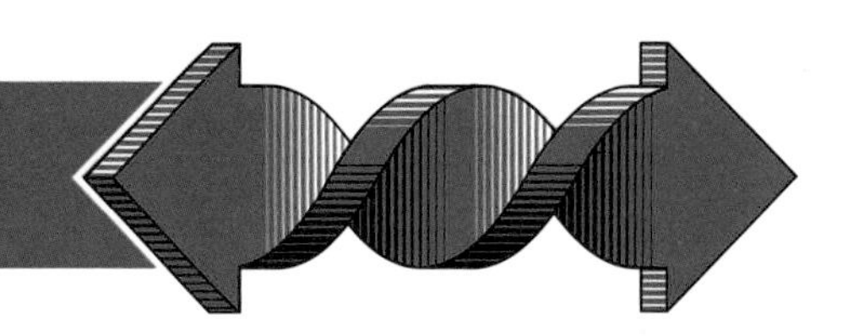

Using a Trundle Wheel

A trundle wheel clicks each time it rolls a distance of 1 m.

Use a trundle wheel to find these distances.

A. from your desk to the classroom door

B. from the door to the end of the hall

C. around the walls of your classroom (the perimeter)

ESTIMATING

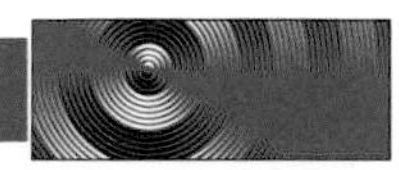

Place 2 pieces of tape where you think they will be 5 m apart.

Use a trundle wheel to check your estimate.

Repeat for these distances.

A. 10 m B. 7 m

The Kilometre

Which of these distances are longer than 1 km?

A. from the front of your school to the back

B. from your school to your home

C. from your school to the next town or city

D. from your school to the parliament buildings in Ottawa

Choosing the Appropriate Unit

Think of centimetres, metres, or kilometres.
Which unit would you use to measure each length?

1.

the distance between two cities

2.

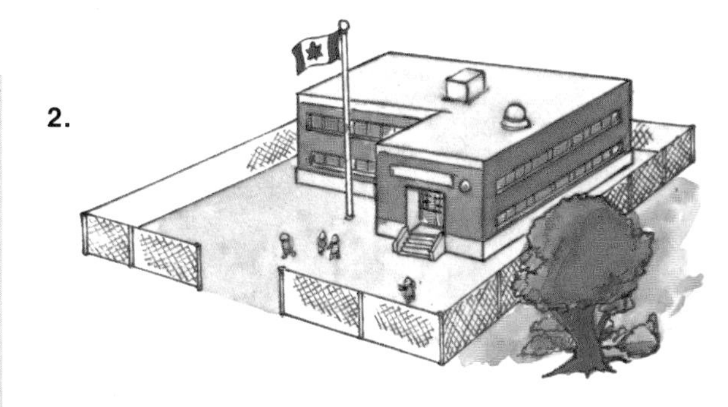

the length of a school yard

3.

the length of a pet rabbit

4.

the distance from your home to school

5.

your height

6.

the height of a building

7.

the distance run in a race

8.

the length of a bicycle

CHECK-UP

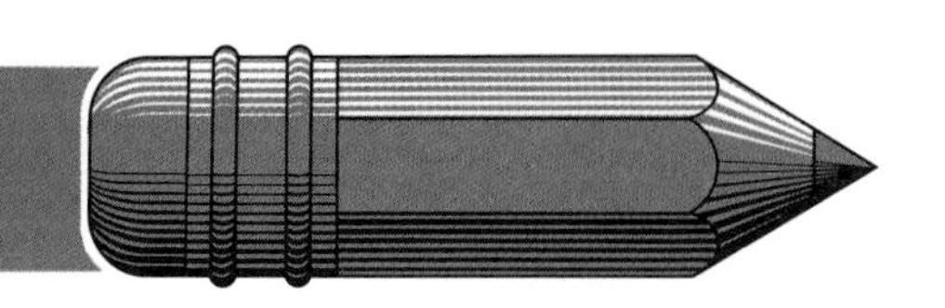

Measure to the nearest centimetre.

1. 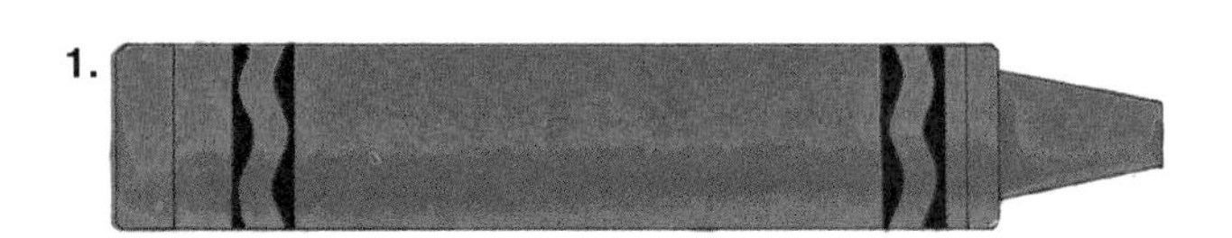

2. 

Use the graph to answer these questions.

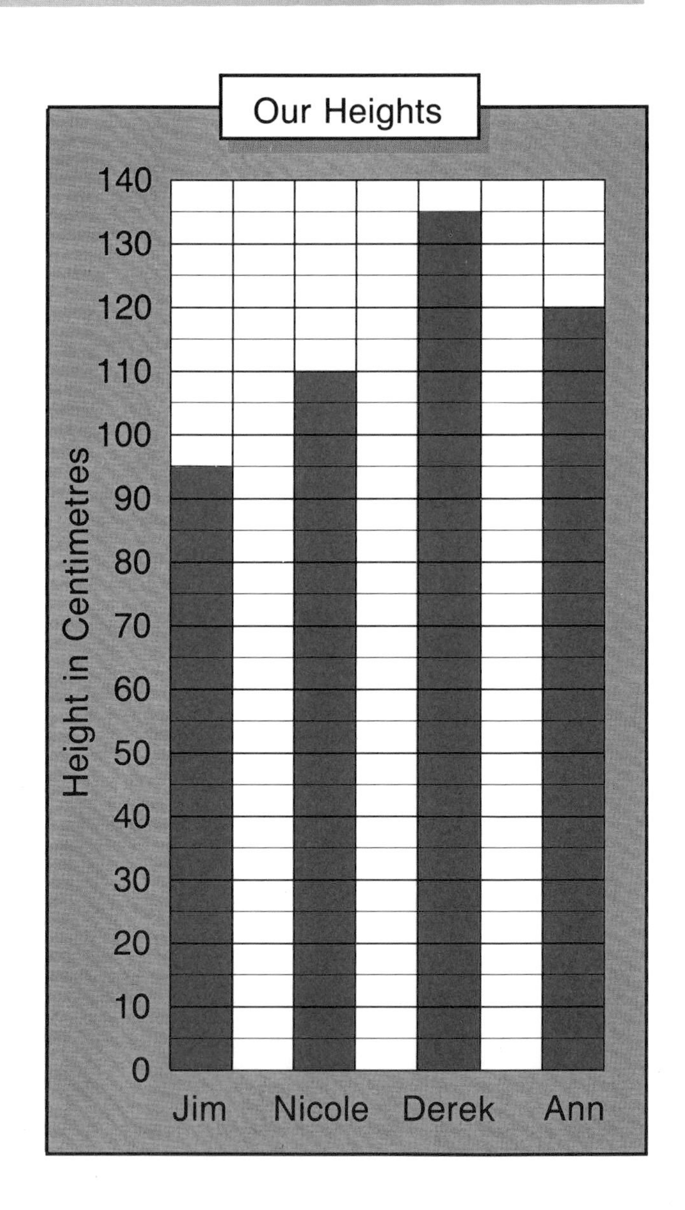

3. Who is tallest?
4. Who is shortest?
5. How much taller is Ann than Nicole?

Find the perimeter of each figure. Use your ruler.

6. 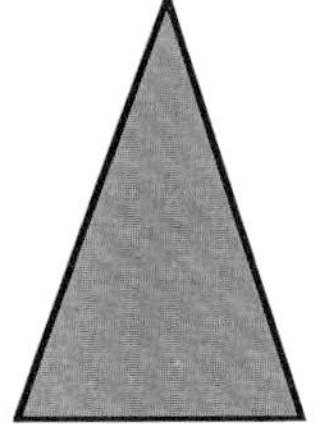

7. 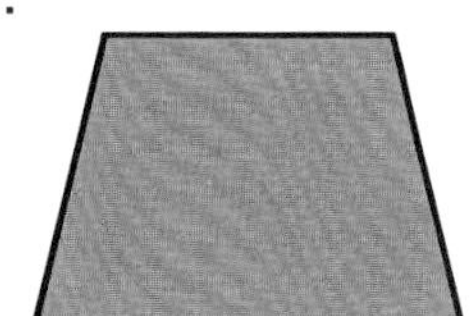

8. Name 2 things longer than 1 m.

9. Name 2 things shorter than 1 m.

10. Think of centimetres, metres, and kilometres.
    Choose the unit you would use to measure each of these.

    A. the length of your school
    B. the distance between cities
    C. the width of a fingernail
    D. the width of your hand

CUMULATIVE REVIEW

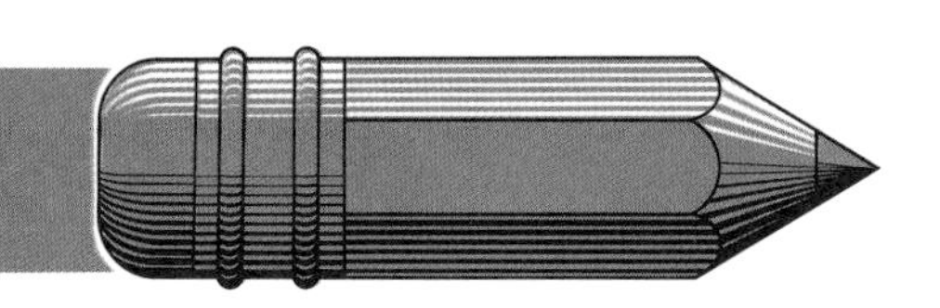

Trade a ten for 10 ones.

1. 6 tens 4 ones = 5 tens ▒ ones
2. 3 tens 7 ones = 2 tens ▒ ones

Subtract.

3. $\begin{array}{r}57\\-\ 3\\\hline\end{array}$	4. $\begin{array}{r}49\\-26\\\hline\end{array}$	5. $\begin{array}{r}85\\-37\\\hline\end{array}$	6. $\begin{array}{r}54\\-45\\\hline\end{array}$	7. $\begin{array}{r}32\\-\ 5\\\hline\end{array}$	8. $\begin{array}{r}70\\-\ 8\\\hline\end{array}$
9. $\begin{array}{r}84\\-42\\\hline\end{array}$	10. $\begin{array}{r}67\\-20\\\hline\end{array}$	11. $\begin{array}{r}46\\-38\\\hline\end{array}$	12. $\begin{array}{r}93\\-56\\\hline\end{array}$	13. $\begin{array}{r}46\\-\ 7\\\hline\end{array}$	14. $\begin{array}{r}80\\-\ 7\\\hline\end{array}$

15. There are 37 children in the school band. There are 21 in the choir. How many more are in the band?

16. Mrs. Lum baked 48 cookies. The children ate 29. How many are left?

Is the object shaped like a cube, sphere, cylinder or rectangular prism?

17. 

18. 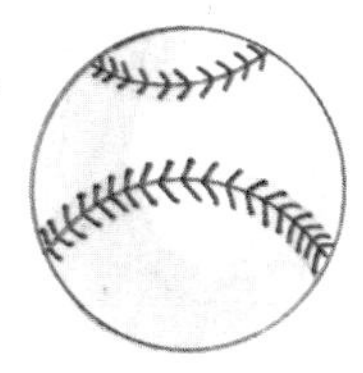

19. 

20. 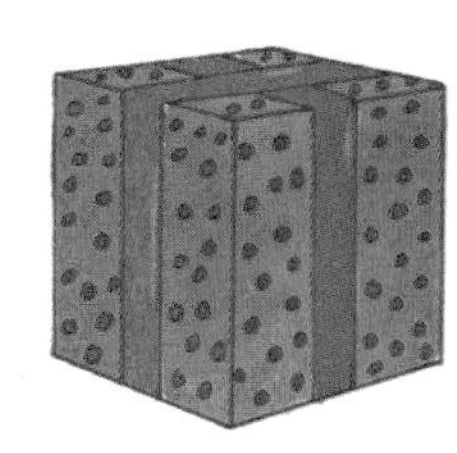

How many faces does each solid have? Use a table to record your answers.

21. 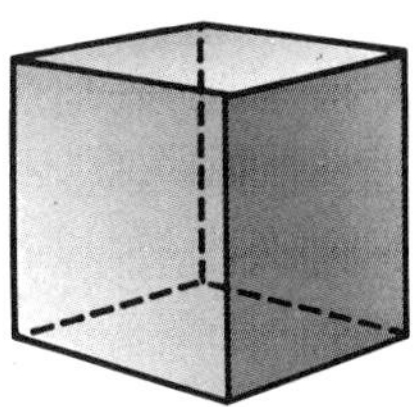

22. 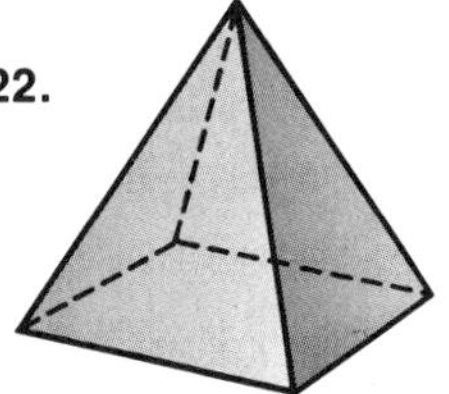

23. 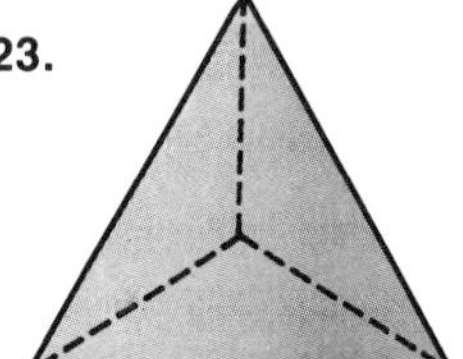

24. 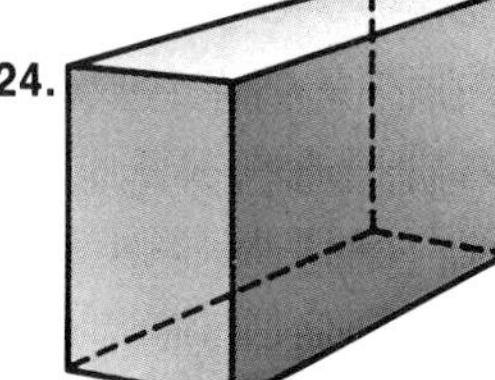

CUMULATIVE REVIEW

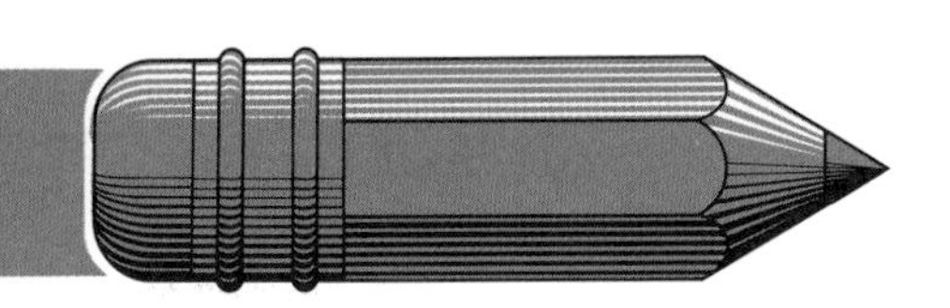

Measure the length of each object to the nearest centimetre.

25.

26.

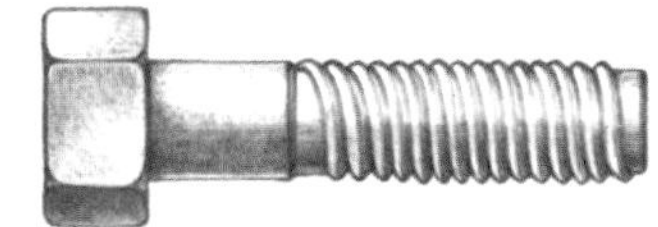

Use your ruler to find the perimeter of each figure.

27.

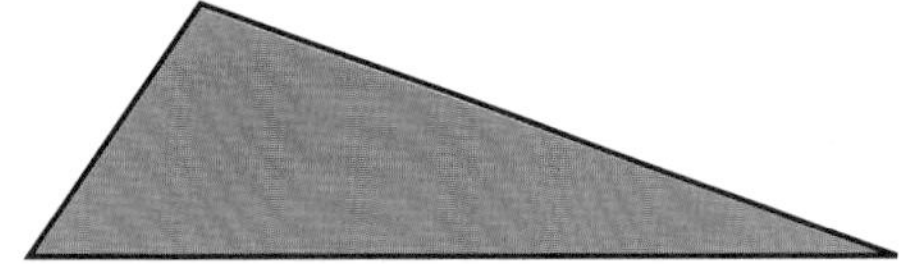

28.

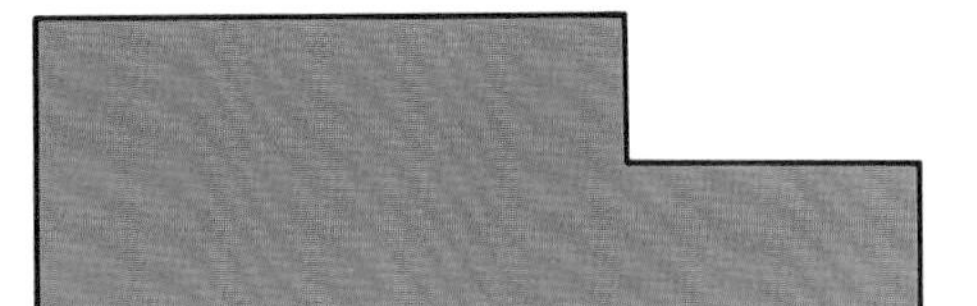

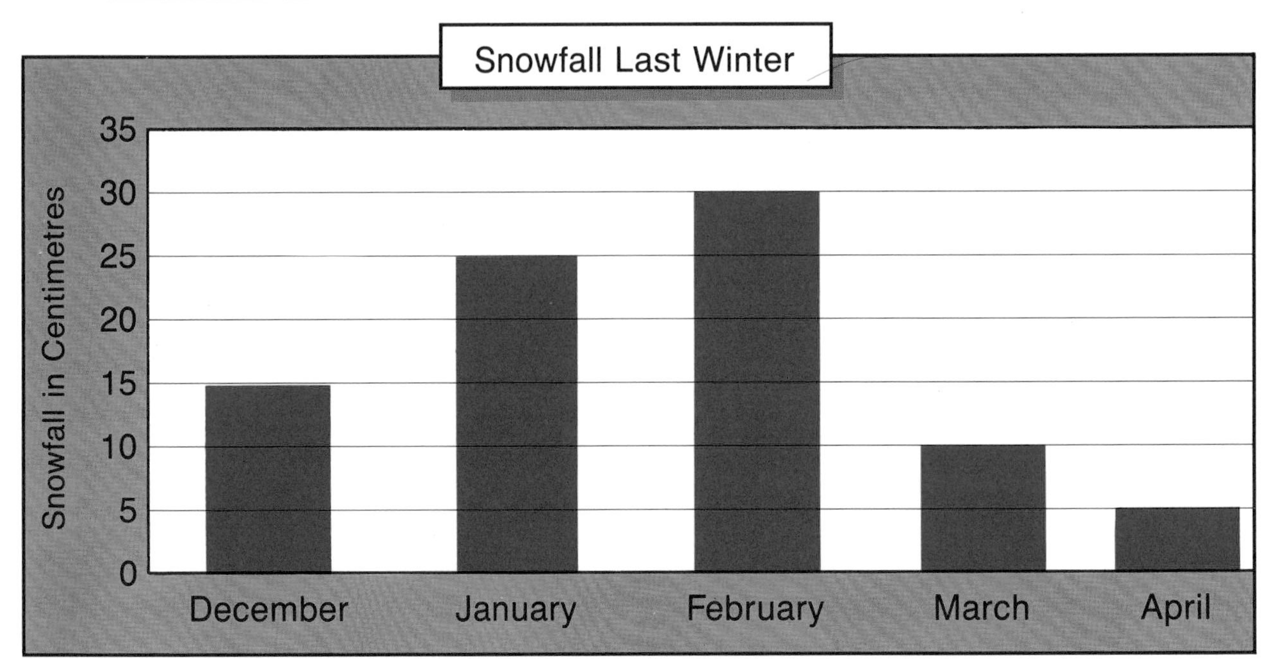

29. Which month had the greatest snowfall?

30. How much more snow fell in December than in March?

31. Which month had 15 cm less snowfall than January?

32. How much snow fell in all 5 months?

PLACE VALUE TO 9999

Canada is the Rocky Mountains,
Canada is Prince Edward Island,
Canada is a country made for love.

Canada is the prairie cowboy,
Canada is the Yukon miner,
Canada is a country full of love.

We have love for our neighbours
of whatever creed or colour,

We have love for our cities
and our valleys and our plains,

We have a voice that is calling
telling all the world we're willing

To welcome them to this great land,
And that's what Canada is.

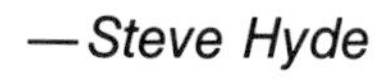

—*Steve Hyde*

Thousands

This shows
one hundred small cubes.

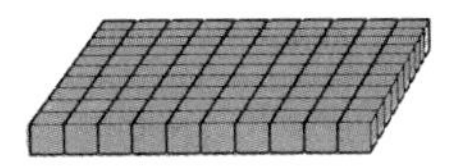

100

How many cubes
are there in
10 hundreds?

Count.

100
200
300
400
500
600
700
800
900
1000

10 hundreds = one thousand

This shows
one thousand small cubes.

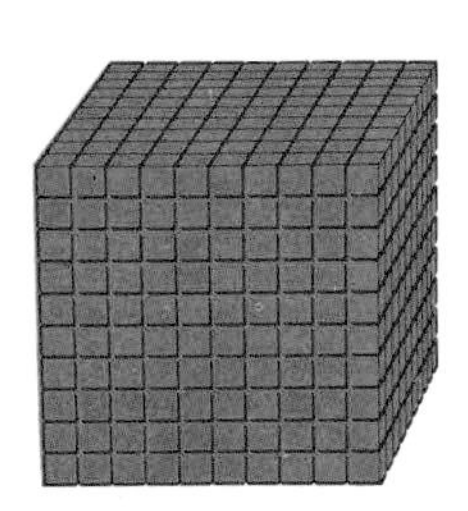

1000

How many small cubes are shown?

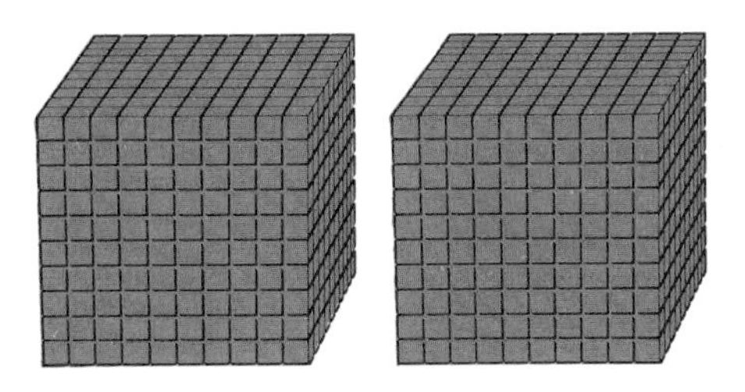

We think: two thousand

We write: **2000**

How many small cubes are in each picture?

1.

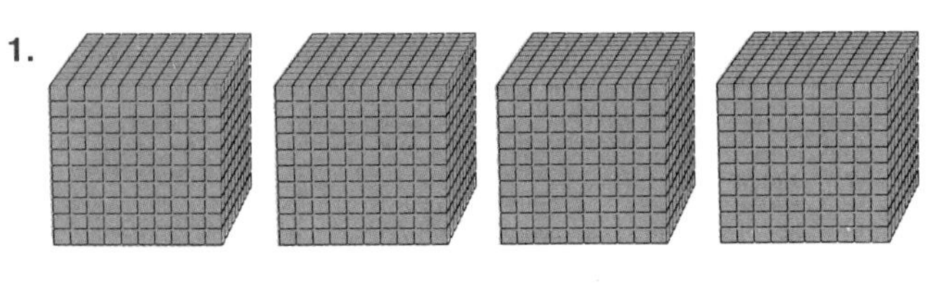

2.

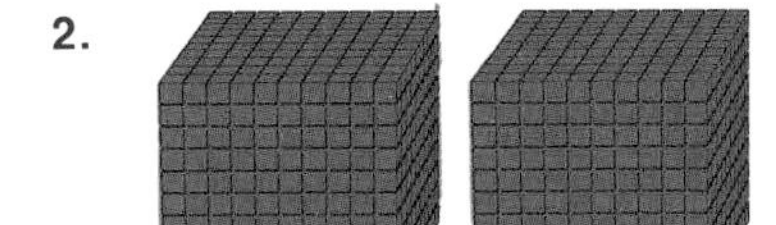

3.

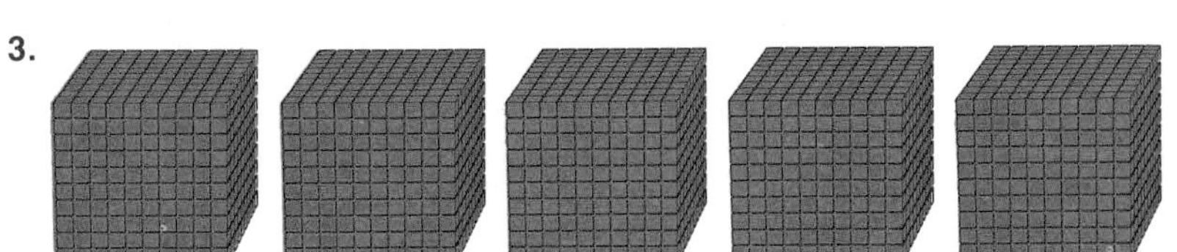

4.

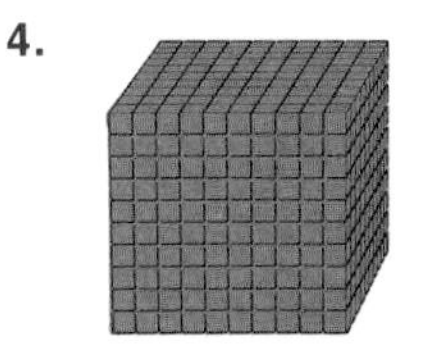

Numerals to 9999

We see:

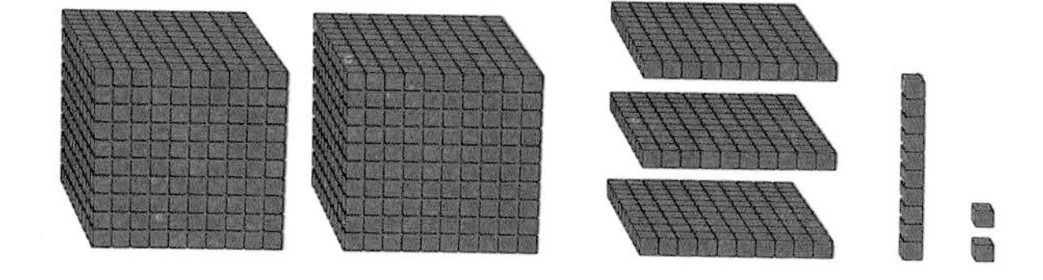

We think:

Thousands	Hundreds	Tens	Ones
2	3	1	2

We write: 2312

How many small cubes are in each picture?

1.

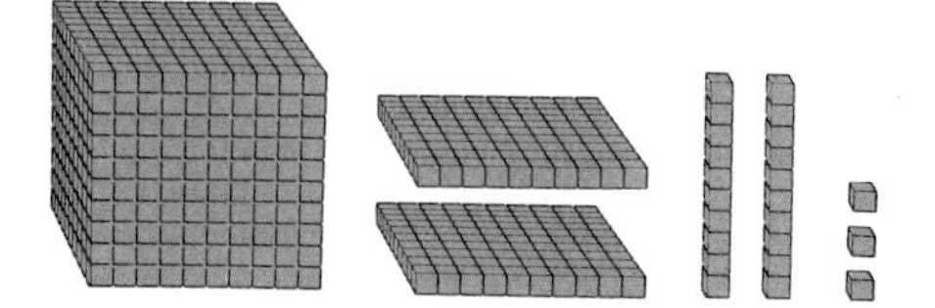

2.

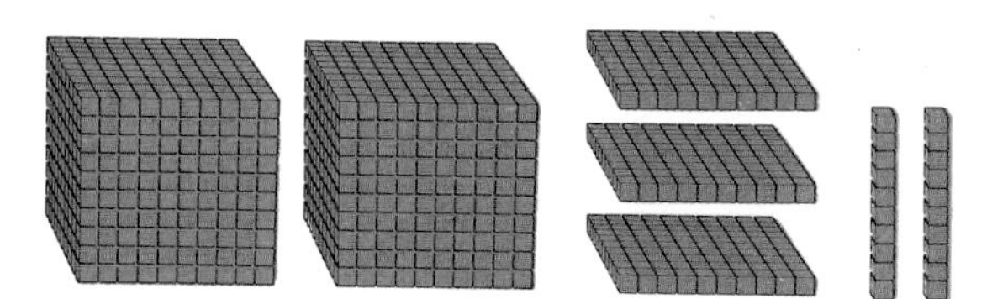

3.

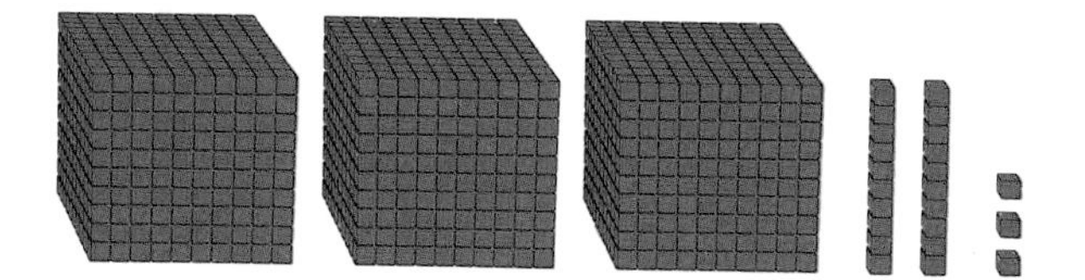

4.

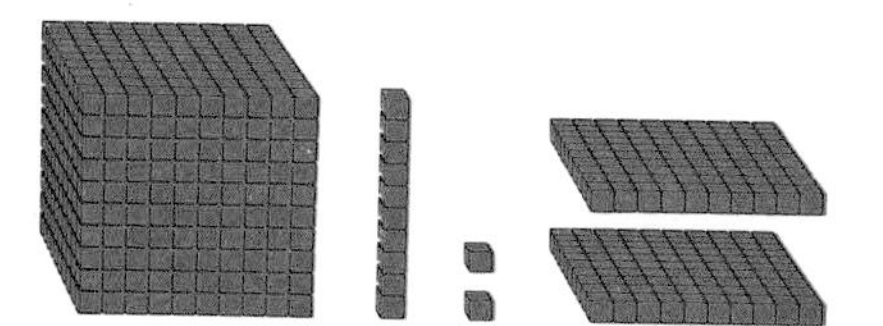

5.

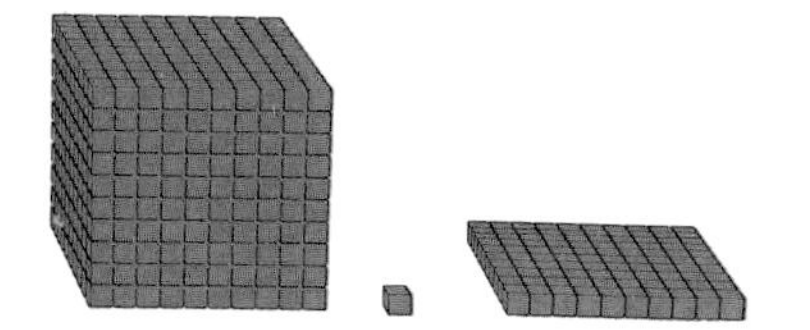

6.

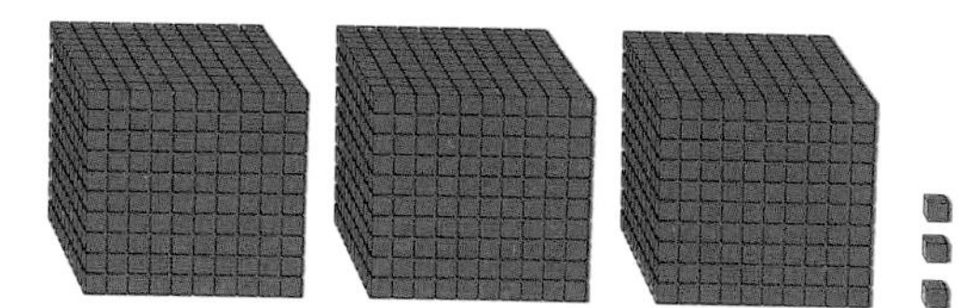

Place Value to 9999

We say: five thousand eight hundred twenty-seven

We think: 5000 + 800 + 20 + 7

We write: 5827

Copy and complete.

1. 6283 = ■ thousands ■ hundreds ■ tens ■ ones

1. 6283 = 6 thousands 2 hundreds 8 tens 3 ones

2. 7120 = ■ thousands ■ hundreds ■ tens ■ ones
3. 8094 = ■ thousands ■ hundreds ■ tens ■ ones
4. 5207 = ■ thousands ■ hundreds ■ tens ■ ones
5. 9003 = ■ thousands ■ hundreds ■ tens ■ ones

Write each number in words.

6. 7816
7. 2187
8. 4091
9. 7308
10. 9006
11. 5090

Write the meaning of each coloured digit.

12. 5841

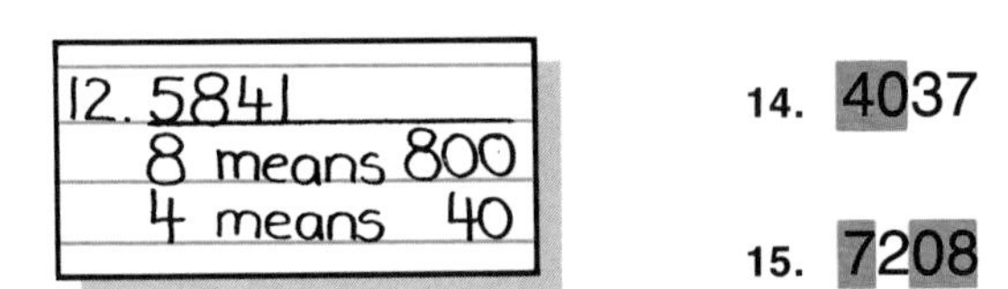

13. 9134
14. 4037
15. 7208

Copy and complete.

16. 6829 = 6000 + ■ + 20 + 9
17. 4307 = 4000 + 300 + ■
18. 5026 = ■ + 20 + 6
19. 9080 = 9000 + ■

Writing Numerals to 9999

An airplane flying from St. John's, Newfoundland to Vancouver, British Columbia would travel five thousand six hundred nineteen kilometres.

We write: 5619

Write the numeral.

1. The population of Bracebridge, Ontario is eight thousand four hundred forty-five.
2. The height of Mount Logan in the Yukon is five thousand nine hundred fifty-one metres.
3. The border between Canada and the United States is eight thousand eight hundred ninety-two kilometres long.

Write the numeral.

4. 7 thousands, 2 hundreds, 7 tens, 6 ones
5. 2 thousands, 7 tens, 6 ones
6. 9 thousands, 9 hundreds, 3 ones
7. 3 thousands, 3 tens
8. 4 thousands, 7 ones
9. six thousand nine hundred seventy-six
10. eight thousand five hundred sixty-seven

ESTIMATING

Estimate: Are there more red circles or orange circles?

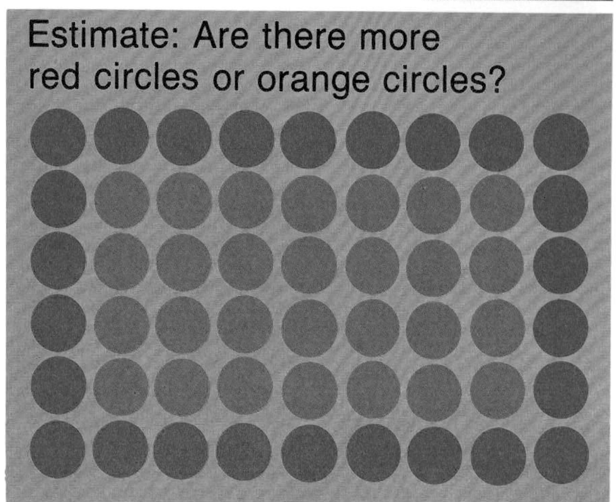

Number Sequences to 9999

Which seats have animals in them?
Write the numbers.

Write the next 6 numbers in each pattern.

1. 1552, 1562, 1572,...
2. 8850, 8860, 8870,...
3. 6202, 6302, 6402,...
4. 9975, 9965, 9955,...
5. 734, 834, 934,...
6. 3021, 2921, 2821,...
7. 8939, 8949, 8959,...
8. 3955, 3945, 3935,...
9. 8577, 8477, 8377,...
10. 461, 561, 661,...

JUST FOR FUN

RIDDLE: What must you do before you get off a Ferris wheel?

For each letter, write the number which completes the pattern.

Write the matching letters.

5007	8997	5197		5001	5197	997	4967		5000	4994	4967		8997	4897

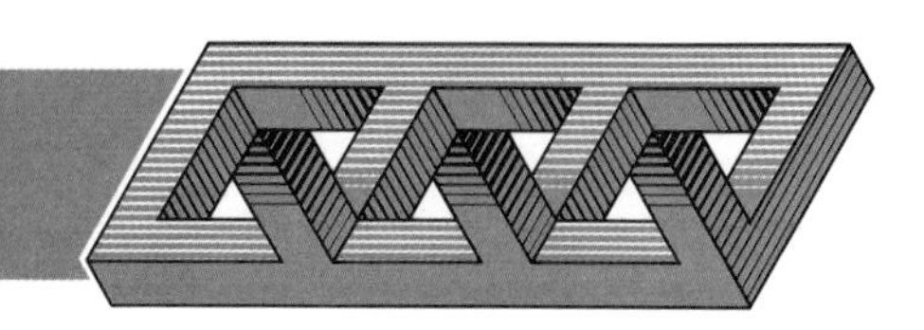

PROBLEM SOLVING

Adding Missing Information

Answer these questions. Use the information to the left and any other information you need.

1. When did Canada celebrate its tenth birthday?
2. When did Canada celebrate its hundredth birthday?
3. When will you celebrate your tenth birthday?
4. When will you celebrate your hundredth birthday?
5. What was the date 10 years before the railway was completed?
6. What was the date 100 years before Marc Garneau's first space flight?
7. When will Canada's flag be 100 years old?
8. When will Canada be 1000 years old?

BITS AND BYTES

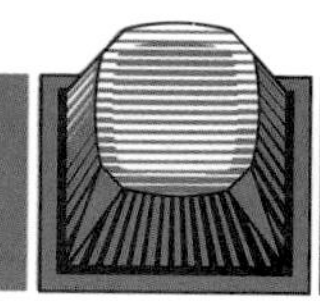
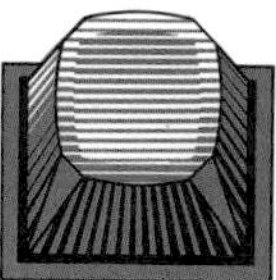
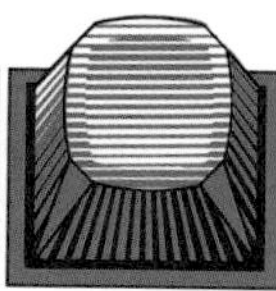

To add on a calculator... press these keys.

8 + 7 = ▒

Use your calculator to complete these.

1. 6 + 9 = ▒
2. 7 + 8 = ▒
3. 9 + 10 = ▒
4. 8 + 9 = ▒

5.	6.	7.	8.	9.	10.
8	7	20	30	8	90
+ 10	+ 10	+ 6	+ 7	+ 50	+ 8

To add on a calculator... press these keys.

6 + 7 + 5 = ▒

Use your calculator to complete these.

11. 3 + 4 + 8 = ▒
12. 4 + 5 + 7 = ▒
13. 3 + 5 + 9 = ▒

14.	15.	16.	17.	18.	19.
800	300	600	700	200	400
20	40	20	80	90	60
+ 4	+ 7	+ 8	+ 5	+ 7	+ 9

Now add these.

20.	21.	22.	23.	24.	25.
2000	4000	3000	8000	8	800
600	700	600	300	60	40
30	10	80	20	300	6000
+ 7	+ 5	+ 3	+ 9	+ 9000	+ 3

Can you add these faster in your head?

Comparing Numbers to 9999

Which car costs more?

Compare the thousands digits.	Compare the hundreds digits.	Compare the tens digits.	The number with the greater digit is greater.
8918 8926 same	8918 8926 same	8918 8926 2 is greater.	$8926 > 8918$

We say: 8926 is greater than 8918 or 8918 is less than 8926

We write: $8926 > 8918$ or $8918 < 8926$

The green car costs more.

Write the number that is greater.

1. 7382
7391
2. 4637
4652
3. 5128
5371
4. 6387
5291
5. 7311
7412

Write the number that is less.

6. 5008
4992
7. 3716
3714
8. 6708
6807
9. 3700
3692
10. 2067
2226

Copy each sentence. Write $>$ or $<$ for each ▒.

11. 5150 ▒ 5299
12. 8660 ▒ 8661
13. 8489 ▒ 8480

EXTENSIONS

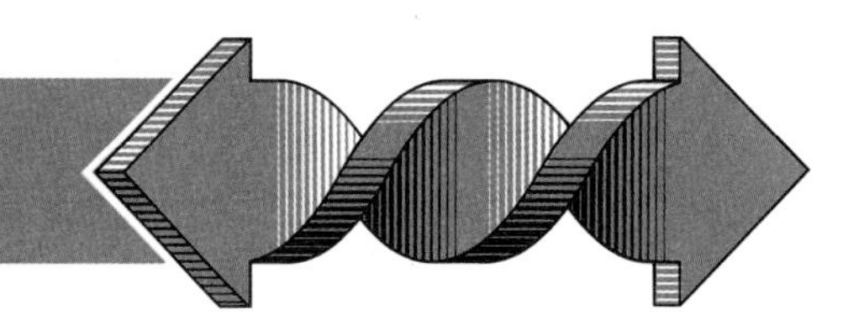

Ordering Numbers to 9999

Which jar has the most jelly beans?

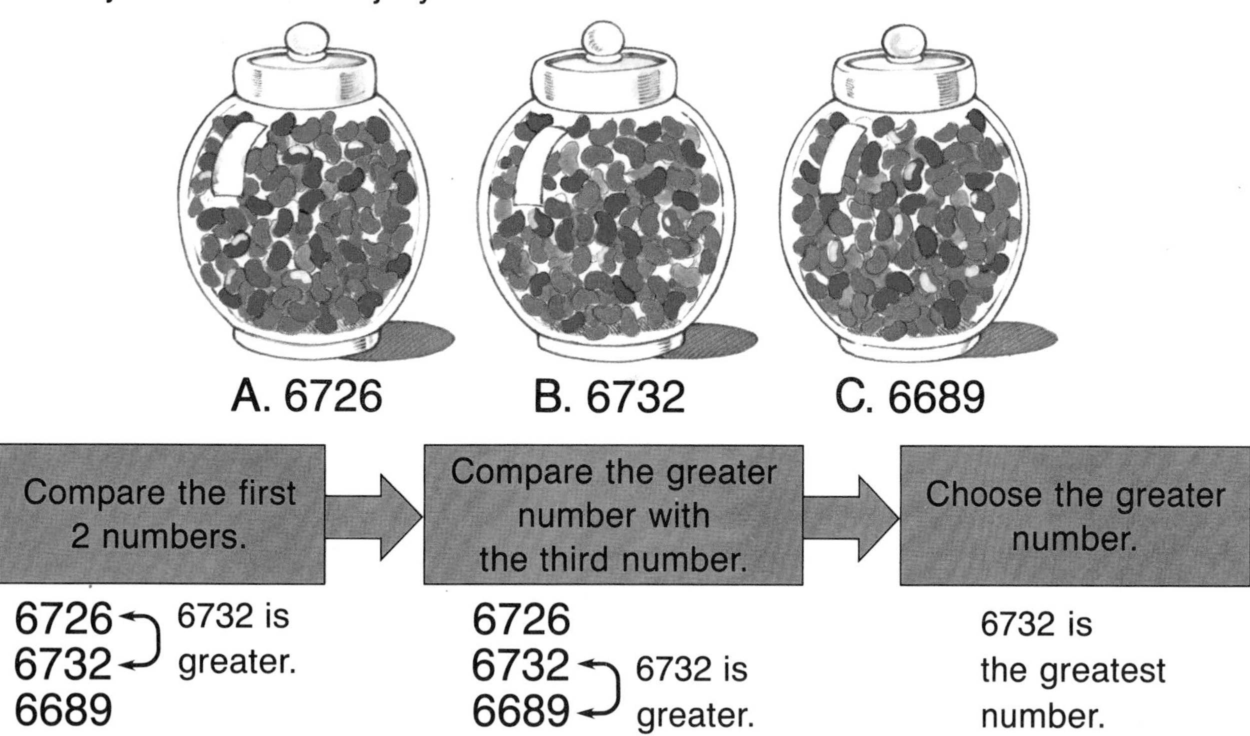

Jar B contains the most jelly beans.

Write the greatest number of the three.

1. 5062
 3216
 7180
2. 3033
 3303
 3003
3. 7216
 7126
 7621

Write the least number of the three.

4. 8008
 9728
 2982
5. 3506
 3605
 5306
6. 2119
 2190
 2281

Write the numbers from least to greatest.

7. 2817, 1909, 2716
8. 8812, 8919, 8099

CHECK-UP

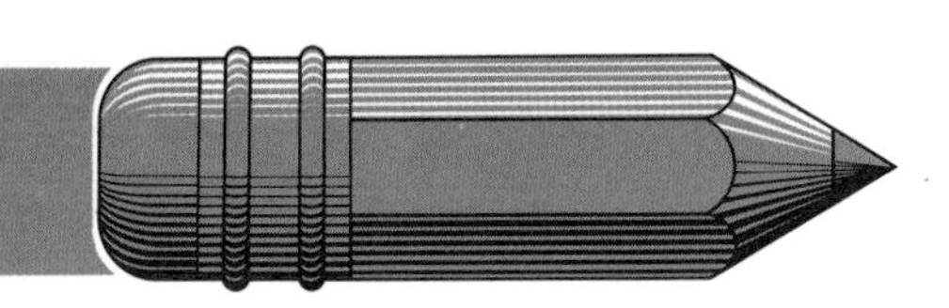

How many small cubes are in each picture?

1.

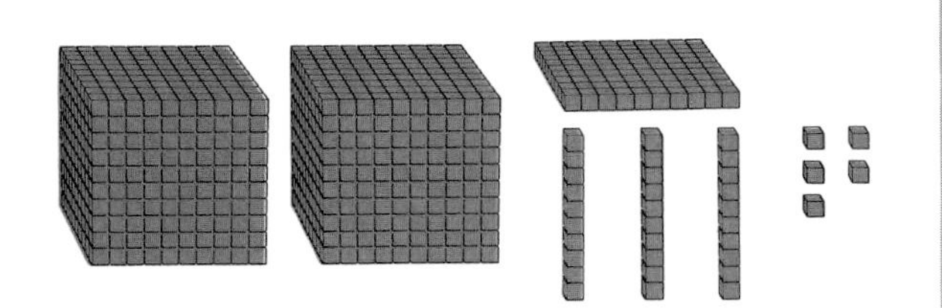

2.

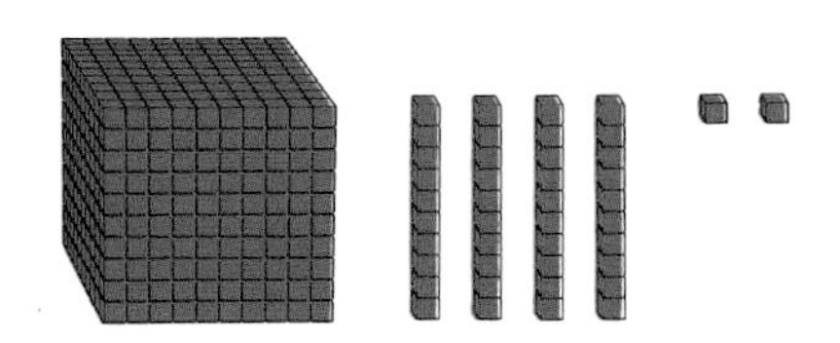

Copy and complete.

3. 9286 = ■ + 200 + 80 + 6
4. 4231 = 4000 + ■ + 30 + 1
5. 5278 = 5000 + 200 + ■ + 8
6. 9630 = 9000 + 600 + ■

Write the numeral.

7. 4 thousands, 9 hundreds, 3 tens, 6 ones
8. two thousand seven hundred sixty-eight

Write the next 5 numbers in each pattern.

9. 2734, 2735, 2736,...
10. 5839, 5838, 5837,...
11. 7614, 7624, 7634,...
12. 4169, 4069, 3969,...

Write the greater number.

13. \$7219
    \$7283
14. 8532
    8618

ADDITION AND SUBTRACTION OF 3-DIGIT NUMBERS

Alligator Pie

Alligator pie, alligator pie,
If I don't get some I think I'm gonna die.
Give away the green grass, give away the sky,
But don't give away my alligator pie.

Alligator stew, alligator stew,
If I don't get some I don't know what I'll do.
Give away my furry hat, give away my shoe,
But don't give away my alligator stew.

Alligator soup, alligator soup,
If I don't get some I think I'm gonna droop.
Give away my hockey-stick, give away my hoop,
But don't give away my alligator soup.

—Dennis Lee

Adding 3-Digit Numbers: Without Trading

Judy has 224 stamps

Hundreds	Tens	Ones

Charles has 165 stamps.

Hundreds	Tens	Ones

How many stamps are there in all?

Add the ones. → Add the tens. → Add the hundreds.

$$\begin{array}{r} 224 \\ +165 \\ \hline 9 \end{array} \qquad \begin{array}{r} 224 \\ +165 \\ \hline 89 \end{array} \qquad \begin{array}{r} 224 \\ +165 \\ \hline 389 \end{array}$$

Judy and Charles have 389 stamps in all.

Add.

1. $\begin{array}{r} 382 \\ +416 \\ \hline \end{array}$
2. $\begin{array}{r} 527 \\ +232 \\ \hline \end{array}$
3. $\begin{array}{r} 107 \\ +361 \\ \hline \end{array}$
4. $\begin{array}{r} 610 \\ +156 \\ \hline \end{array}$
5. $\begin{array}{r} 703 \\ +285 \\ \hline \end{array}$
6. $\begin{array}{r} 613 \\ +265 \\ \hline \end{array}$
7. $\begin{array}{r} 345 \\ +253 \\ \hline \end{array}$
8. $\begin{array}{r} 529 \\ +170 \\ \hline \end{array}$
9. $\begin{array}{r} 611 \\ +\ 68 \\ \hline \end{array}$
10. $\begin{array}{r} 413 \\ +376 \\ \hline \end{array}$
11. $\begin{array}{r} 440 \\ +350 \\ \hline \end{array}$
12. $\begin{array}{r} 128 \\ +651 \\ \hline \end{array}$
13. $\begin{array}{r} 517 \\ +281 \\ \hline \end{array}$
14. $\begin{array}{r} 261 \\ +127 \\ \hline \end{array}$
15. $\begin{array}{r} 624 \\ +335 \\ \hline \end{array}$
16. $\begin{array}{r} 521 \\ +336 \\ \hline \end{array}$
17. $\begin{array}{r} 446 \\ +221 \\ \hline \end{array}$
18. $\begin{array}{r} 381 \\ +407 \\ \hline \end{array}$

Add.

1. 173 + 215
2. 735 + 253
3. 617 + 172
4. 506 + 323
5. 45 + 723
6. 519 + 260
7. 802 + 163
8. 571 + 406
9. 35 + 762
10. 413 + 72
11. 321 + 672
12. 262 + 517
13. 372 + 26
14. 427 + 202
15. 305 + 594
16. 165 + 732
17. 162 + 827
18. 555 + 21
19. 102 + 756
20. 93 + 105
21. 820 + 136
22. 551 + 237
23. 212 + 614
24. 282 + 706
25. 279 + 620
26. 117 + 682
27. 423 + 504
28. 371 + 26
29. 6 + 253
30. 527 + 241

31. Ling needs 243 more stickers to complete her collection. She already has 612 stickers. How many stickers make up the whole collection?

32. Ruth has 276 stamps. Jill has 200 more than Ruth. How many stamps does Jill have?

PROBLEM SOLVING

When Alli Gator added 224 + 227, he said his answer was four hundred forty-eleven.

Explain why Alli's answer is wrong.

Adding 3-Digit Numbers: Trading Ones

Mr. Brown's class raised \$127.

Mrs. Cook's class raised \$146.

How much money did both classes raise?

Add the ones. Trade 10 ones for a ten.	Add the tens.	Add the hundreds.
$\begin{array}{r} {}^{1} \\ 127 \\ +146 \\ \hline 3 \end{array}$	$\begin{array}{r} {}^{1} \\ 127 \\ +146 \\ \hline 73 \end{array}$	$\begin{array}{r} 127 \\ +146 \\ \hline 273 \end{array}$

The two classes raised \$273.

Add.

1. $\begin{array}{r} 237 \\ +148 \\ \hline \end{array}$
2. $\begin{array}{r} 159 \\ +226 \\ \hline \end{array}$
3. $\begin{array}{r} 446 \\ +228 \\ \hline \end{array}$
4. $\begin{array}{r} 509 \\ +47 \\ \hline \end{array}$
5. $\begin{array}{r} 228 \\ +336 \\ \hline \end{array}$
6. $\begin{array}{r} 427 \\ +257 \\ \hline \end{array}$
7. $\begin{array}{r} 82 \\ +102 \\ \hline \end{array}$
8. $\begin{array}{r} 409 \\ +208 \\ \hline \end{array}$
9. $\begin{array}{r} 637 \\ +25 \\ \hline \end{array}$
10. $\begin{array}{r} 239 \\ +527 \\ \hline \end{array}$
11. $\begin{array}{r} 309 \\ +502 \\ \hline \end{array}$
12. $\begin{array}{r} 469 \\ +217 \\ \hline \end{array}$
13. $\begin{array}{r} 886 \\ +109 \\ \hline \end{array}$
14. $\begin{array}{r} 86 \\ +307 \\ \hline \end{array}$
15. $\begin{array}{r} 521 \\ +244 \\ \hline \end{array}$
16. $\begin{array}{r} 628 \\ +168 \\ \hline \end{array}$
17. $\begin{array}{r} 225 \\ +148 \\ \hline \end{array}$
18. $\begin{array}{r} 365 \\ +127 \\ \hline \end{array}$

Add.

1. $\begin{array}{r} 345 \\ +227 \\ \hline \end{array}$ 2. $\begin{array}{r} 523 \\ +158 \\ \hline \end{array}$ 3. $\begin{array}{r} 256 \\ +137 \\ \hline \end{array}$ 4. $\begin{array}{r} 907 \\ +\ 38 \\ \hline \end{array}$ 5. $\begin{array}{r} 219 \\ +358 \\ \hline \end{array}$ 6. $\begin{array}{r} 429 \\ +335 \\ \hline \end{array}$

7. $\begin{array}{r} 387 \\ +207 \\ \hline \end{array}$ 8. $\begin{array}{r} 564 \\ +127 \\ \hline \end{array}$ 9. $\begin{array}{r} 618 \\ +268 \\ \hline \end{array}$ 10. $\begin{array}{r} 729 \\ +108 \\ \hline \end{array}$ 11. $\begin{array}{r} 887 \\ +103 \\ \hline \end{array}$ 12. $\begin{array}{r} 223 \\ +446 \\ \hline \end{array}$

13. $\begin{array}{r} 517 \\ +265 \\ \hline \end{array}$ 14. $\begin{array}{r} 727 \\ +142 \\ \hline \end{array}$ 15. $\begin{array}{r} 906 \\ +\ 86 \\ \hline \end{array}$ 16. $\begin{array}{r} 68 \\ +729 \\ \hline \end{array}$

17. $\begin{array}{r} 602 \\ +\ \ 9 \\ \hline \end{array}$ 18. $\begin{array}{r} 317 \\ +569 \\ \hline \end{array}$ 19. $\begin{array}{r} 225 \\ +545 \\ \hline \end{array}$

20. $\begin{array}{r} 366 \\ +224 \\ \hline \end{array}$ 21. $\begin{array}{r} 538 \\ +228 \\ \hline \end{array}$ 22. $\begin{array}{r} 650 \\ +240 \\ \hline \end{array}$

23. $\begin{array}{r} 700 \\ +163 \\ \hline \end{array}$ 24. $\begin{array}{r} 217 \\ +156 \\ \hline \end{array}$ 25. $\begin{array}{r} 339 \\ +549 \\ \hline \end{array}$

26. $\begin{array}{r} 553 \\ +\ 39 \\ \hline \end{array}$ 27. $\begin{array}{r} 707 \\ +103 \\ \hline \end{array}$ 28. $\begin{array}{r} 446 \\ +225 \\ \hline \end{array}$

29. Last year the telephone book had 642 pages.
This year it has 29 pages more.
How many pages are there this year?

30. Last week Kim sold 235 newspapers.
This week she sold 15 more newspapers.
How many did she sell this week?

31. Lester B. Pearson school has 328 boys and 339 girls.
It has 34 teachers.
How many students are in that school?

32. Kevin let out 118 m of kite string.
Later, he let out the remaining 24 m.
How much string was attached to the kite?

Trading 10 Tens for a Hundred

Brad had 2 hundreds and 12 tens.

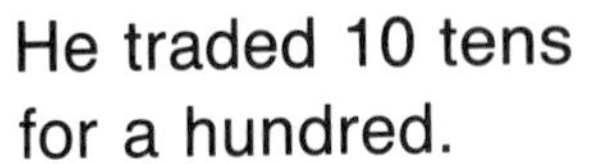

He traded 10 tens for a hundred.

Now he has 3 hundreds and 2 tens.

Trade 10 tens for a hundred. Copy and complete.

1.

Hundreds	Tens
2	17
▒	▒

1. Hundreds	Tens
2	17
3	7

2.

Hundreds	Tens
5	19
▒	▒

3.

Hundreds	Tens
3	12
▒	▒

4.

Hundreds	Tens	Ones
3	6	5
+4	8	3
▒	▒	▒
	trade	
▒	▒	▒

4. Hundreds	Tens	Ones
3	6	5
+4	8	3
7	14	8
	trade	
8	4	8

5.

Hundreds	Tens	Ones
2	3	7
+1	9	2
▒	▒	▒
	trade	
▒	▒	▒

6.

Hundreds	Tens	Ones
5	4	3
+2	7	1
▒	▒	▒
	trade	
▒	▒	▒

7.

Hundreds	Tens	Ones
4	7	6
+3	5	2
▒	▒	▒
	trade	
▒	▒	▒

8.

Hundreds	Tens	Ones
7	9	2
+1	9	5
▒	▒	▒
	trade	
▒	▒	▒

Adding 3-Digit Numbers: Trading Tens

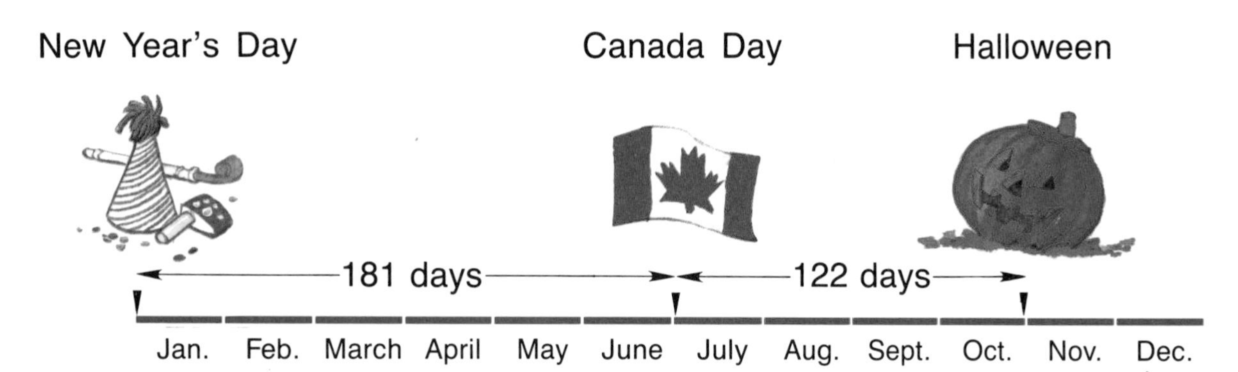

Canada Day is 181 days after New Year's Day.
Halloween is 122 days after Canada Day.
How many days is Halloween after New Year's Day?

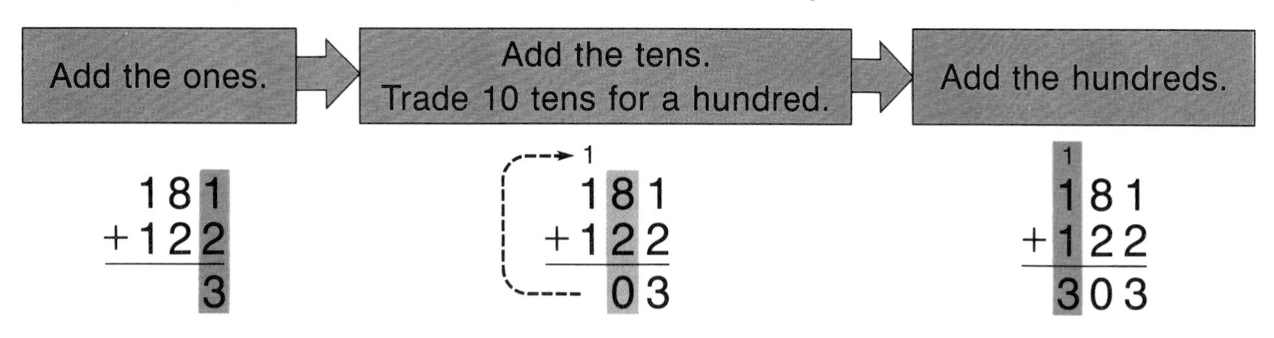

Halloween is 303 days after New Year's Day.

Add.

1. 371 + 284	2. 526 + 183	3. 428 + 391	4. 647 + 181	5. 562 + 284	6. 372 + 294
7. 552 + 361	8. 483 + 294	9. 621 + 273	10. 521 + 293	11. 670 + 298	12. 236 + 571
13. 683 + 184	14. 372 + 461	15. 766 + 190	16. 328 + 513	17. 791 + 128	18. 673 + 94

Adding 3-Digit Numbers: Trading Tens and Ones

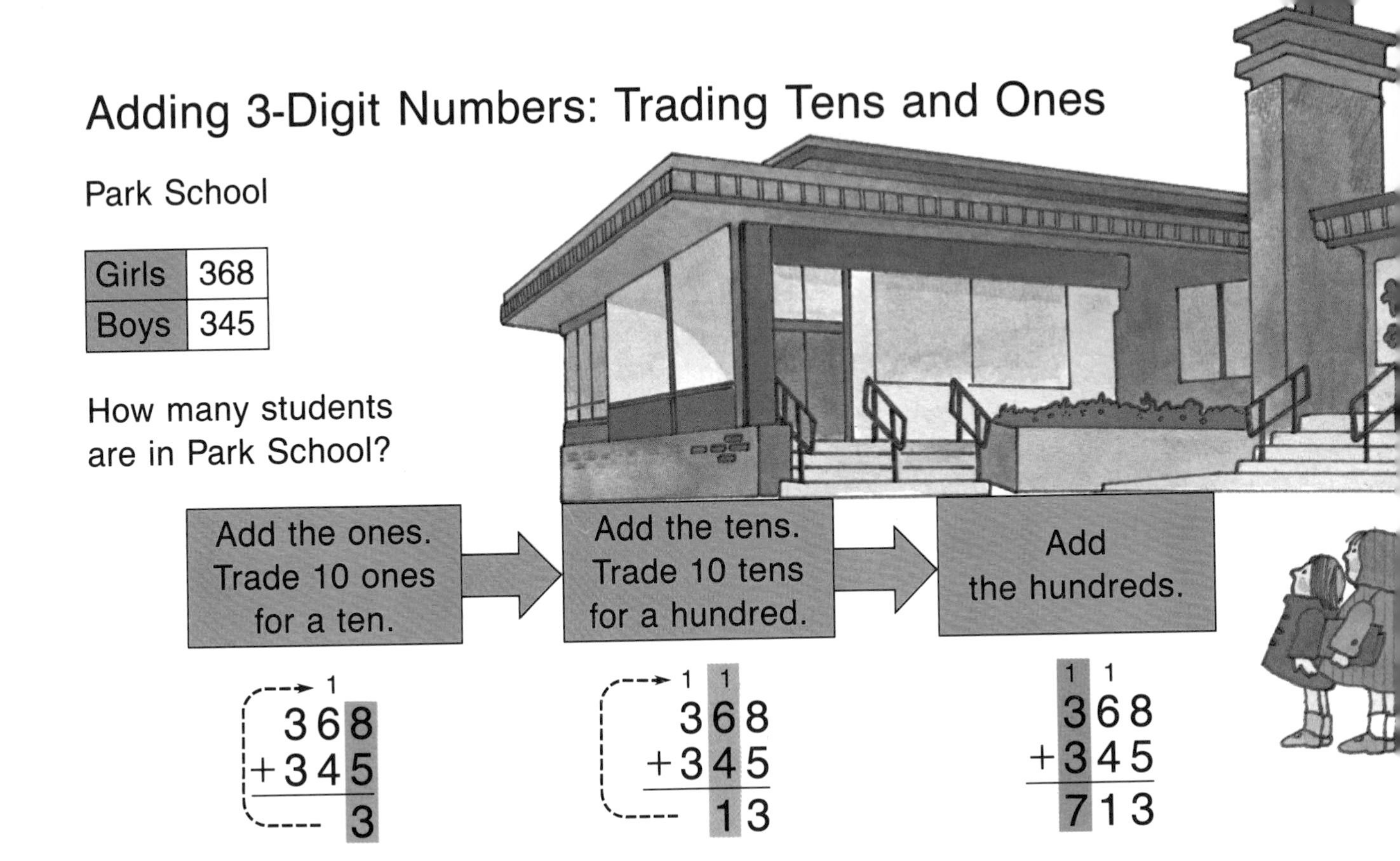

Park School

Girls	368
Boys	345

How many students are in Park School?

Add the ones. Trade 10 ones for a ten. → Add the tens. Trade 10 tens for a hundred. → Add the hundreds.

$$\begin{array}{r} {}^{1} \\ 368 \\ +345 \\ \hline 3 \end{array} \qquad \begin{array}{r} {}^{1}{}^{1} \\ 368 \\ +345 \\ \hline 13 \end{array} \qquad \begin{array}{r} {}^{1}{}^{1} \\ 368 \\ +345 \\ \hline 713 \end{array}$$

There are 713 students in Park School.

Add.

1. 547 + 368
2. 387 + 293
3. 429 + 392
4. 526 + 297
5. 695 + 287
6. 98 + 829
7. 507 + 489
8. 629 + 293
9. 705 + 197
10. 844 + 77
11. 663 + 229
12. 441 + 372
13. 239 + 486
14. 186 + 477
15. 329 + 518
16. 98 + 672
17. 409 + 208
18. 445 + 423
19. 323 + 521
20. 697 + 188
21. 237 + 563
22. 591 + 187
23. 629 + 183
24. 552 + 278

Add.

1. $\begin{array}{r} 627 \\ +184 \\ \hline \end{array}$
2. $\begin{array}{r} 529 \\ +382 \\ \hline \end{array}$
3. $\begin{array}{r} 209 \\ +626 \\ \hline \end{array}$
4. $\begin{array}{r} 582 \\ +329 \\ \hline \end{array}$
5. $\begin{array}{r} 477 \\ +377 \\ \hline \end{array}$
6. $\begin{array}{r} 288 \\ +319 \\ \hline \end{array}$
7. $\begin{array}{r} 703 \\ +126 \\ \hline \end{array}$
8. $\begin{array}{r} 512 \\ +318 \\ \hline \end{array}$
9. $\begin{array}{r} 756 \\ +129 \\ \hline \end{array}$
10. $\begin{array}{r} 607 \\ +109 \\ \hline \end{array}$
11. $\begin{array}{r} 752 \\ +219 \\ \hline \end{array}$
12. $\begin{array}{r} 214 \\ +732 \\ \hline \end{array}$
13. $\begin{array}{r} 625 \\ +215 \\ \hline \end{array}$
14. $\begin{array}{r} 529 \\ +385 \\ \hline \end{array}$
15. $\begin{array}{r} 425 \\ +535 \\ \hline \end{array}$
16. $\begin{array}{r} 219 \\ +686 \\ \hline \end{array}$
17. $\begin{array}{r} 527 \\ +384 \\ \hline \end{array}$
18. $\begin{array}{r} 822 \\ +\ 86 \\ \hline \end{array}$
19. $\begin{array}{r} 425 \\ +475 \\ \hline \end{array}$
20. $\begin{array}{r} 381 \\ +529 \\ \hline \end{array}$
21. $\begin{array}{r} 427 \\ +\ 83 \\ \hline \end{array}$
22. $\begin{array}{r} 256 \\ +512 \\ \hline \end{array}$
23. $\begin{array}{r} 782 \\ +119 \\ \hline \end{array}$
24. $\begin{array}{r} 279 \\ +382 \\ \hline \end{array}$

25. Angelo's reader has 368 pages.
Jessica's science book has
56 pages more.
How many pages are in Jessica's
science book?

26. This year there are 713 students
in Park School.
Last year there were 69 more.
How many students were in
Park School last year?

27. Park School has 768 old math books
and 259 French books.
It has 139 new math books.
How many math books do they have
in all?

28. The school raised \$389 in September
and \$426 in October.
How much was raised in both
months?

Adding 3-Digit Numbers: 4-Digit Sums

Ninety years ago the bison
was almost extinct.
There were about 850 bison in
Canada and 950 outside Canada.

How many bison were there in all?

Add the ones.	Add the tens. Trade.	Add the hundreds.	Write 10 hundreds as a thousand.
850 + 950 = 0	$\overset{1}{8}50$ + 950 = 00	$\overset{1}{8}50$ + 950 = 1800 (18 hundreds)	850 + 950 = 1800

There were about 1800 bison in all.

Add.

1. 578 + 720	2. 237 + 926	3. 581 + 296	4. 398 + 965	5. 887 + 426	6. 563 + 729
7. 473 + 609	8. 538 + 727	9. 399 + 686	10. 448 + 729	11. 307 + 809	12. 677 + 366
13. 433 + 728	14. 292 + 98	15. 680 + 720	16. 998 + 89	17. 326 + 623	18. 492 + 924
19. 868 + 709	20. 687 + 19	21. 693 + 387	22. 557 + 727	23. 816 + 86	24. 637 + 509

Add.

1. $\begin{array}{r} 743 \\ +289 \\ \hline \end{array}$
2. $\begin{array}{r} 547 \\ +655 \\ \hline \end{array}$
3. $\begin{array}{r} 396 \\ +867 \\ \hline \end{array}$
4. $\begin{array}{r} 446 \\ +883 \\ \hline \end{array}$
5. $\begin{array}{r} 909 \\ +807 \\ \hline \end{array}$
6. $\begin{array}{r} 888 \\ +777 \\ \hline \end{array}$
7. $\begin{array}{r} 336 \\ +597 \\ \hline \end{array}$
8. $\begin{array}{r} 485 \\ +733 \\ \hline \end{array}$
9. $\begin{array}{r} 659 \\ +748 \\ \hline \end{array}$
10. $\begin{array}{r} 709 \\ +866 \\ \hline \end{array}$
11. $\begin{array}{r} 337 \\ +954 \\ \hline \end{array}$
12. $\begin{array}{r} 389 \\ +692 \\ \hline \end{array}$
13. $\begin{array}{r} 627 \\ +909 \\ \hline \end{array}$
14. $\begin{array}{r} 767 \\ +593 \\ \hline \end{array}$
15. $\begin{array}{r} 444 \\ +76 \\ \hline \end{array}$
16. $\begin{array}{r} 997 \\ +27 \\ \hline \end{array}$
17. $\begin{array}{r} 353 \\ +778 \\ \hline \end{array}$
18. $\begin{array}{r} 918 \\ +86 \\ \hline \end{array}$

19. Ten years ago there were about 6000 bison living in parks. Now there are about 950 more. About how many bison are there now?

20. One herd contained 829 bison. Another herd contained 927 bison. How many bison were there in all?

ESTIMATING

Estimate: Are there more apples on the bison's side of the fence or on the other side?

Adding with Trading: 3 Addends

Lynne collected 287 cards.
Mark collected 376 cards.
Heather collected 398 cards.
How many cards are there in all?

Add the ones. Trade 20 ones for 2 tens.	→	Add the tens. Trade 20 tens for 2 hundreds.	→	Add the hundreds.

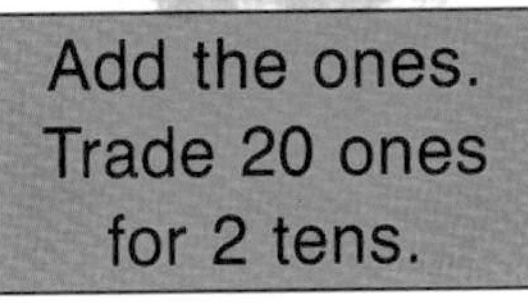

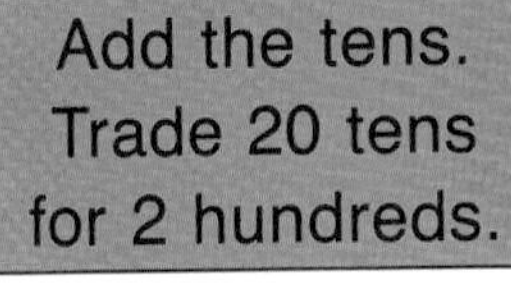


```
  2 2
  287
  376
+ 398
 1061
```

There are 1061 cards in all.

Add.

1.	2.	3.	4.	5.	6.
576 357 +289	109 760 +847	667 389 +946	387 946 +707	221 568 +735	606 909 +321

7.	8.	9.	10.	11.	12.
472 98 +361	331 22 +435	697 388 +429	554 380 +229	119 256 +309	67 551 +735

CARD COLLECTING CONTEST WINNERS

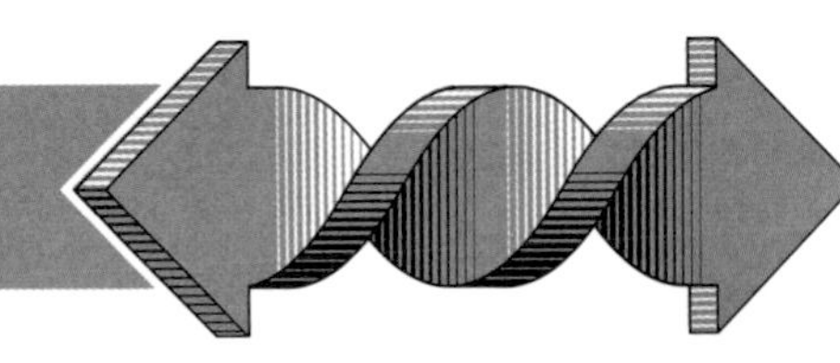

Tara: 467 cards   Roy: 399 cards   Janine: 487 cards

Louis: 489 cards   Tim: 407 cards   Laura: 529 cards

1. How many cards do the three girls have in all?
2. Do Tara and Laura together have more cards than Roy and Janine?
3. Which child collected 90 cards more than Roy?
4. Name 3 children who collected a total of more than 1400 cards.
5. How many cards do the 6 winners have in all?
6. Which 2 children collected a total of 806 cards?

Subtracting 3-Digit Numbers: Without Trading

One year there was rain on 122 days.
There were 365 days in the year.

How many days were without rain?

Subtract the ones.	Subtract the tens.	Subtract the hundreds.
$\begin{array}{r} 365 \\ -122 \\ \hline 3 \end{array}$	$\begin{array}{r} 365 \\ -122 \\ \hline 43 \end{array}$	$\begin{array}{r} 365 \\ -122 \\ \hline 243 \end{array}$

There were 243 days without rain.

Subtract.

1. $\begin{array}{r} 587 \\ -253 \\ \hline \end{array}$	2. $\begin{array}{r} 476 \\ -352 \\ \hline \end{array}$	3. $\begin{array}{r} 695 \\ -273 \\ \hline \end{array}$	4. $\begin{array}{r} 887 \\ -671 \\ \hline \end{array}$	5. $\begin{array}{r} 890 \\ -720 \\ \hline \end{array}$	6. $\begin{array}{r} 535 \\ -205 \\ \hline \end{array}$
7. $\begin{array}{r} 394 \\ -381 \\ \hline \end{array}$	8. $\begin{array}{r} 558 \\ -427 \\ \hline \end{array}$	9. $\begin{array}{r} 649 \\ -35 \\ \hline \end{array}$	10. $\begin{array}{r} 804 \\ -602 \\ \hline \end{array}$	11. $\begin{array}{r} 596 \\ -204 \\ \hline \end{array}$	12. $\begin{array}{r} 999 \\ -387 \\ \hline \end{array}$
13. $\begin{array}{r} 778 \\ -772 \\ \hline \end{array}$	14. $\begin{array}{r} 898 \\ -673 \\ \hline \end{array}$	15. $\begin{array}{r} 409 \\ -403 \\ \hline \end{array}$	16. $\begin{array}{r} 677 \\ -574 \\ \hline \end{array}$	17. $\begin{array}{r} 986 \\ -906 \\ \hline \end{array}$	18. $\begin{array}{r} 489 \\ -370 \\ \hline \end{array}$

Subtract.

1. $\begin{array}{r} 567 \\ -463 \\ \hline \end{array}$
2. $\begin{array}{r} 289 \\ -274 \\ \hline \end{array}$
3. $\begin{array}{r} 683 \\ -583 \\ \hline \end{array}$
4. $\begin{array}{r} 628 \\ -207 \\ \hline \end{array}$
5. $\begin{array}{r} 539 \\ -128 \\ \hline \end{array}$
6. $\begin{array}{r} 238 \\ -127 \\ \hline \end{array}$
7. $\begin{array}{r} 599 \\ -398 \\ \hline \end{array}$
8. $\begin{array}{r} 409 \\ -206 \\ \hline \end{array}$
9. $\begin{array}{r} 667 \\ -332 \\ \hline \end{array}$
10. $\begin{array}{r} 697 \\ -581 \\ \hline \end{array}$
11. $\begin{array}{r} 800 \\ -300 \\ \hline \end{array}$
12. $\begin{array}{r} 428 \\ -317 \\ \hline \end{array}$
13. $\begin{array}{r} 226 \\ -105 \\ \hline \end{array}$
14. $\begin{array}{r} 888 \\ -603 \\ \hline \end{array}$
15. $\begin{array}{r} 992 \\ -720 \\ \hline \end{array}$

16. There was snow on 65 days out of 365 days.
How many days were there without snow?

17. The first year there were 103 days of rain and 30 days of snow.
The second year there were 116 days of rain and 32 days of snow.
How many more days of rain were there in the second year than in the first?

18. A leap year has 366 days.
There was sunshine on 252 days in a leap year.
How many days were without sunshine that year?

PROBLEM SOLVING

One leap year there were 106 days with rain.
On 35 other days, it snowed.
How many days were without rain or snow?

Subtracting 3-Digit Numbers: Trading a Ten for 10 Ones

Mrs. Higgins had $323.
She spent $119.
How much money
did she have left?

Mrs. Higgins had $323.

She traded a ten for 10 ones.
She spent $119.

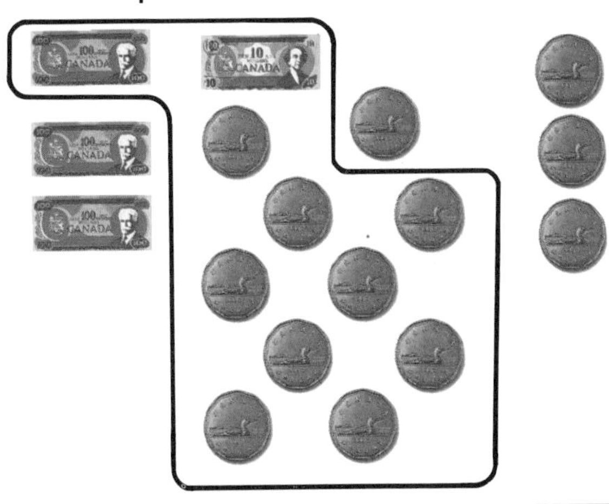

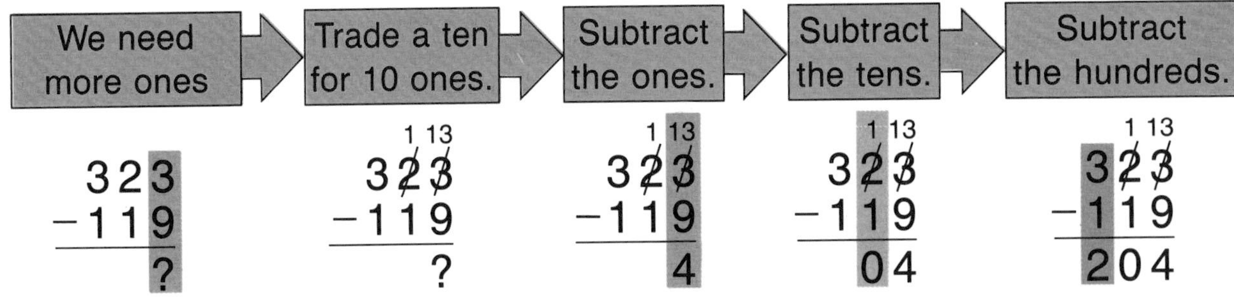

Mrs. Higgins had $204 left.

Subtract.

1.	2.	3.	4.	5.	6.
728 − 519	897 − 348	256 − 139	492 − 284	867 − 349	653 − 237

7.	8.	9.	10.	11.	12.
392 − 138	451 − 237	395 − 86	722 − 306	557 − 332	636 − 207

Subtract.

1. $583 - 265$	2. $392 - 147$	3. $586 - 267$	4. $428 - 219$	5. $637 - 129$	6. $525 - 106$
7. $673 - 246$	8. $722 - 318$	9. $437 - 119$	10. $388 - 229$	11. $471 - 237$	12. $526 - 19$
13. $739 - 226$	14. $647 - 129$	15. $283 - 76$	16. $633 - 424$	17. $846 - 518$	18. $731 - 309$
19. $444 - 335$	20. $666 - 448$	21. $723 - 618$	22. $592 - 177$	23. $611 - 208$	24. $709 - 306$

25. Last year Ann saved $129.
This year she saved $244.
How much more did she save this year?

27. The ABC movie theatre has 982 seats.
727 seats are taken.
How many seats are left?

26. In one leap year, Labor Day was on the 247th day of the year.
Halloween was on the 305th day of the year.
How many days came after Labor Day that year?

PROBLEM SOLVING

In a leap year, Remembrance Day is the 316th day of the year.
Christmas is the 360th day of the year.
How many days from Remembrance Day to Christmas?

316 days

360 days

New Year's Day
Jan. 1

Remembrance Day
Nov. 11

Christmas Day
Dec. 25

Subtracting 3-Digit Numbers: Trading Hundreds

One year there were 193 school days. How many days were not school days?

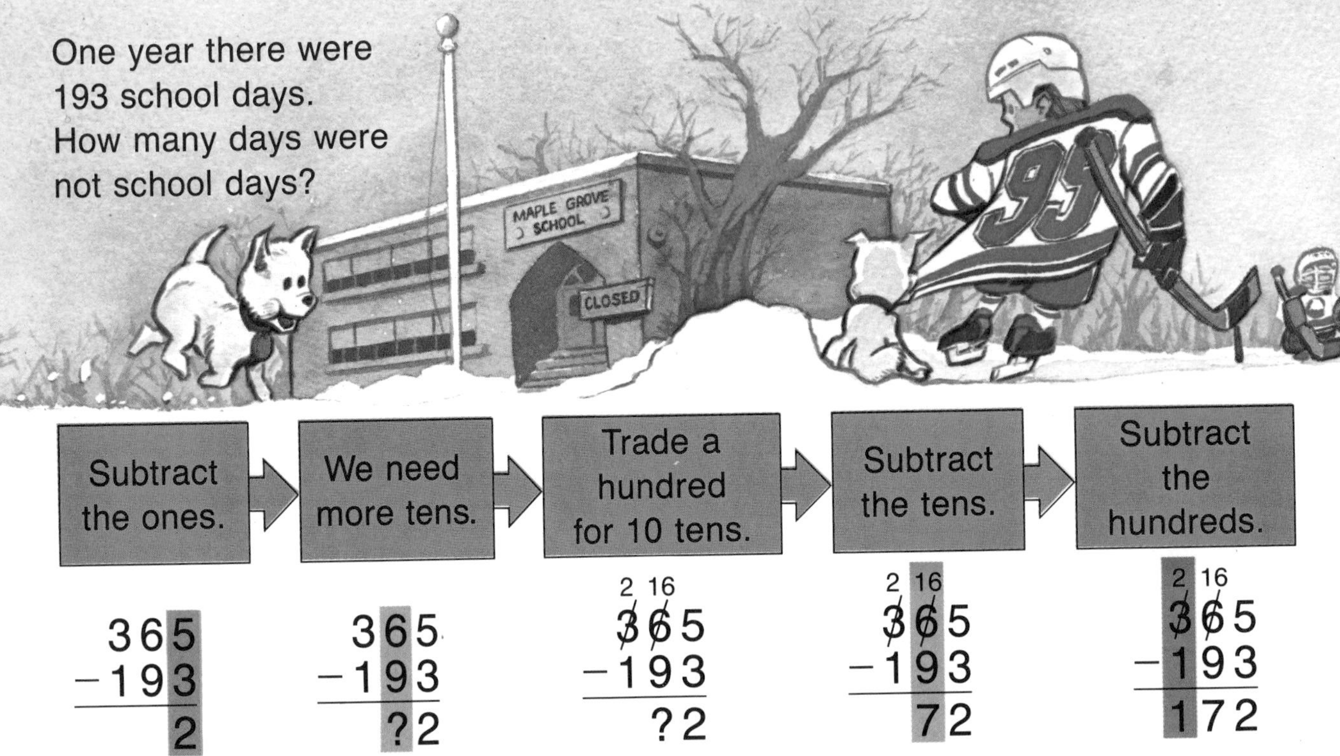

Subtract the ones. → We need more tens. → Trade a hundred for 10 tens. → Subtract the tens. → Subtract the hundreds.

$$\begin{array}{r} 365 \\ -193 \\ \hline 2 \end{array} \qquad \begin{array}{r} 365 \\ -193 \\ \hline ?2 \end{array} \qquad \begin{array}{r} {}^{2\,16} \\ \not{3}\not{6}5 \\ -193 \\ \hline ?2 \end{array} \qquad \begin{array}{r} {}^{2\,16} \\ \not{3}\not{6}5 \\ -193 \\ \hline 72 \end{array} \qquad \begin{array}{r} {}^{2\,16} \\ \not{3}\not{6}5 \\ -193 \\ \hline 172 \end{array}$$

172 days were not school days.

Subtract.

1. $348 - 163$	2. $629 - 357$	3. $837 - 256$	4. $549 - 167$	5. $888 - 692$	6. $379 - 286$
7. $746 - 360$	8. $577 - 294$	9. $860 - 580$	10. $428 - 235$	11. $739 - 686$	12. $440 - 280$
13. $469 - 184$	14. $326 - 182$	15. $727 - 633$	16. $826 - 105$	17. $688 - 392$	18. $557 - 365$
19. $365 - 183$	20. $782 - 590$	21. $826 - 345$	22. $706 - 503$	23. $588 - 294$	24. $853 - 382$

Subtracting 3-Digit Numbers: One Trade

Subtract.

1. 864 − 259	2. 843 − 561	3. 447 − 362	4. 684 − 566	5. 793 − 577	6. 927 − 684
7. 156 − 84	8. 569 − 288	9. 783 − 259	10. 327 − 273	11. 659 − 595	12. 780 − 390
13. 538 − 386	14. 446 − 338	15. 509 − 307	16. 736 − 482	17. 597 − 329	18. 480 − 190
19. 666 − 382	20. 597 − 148	21. 367 − 286	22. 729 − 607	23. 837 − 787	24. 627 − 536

25. There were 284 pairs of skates at a skate sale.
159 pairs were sold.
How many pairs were left?

26. 92 of the 284 pairs of skates were white.
How many pairs were not white?

PROBLEM SOLVING 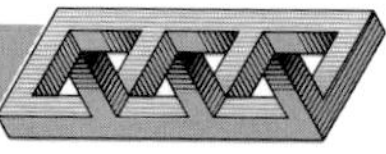

A hockey league bought
1000 hockey sticks.
569 were right-handed sticks.
How many were
left-handed sticks?

Hint: Think of 999 instead of 1000.

Subtracting: Two Trades

One year Wayne Gretzky earned 164 points.
The next year he scored 212 points.
How many more points did he earn in the second year?

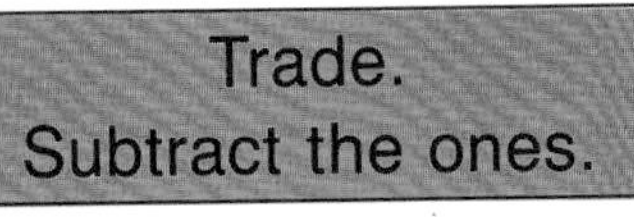

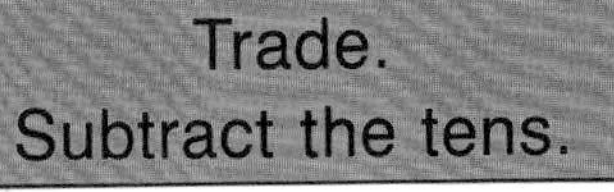

Subtract the hundreds.

Trade. Subtract the ones.	Trade. Subtract the tens.	Subtract the hundreds.
$\begin{array}{r} \scriptstyle 0\,12 \\ 2\not{1}\not{2} \\ -164 \\ \hline 8 \end{array}$	$\begin{array}{r} \scriptstyle 1\,10\,12 \\ \not{2}\not{1}\not{2} \\ -164 \\ \hline 48 \end{array}$	$\begin{array}{r} \scriptstyle 1\,10\,12 \\ \not{2}\not{1}\not{2} \\ -164 \\ \hline 48 \end{array}$

He scored 48 more points.

Subtract.

1. $\begin{array}{r} 358 \\ -179 \\ \hline \end{array}$
2. $\begin{array}{r} 765 \\ -587 \\ \hline \end{array}$
3. $\begin{array}{r} 692 \\ -293 \\ \hline \end{array}$
4. $\begin{array}{r} 217 \\ -188 \\ \hline \end{array}$
5. $\begin{array}{r} 359 \\ -173 \\ \hline \end{array}$
6. $\begin{array}{r} 536 \\ -358 \\ \hline \end{array}$
7. $\begin{array}{r} 231 \\ -87 \\ \hline \end{array}$
8. $\begin{array}{r} 697 \\ -598 \\ \hline \end{array}$
9. $\begin{array}{r} 488 \\ -289 \\ \hline \end{array}$
10. $\begin{array}{r} 357 \\ -261 \\ \hline \end{array}$
11. $\begin{array}{r} 612 \\ -319 \\ \hline \end{array}$
12. $\begin{array}{r} 287 \\ -169 \\ \hline \end{array}$
13. $\begin{array}{r} 845 \\ -357 \\ \hline \end{array}$
14. $\begin{array}{r} 668 \\ -95 \\ \hline \end{array}$
15. $\begin{array}{r} 822 \\ -657 \\ \hline \end{array}$
16. $\begin{array}{r} 221 \\ -106 \\ \hline \end{array}$
17. $\begin{array}{r} 477 \\ -377 \\ \hline \end{array}$
18. $\begin{array}{r} 826 \\ -362 \\ \hline \end{array}$
19. $\begin{array}{r} 791 \\ -92 \\ \hline \end{array}$
20. $\begin{array}{r} 665 \\ -372 \\ \hline \end{array}$
21. $\begin{array}{r} 569 \\ -388 \\ \hline \end{array}$
22. $\begin{array}{r} 784 \\ -526 \\ \hline \end{array}$
23. $\begin{array}{r} 817 \\ -578 \\ \hline \end{array}$
24. $\begin{array}{r} 927 \\ -758 \\ \hline \end{array}$

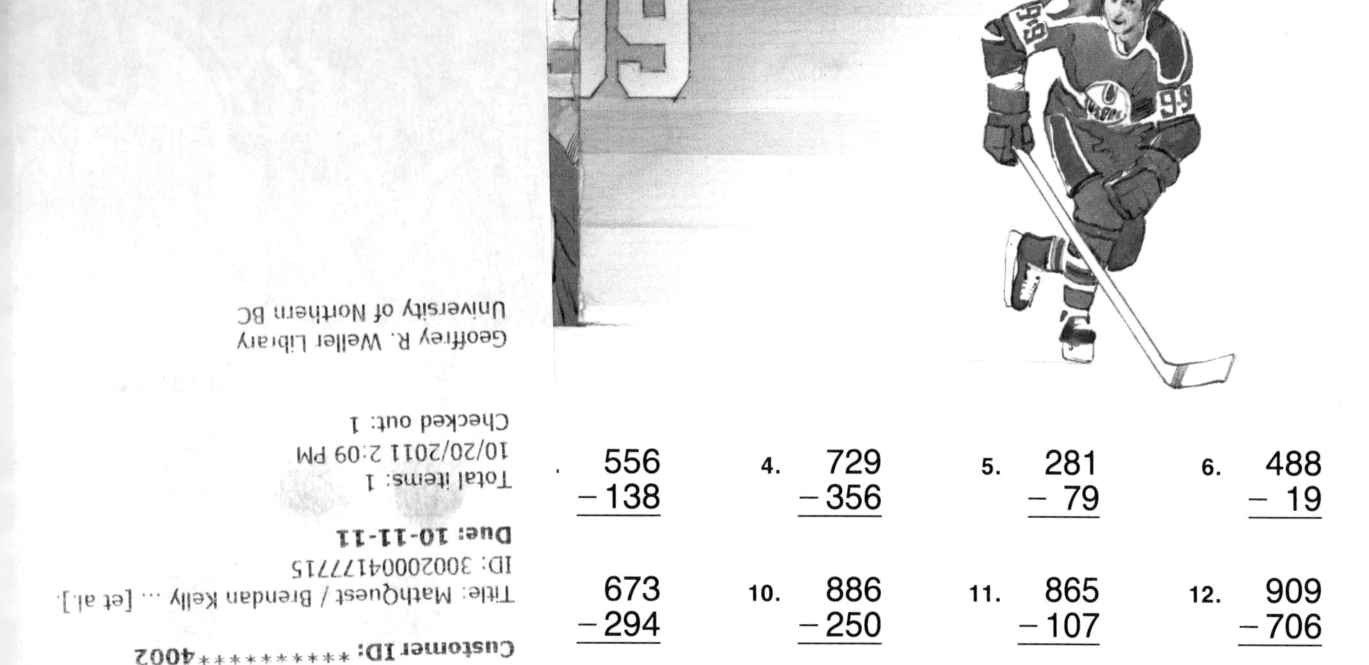

$\begin{array}{r} 556 \\ -138 \\ \hline \end{array}$	4. $\begin{array}{r} 729 \\ -356 \\ \hline \end{array}$	5. $\begin{array}{r} 281 \\ -79 \\ \hline \end{array}$	6. $\begin{array}{r} 488 \\ -19 \\ \hline \end{array}$
$\begin{array}{r} 673 \\ -294 \\ \hline \end{array}$	10. $\begin{array}{r} 886 \\ -250 \\ \hline \end{array}$	11. $\begin{array}{r} 865 \\ -107 \\ \hline \end{array}$	12. $\begin{array}{r} 909 \\ -706 \\ \hline \end{array}$
$\begin{array}{r} 641 \\ -387 \\ \hline \end{array}$	16. $\begin{array}{r} 294 \\ -107 \\ \hline \end{array}$	17. $\begin{array}{r} 383 \\ -127 \\ \hline \end{array}$	18. $\begin{array}{r} 456 \\ -328 \\ \hline \end{array}$
$\begin{array}{r} 493 \\ -339 \\ \hline \end{array}$	22. $\begin{array}{r} 278 \\ -156 \\ \hline \end{array}$	23. $\begin{array}{r} 983 \\ -709 \\ \hline \end{array}$	24. $\begin{array}{r} 710 \\ -306 \\ \hline \end{array}$

25. This table shows the number of goals and assists earned one year by the top 3 players.
Fill in the missing numbers.

	Gretzky	Bossy	Stastny
Goals	92	64	▒
Assists	120	▒	93
Total	212	147	139

How much greater is Gretzky's total than Stastny's total?

PROBLEM SOLVING

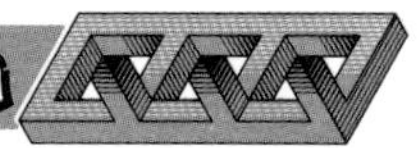

When Wayne Gretzky scored 212 points, he beat his old record by 48 points.
What was his old record?

Subtracting from Zeros

HOW DO YOU SUBTRACT FROM A NUMBER WITH ZERO TENS?

When you need more tens. . . → trade a hundred for 10 tens.

$$\begin{array}{r} 509 \\ -\,236 \\ \hline \end{array} \qquad \begin{array}{r} {\scriptstyle 4\ 10} \\ \not{5}\not{0}9 \\ -\,236 \\ \hline \end{array}$$

When you need more ones... → trade a ten for 10 ones.

$$\begin{array}{r} 307 \\ -\,119 \\ \hline \end{array} \qquad \begin{array}{r} {\scriptstyle 2\ 9\ 17} \\ \not{3}\not{0}\not{7} \\ -\,119 \\ \hline \end{array}$$

Copy each question.
Show how to trade for subtraction.
Do not subtract.

1.	2.	Example	3.	4.
607 − 209	706 − 523	1. (5 9 17) 607 − 209 2. (6 10) 706 − 523	607 − 314	507 − 183

5.	6.	7.	8.	9.	10.
800 − 128	500 − 235	600 − 150	320 − 180	908 − 379	606 − 537

11.	12.	13.	14.	15.	16.
305 − 109	607 − 413	808 − 521	903 − 588	800 − 730	700 − 607

Subtracting from Zeros

There are 305 floats in the parade. Shannon has seen 127 floats go by. How many floats are yet to come?

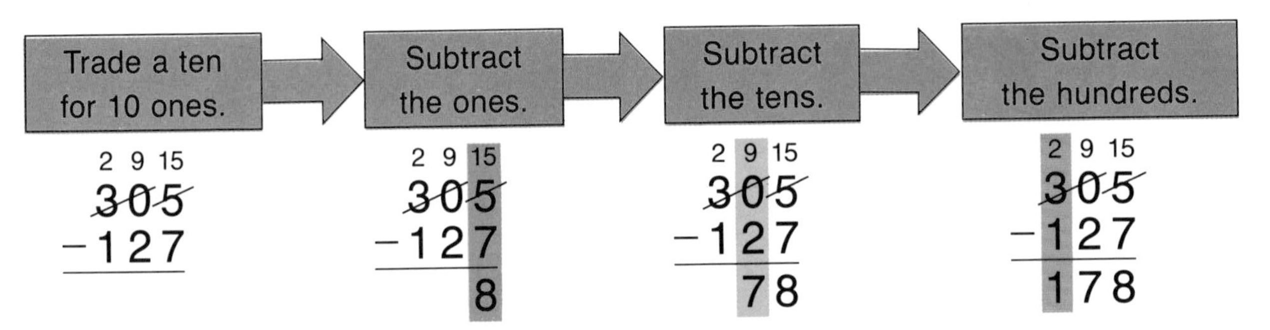

There are 178 floats yet to come.

Subtract.

1. $603 - 227$	2. $506 - 298$	3. $602 - 386$	4. $709 - 356$	5. $507 - 324$	6. $660 - 380$
7. $740 - 428$	8. $690 - 236$	9. $520 - 150$	10. $370 - 184$	11. $906 - 352$	12. $604 - 473$
13. $800 - 304$	14. $701 - 388$	15. $409 - 366$	16. $906 - 312$	17. $103 - 87$	18. $805 - 798$
19. $105 - 36$	20. $207 - 52$	21. $500 - 128$	22. $700 - 360$	23. $707 - 84$	24. $508 - 99$

Subtract.

1. $\begin{array}{r} 307 \\ -126 \\ \hline \end{array}$
2. $\begin{array}{r} 504 \\ -322 \\ \hline \end{array}$
3. $\begin{array}{r} 707 \\ -527 \\ \hline \end{array}$
4. $\begin{array}{r} 301 \\ -65 \\ \hline \end{array}$
5. $\begin{array}{r} 907 \\ -169 \\ \hline \end{array}$
6. $\begin{array}{r} 603 \\ -207 \\ \hline \end{array}$
7. $\begin{array}{r} 806 \\ -238 \\ \hline \end{array}$
8. $\begin{array}{r} 705 \\ -691 \\ \hline \end{array}$
9. $\begin{array}{r} 807 \\ -666 \\ \hline \end{array}$
10. $\begin{array}{r} 903 \\ -569 \\ \hline \end{array}$
11. $\begin{array}{r} 570 \\ -246 \\ \hline \end{array}$
12. $\begin{array}{r} 730 \\ -158 \\ \hline \end{array}$

13. There are 28 drummers in a band of 200.
How many in the band are not drummers?

14. There are 15 marching bands in the parade.
There are 208 floats pulled by tractors.
There are 99 floats pulled by motorcycles.
How many more floats are pulled by tractors than by motorcycles?

15. Mr. Dawson had $800.
He bought an old piano for $673.
How much money did he have left?

PROBLEM SOLVING

There are 200 people in the band. The table shows how many musicians there are.

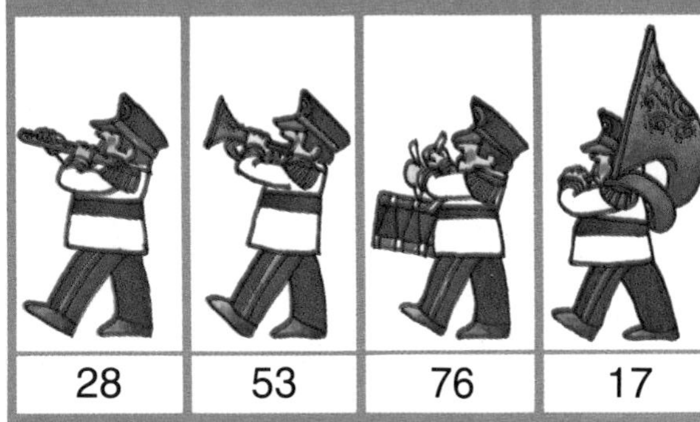

28	53	76	17

How many of the people in the band are not musicians?

JUST FOR FUN

RIDDLE: What kind of cracker isn't good for Polly?

Subtract.

A. 527 − 106

B. 689 − 408

C. 426 − 382

D. 309 − 117

E. 531 − 418

F. 687 − 298

G. 585 − 369

H. 466 − 158

I. 727 − 527

J. 409 − 216

K. 500 − 196

L. 329 − 118

M. 716 − 428

N. 296 − 97

O. 609 − 136

P. 710 − 384

Q. 695 − 507

R. 396 − 288

S. 830 − 529

T. 970 − 582

U. 900 − 63

To answer the riddle write the letters of the questions with these answers.

___ ___ ___ ___ ___ ___ ___ ___ ___ ___ ___ ___

421 389 200 108 113 44 108 421 44 304 113 108

BITS AND BYTES

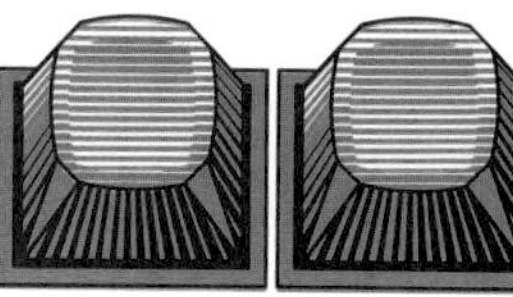

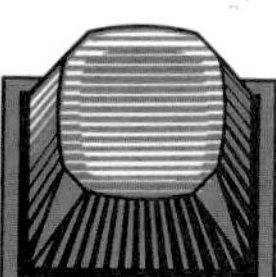

To find the missing number...

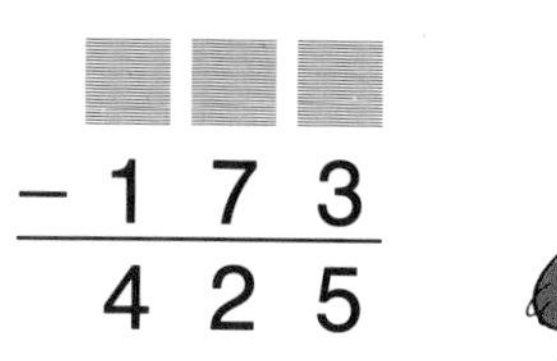

add the other two numbers.

$$\begin{array}{r} 598 \\ -\,173 \\ \hline 425 \end{array}$$

173
+425
598

Use your calculator to find the missing numbers.

1. □□□ − 387 = 202
2. □□□ − 521 = 238
3. □□□ − 693 = 127
4. □□□ − 526 = 283
5. □□□ − 435 = 382
6. □□□ − 298 = 357
7. □□□ − 426 = 288
8. □□□ − 127 = 806
9. □□□ − 356 = 449
10. □□□ − 387 = 293
11. □□□ − 721 = 97
12. □□□ − 686 = 145
13. □□□ − 499 = 223
14. □□□ − 69 = 357
15. □□□ − 185 = 492

16. □ − 327 = 451
17. □ − 587 = 123
18. □ − 421 = 87
19. □ − 289 = 564
20. □ − 88 = 427
21. □ − 177 = 369
22. □ − 809 = 126
23. □ − 243 = 756
24. □ − 392 = 245

Applications

1. When Juno was a pup, he was
   29 cm tall.
   He is now 67 cm tall.
   How much taller is he now?

2. Juno was 48 cm long.
   He is now 54 cm longer.
   How long is he now?

3. Juno's lunch now costs 98¢.
   It used to cost 29¢.
   How much more does it cost now?

Juno ate and ate.
The more he ate,
the bigger he grew.

Juno is now a full-grown sled dog.
He can run a distance of 1 km in
109 seconds.

4. How long does it take Juno to run
   2 km?

Sled dogs like Juno love to run in
the snow.
They are strong enough to pull sleds
when they are 2 years old.

5. Juno was born in 1983.
   How old is he now?

6. In what year will Juno be 8 years old?

It's fun being pulled by Juno!

CHECK-UP

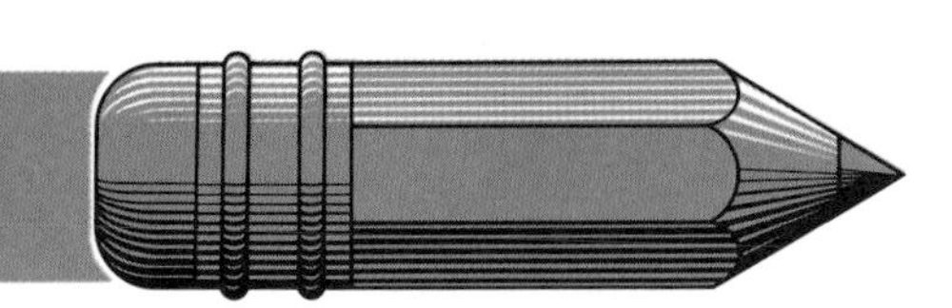

Add.

1. $\begin{array}{r} 347 \\ +532 \\ \hline \end{array}$
2. $\begin{array}{r} 578 \\ +209 \\ \hline \end{array}$
3. $\begin{array}{r} 572 \\ +263 \\ \hline \end{array}$
4. $\begin{array}{r} 613 \\ +525 \\ \hline \end{array}$
5. $\begin{array}{r} 283 \\ +239 \\ \hline \end{array}$
6. $\begin{array}{r} 465 \\ +749 \\ \hline \end{array}$
7. $\begin{array}{r} 701 \\ +163 \\ \hline \end{array}$
8. $\begin{array}{r} 438 \\ +239 \\ \hline \end{array}$
9. $\begin{array}{r} 466 \\ +250 \\ \hline \end{array}$
10. $\begin{array}{r} 415 \\ +923 \\ \hline \end{array}$
11. $\begin{array}{r} 437 \\ +163 \\ \hline \end{array}$
12. $\begin{array}{r} 839 \\ +382 \\ \hline \end{array}$

Subtract.

13. $\begin{array}{r} 327 \\ -215 \\ \hline \end{array}$
14. $\begin{array}{r} 746 \\ -319 \\ \hline \end{array}$
15. $\begin{array}{r} 484 \\ -193 \\ \hline \end{array}$
16. $\begin{array}{r} 828 \\ -439 \\ \hline \end{array}$
17. $\begin{array}{r} 809 \\ -215 \\ \hline \end{array}$
18. $\begin{array}{r} 607 \\ -219 \\ \hline \end{array}$
19. $\begin{array}{r} 968 \\ -752 \\ \hline \end{array}$
20. $\begin{array}{r} 835 \\ -428 \\ \hline \end{array}$
21. $\begin{array}{r} 864 \\ -273 \\ \hline \end{array}$
22. $\begin{array}{r} 534 \\ -275 \\ \hline \end{array}$
23. $\begin{array}{r} 305 \\ -132 \\ \hline \end{array}$
24. $\begin{array}{r} 503 \\ -415 \\ \hline \end{array}$

25. Janet collected 581 stickers.
    Mario collected 418 stickers.
    How many did they collect altogether?

26. Sasha collected 472 stickers.
    She gave 250 to her sister.
    How many did she have left?

27. There are 365 days in a year.
    Summer, fall and winter last 273 days.
    How long is spring?

28. Lockeport has 929 people.
    557 of those people are children.
    Bradwell has 168 people. How many
    people live in the two towns?

GEOMETRIC FIGURES 9

SHAPES

A square was sitting quietly
Outside his rectangular shack
When a triangle came down — *kerplunk!* —
And struck him in the back.
"I must go to the hospital,"
Cried the wounded square,
So a passing rolling circle
Picked him up and took him there.

— Shel Silverstein

from *A Light in the Attic*

Solids and Figures

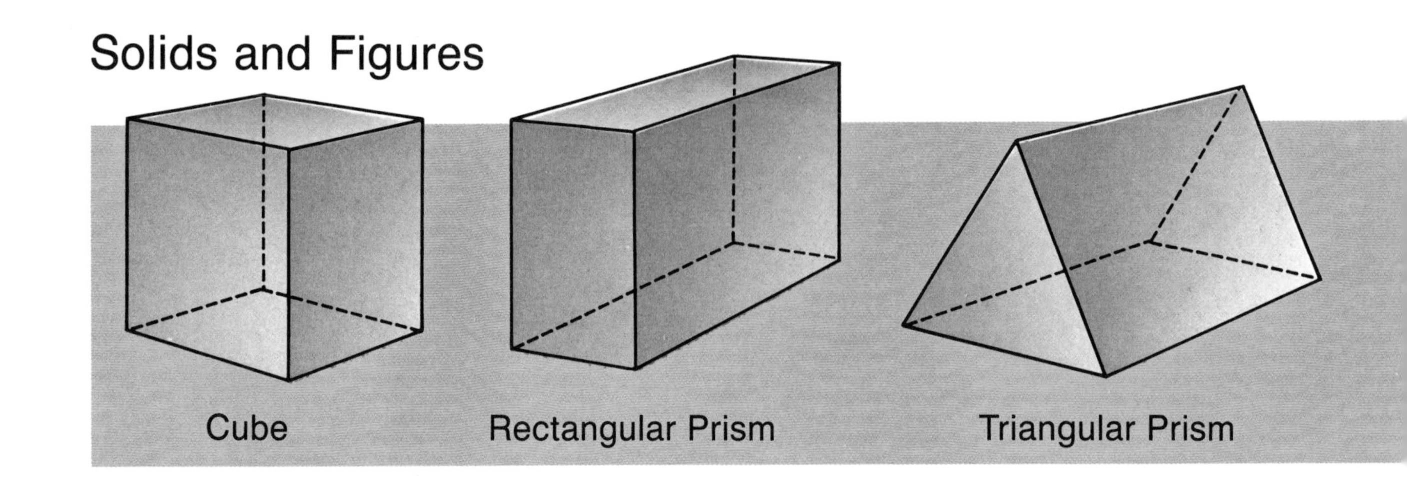

Each drawing was made by tracing a face of a solid.
For each drawing, name 2 solids that might have been used.

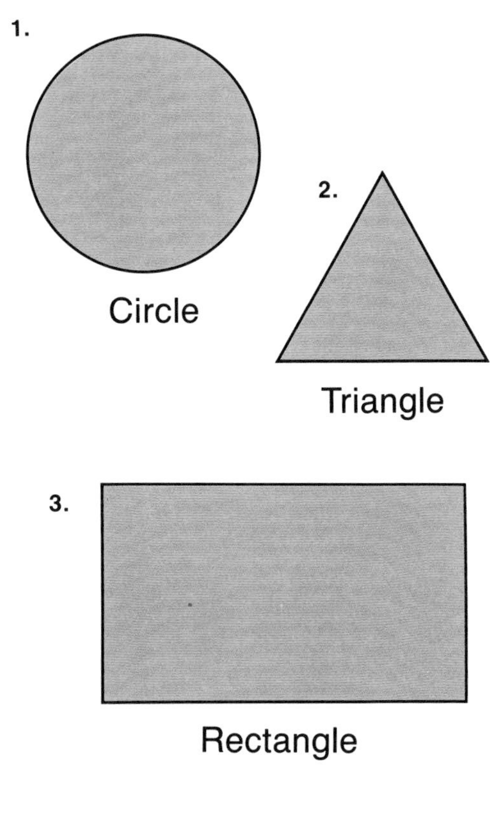

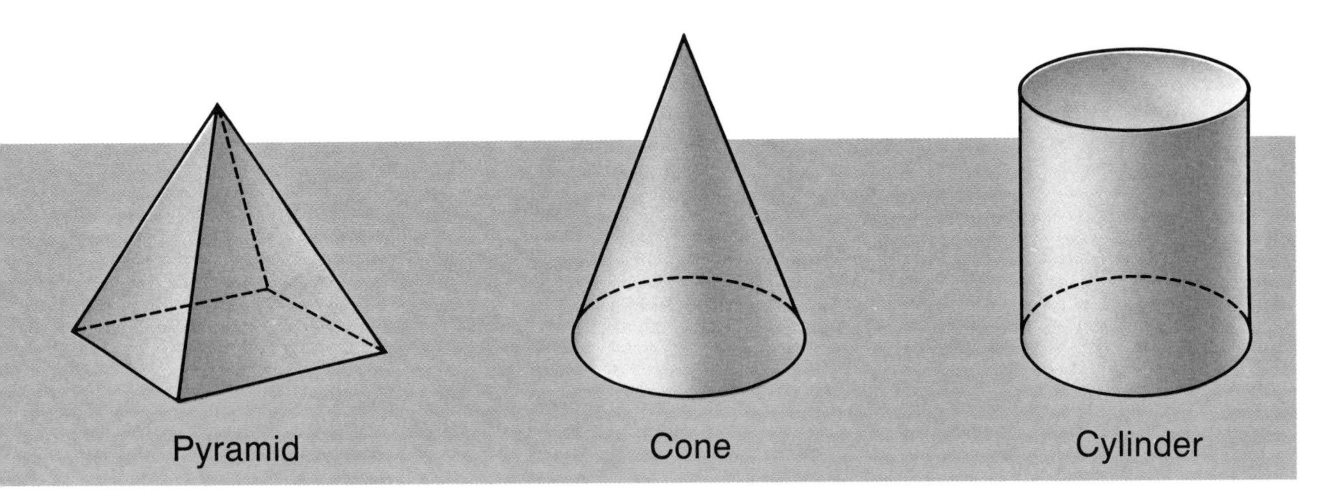

Each drawing shows 2 faces of the same solid.
Which solid was used?

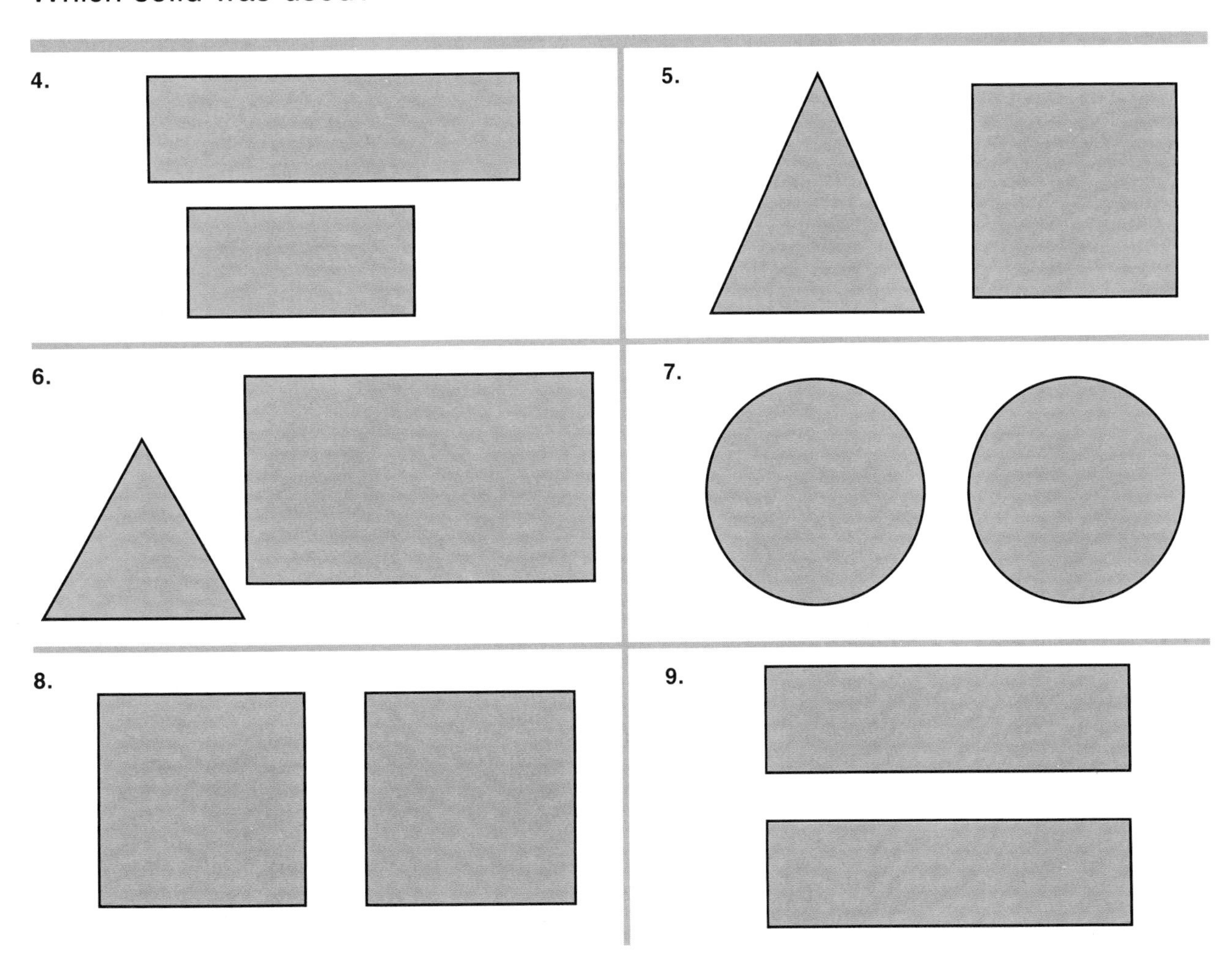

Making Prints

1. Draw the footprint of each robot. Record the shape in a table.

2. Robots with the same footprint are brothers. Name robots which are brothers.

Robot	Shape of Footprint
CRIP	square
KAL	
GORK	
TROK	
DRON	
PIM	

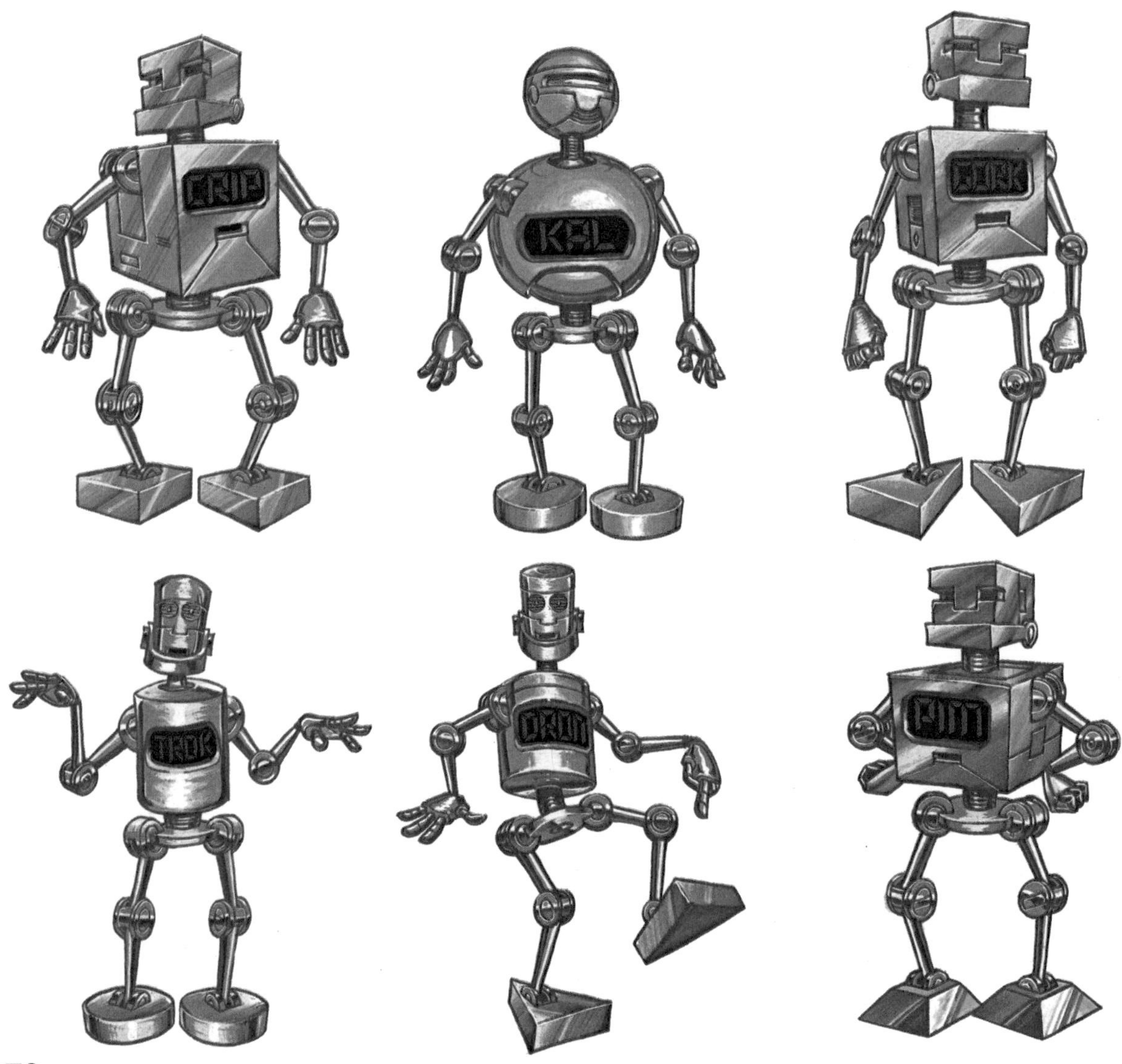

PROBLEM SOLVING

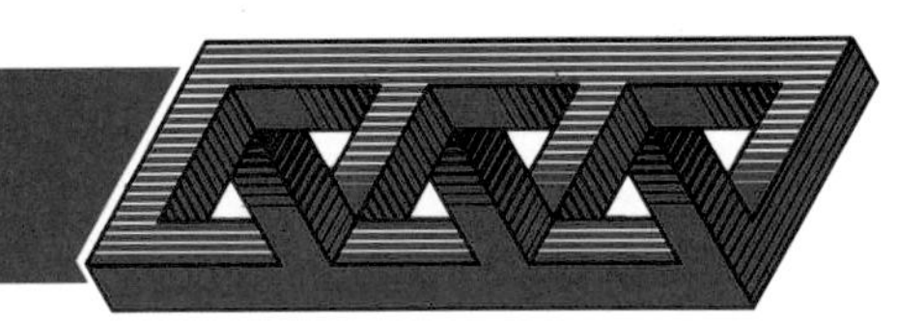

Identifying Likenesses and Differences

1. First Mystery

This robot was seen last night.

He left these footprints.

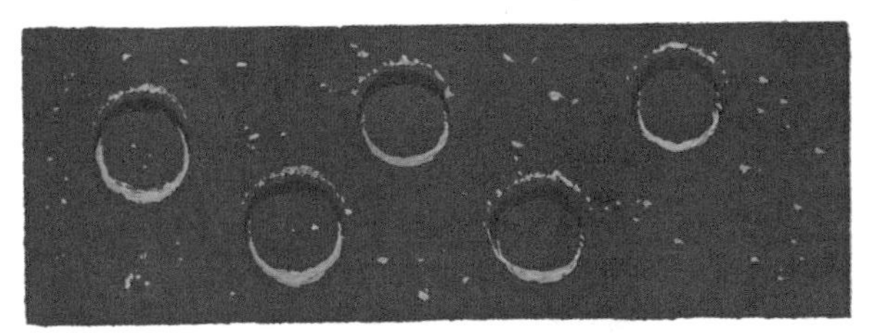

Which robot was it?

2. Second Mystery

This robot was seen last night.

He left these footprints.

Which robot was it?

3. Third Mystery

This robot was also seen last night.

He left these footprints.

Which robot was it?

JUST FOR FUN

Decoding Symbols

Use this code to read the letter.

Symbol	○	□	△	▭	⯃	∠
Meaning	Circle	Square	Triangle	Rectangle	Octagon	Angle

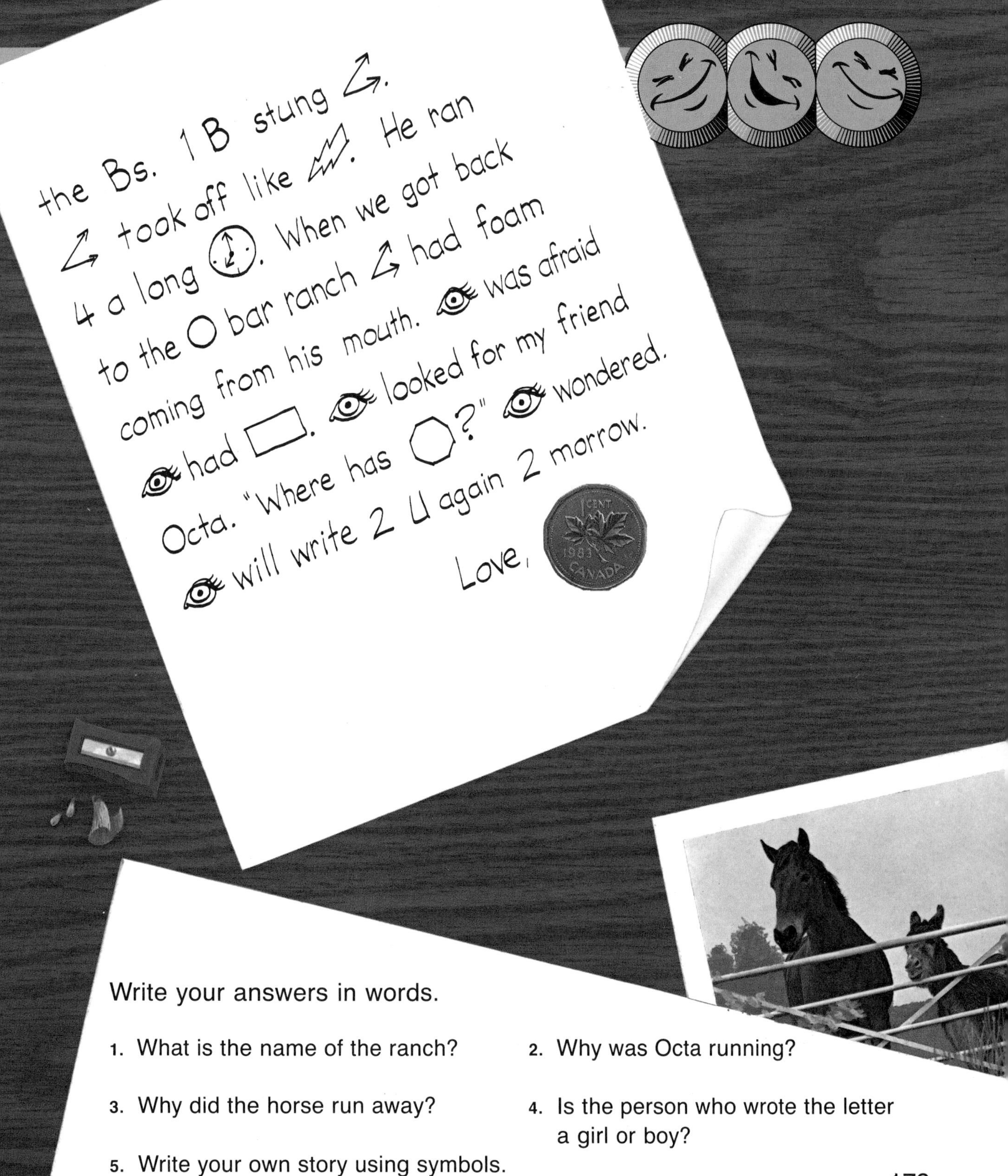

Write your answers in words.

1. What is the name of the ranch?
2. Why was Octa running?
3. Why did the horse run away?
4. Is the person who wrote the letter a girl or boy?
5. Write your own story using symbols.

PROBLEM SOLVING

Identifying Likenesses and Differences

Study these pictures carefully. Then answer each question.

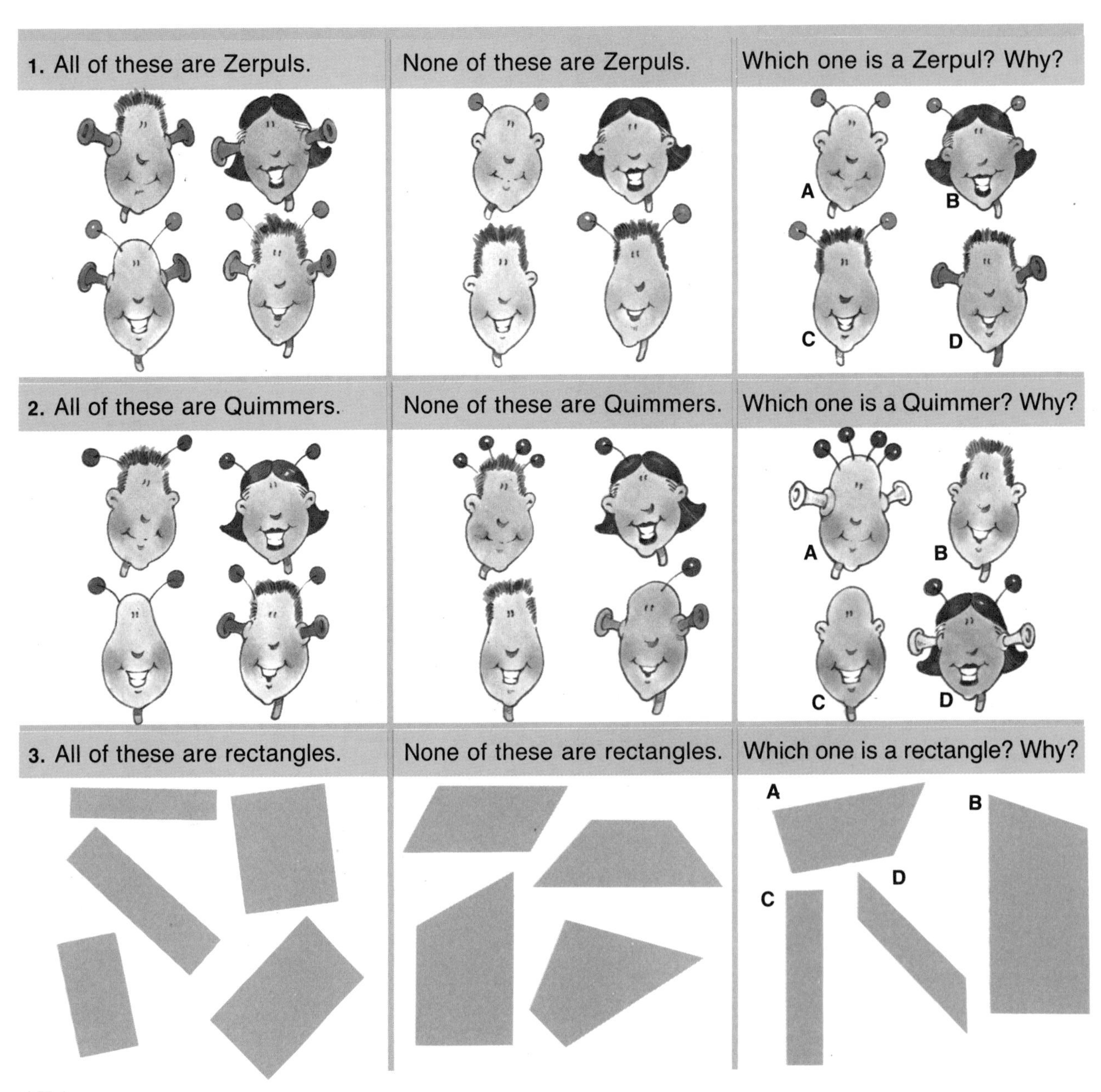

Two big eyes
And 8 long arms
Keep me safe
From deep sea harm.
Can you guess
Without a fuss
Why they call me
Octopus?

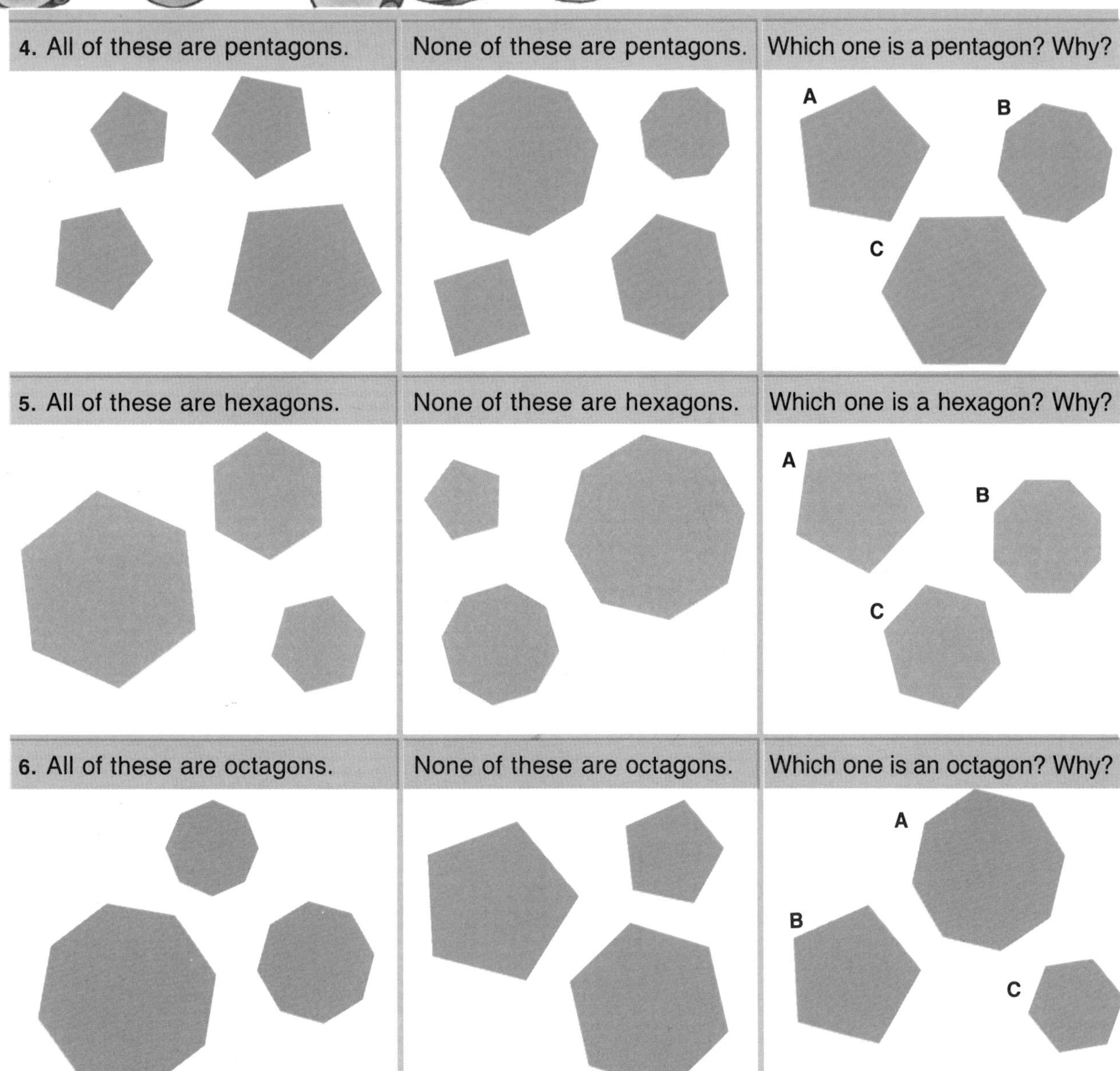
4. All of these are pentagons.
None of these are pentagons.
Which one is a pentagon? Why?
A
B
C
5. All of these are hexagons.
None of these are hexagons.
Which one is a hexagon? Why?
A
B
C
6. All of these are octagons.
None of these are octagons.
Which one is an octagon? Why?
A
B
C

Familiar Figures

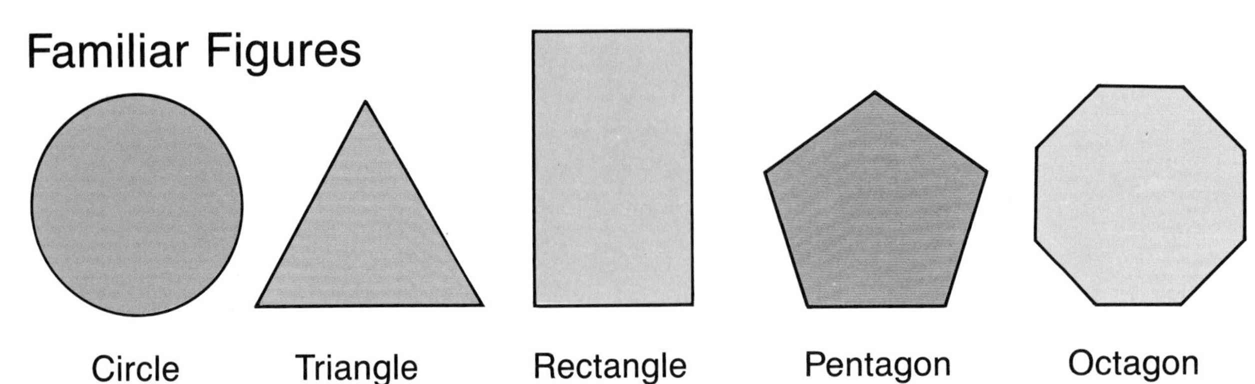

Look at each picture.
Find something shaped like one of the figures above.
Write the name of the figure.

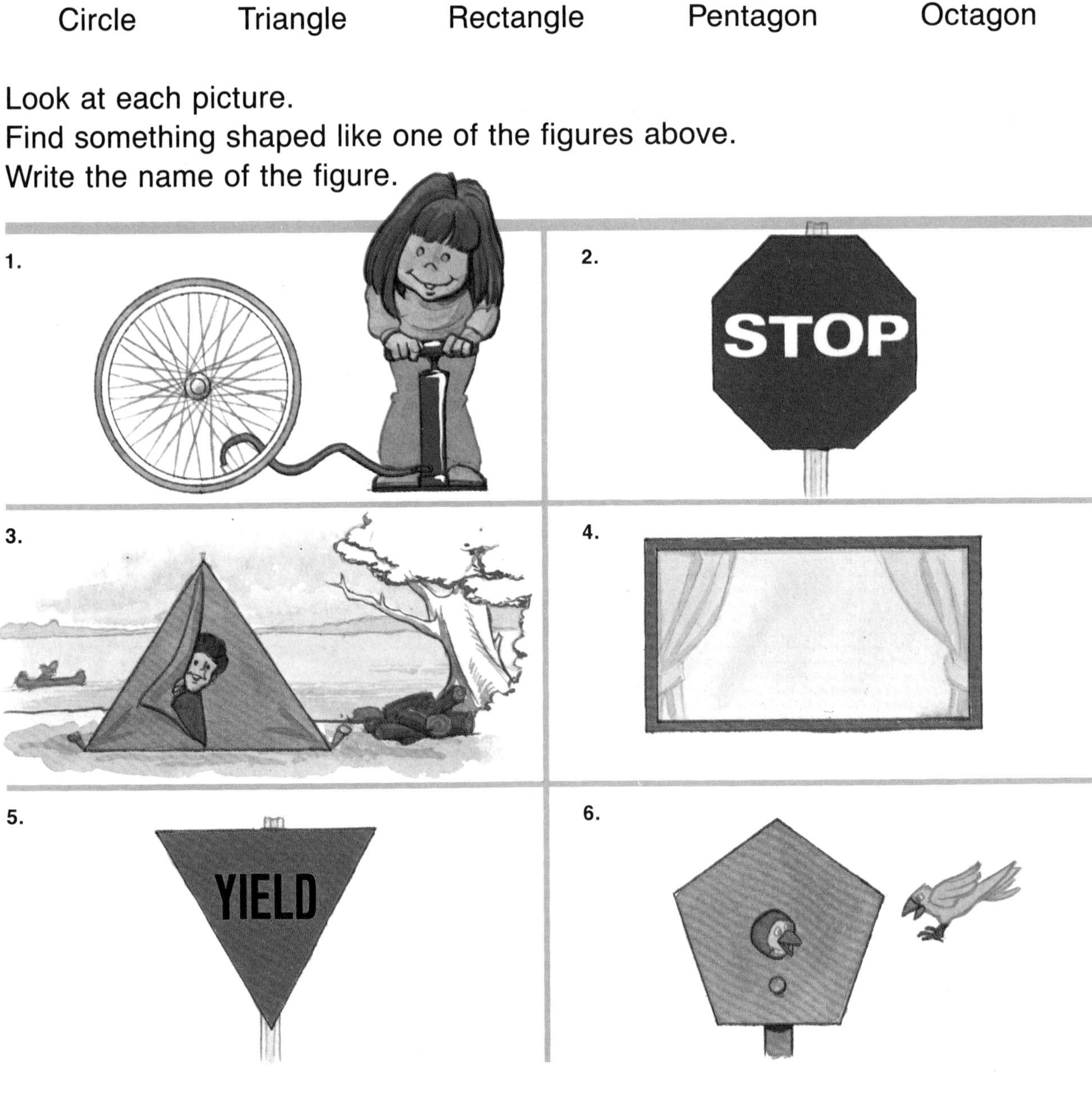

Making a Bar Graph

1. How many circles are shown in the picture?
2. How many squares are shown?
3. How many triangles are shown?
4. How many pentagons are shown?
5. How many octagons are shown?
6. Record the answers in a bar graph. Use centimetre graph paper.

Make a title for your graph.

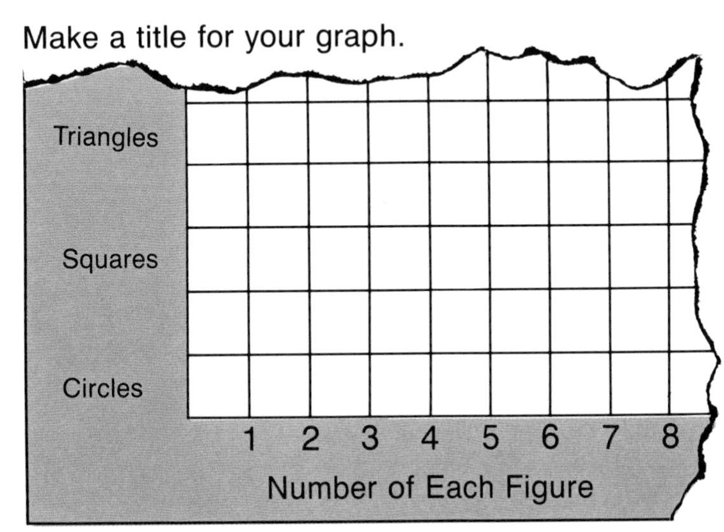

PROBLEM SOLVING

Welcome to

Solving Multi-step Problems

In Shapeland, the cost of an item is the total cost of its parts.

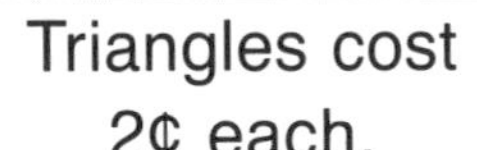

Triangles cost 2¢ each.

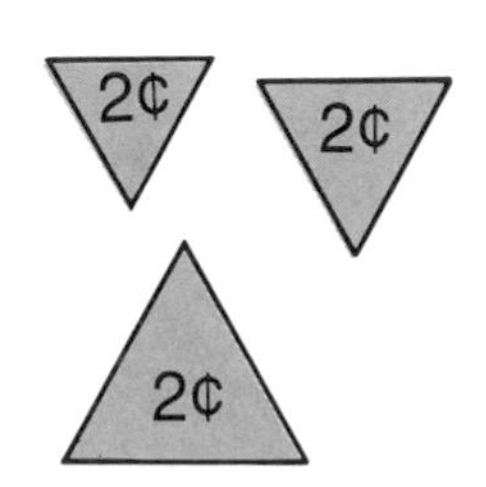

Squares and rectangles cost 3¢ each.

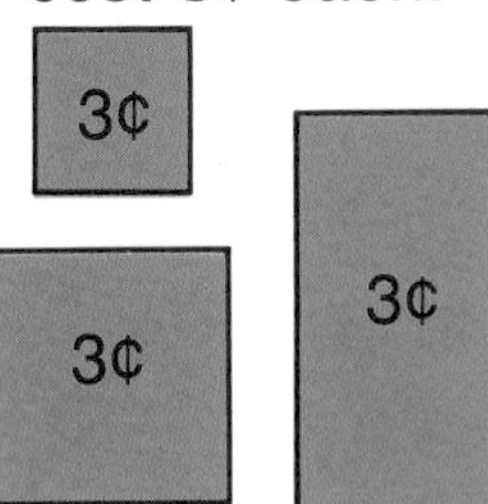

Circles cost 4¢ each.

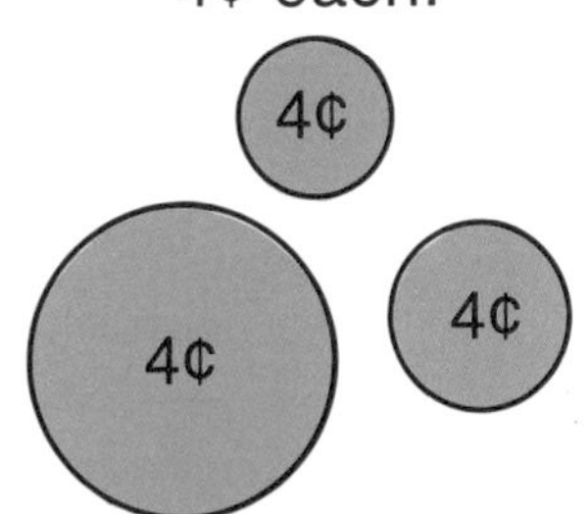

We can find the cost of this figure.

Cost of:

1 square ⟶	3¢
1 circle ⟶	4¢
4 rectangles (3¢ + 3¢ + 3¢ + 3¢) ⟶	12¢
Total cost ⟶	19¢

Find the cost of each figure.

1.

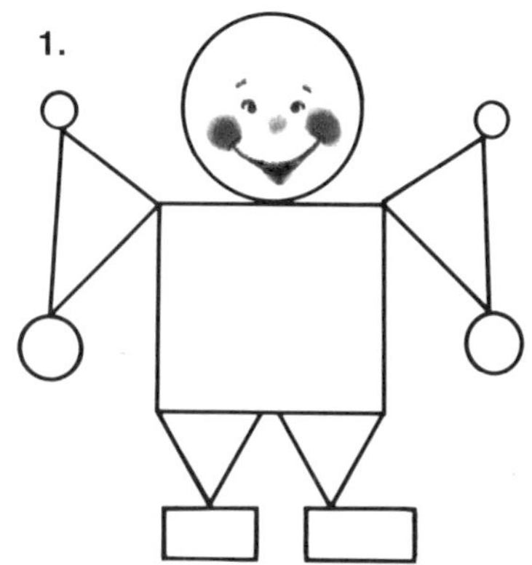

2.

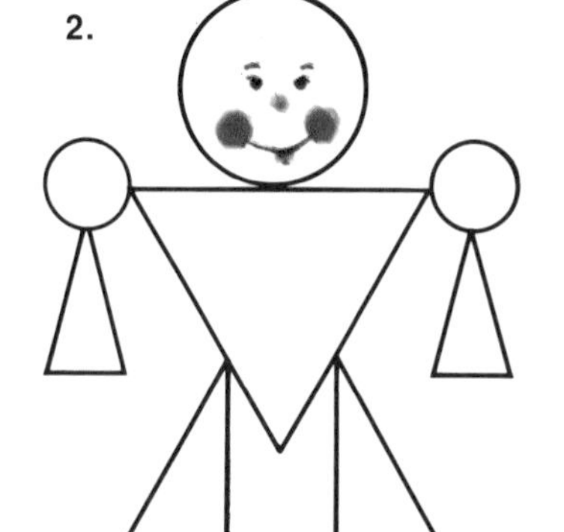

3.

4. Draw your own Shapeland figure. Find its cost.

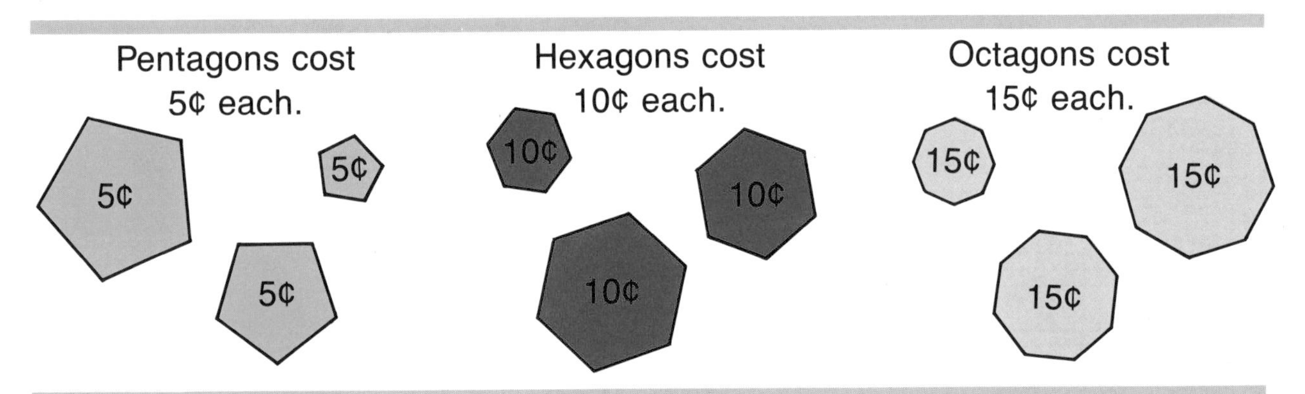

How much would each of these figures cost?

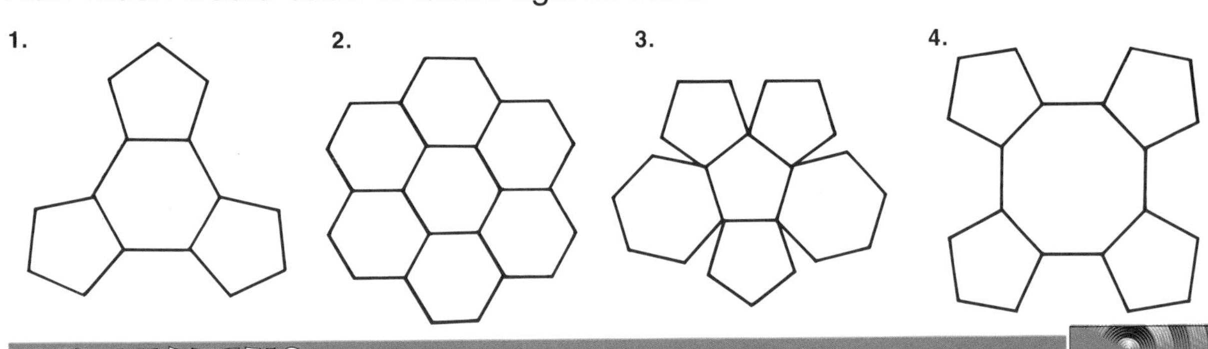

ESTIMATING

Estimate: How many geometric figures are there in all?

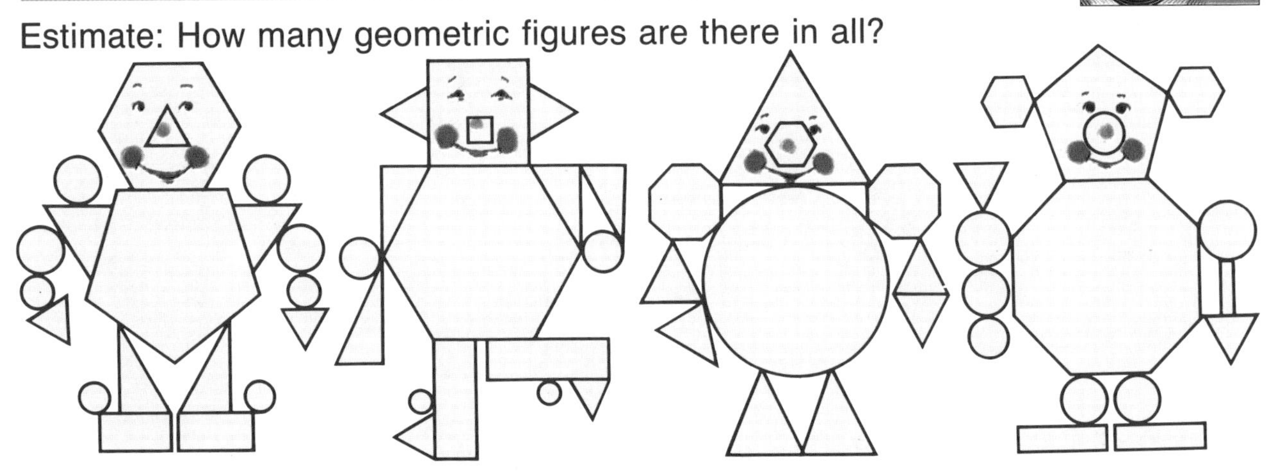

Matching Rectangles

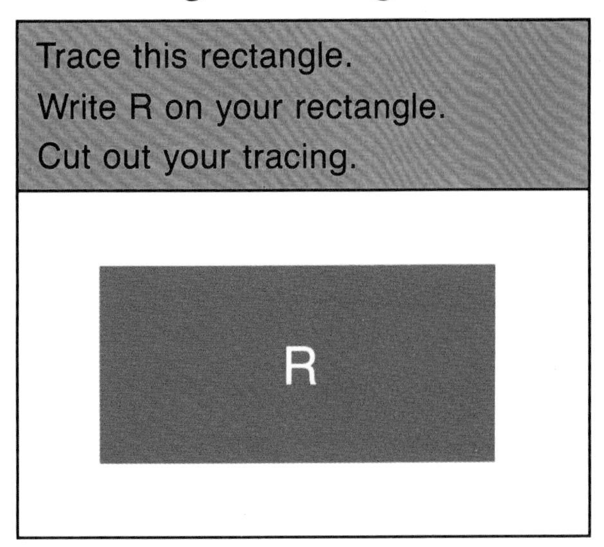

Trace this rectangle.
Write R on your rectangle.
Cut out your tracing.

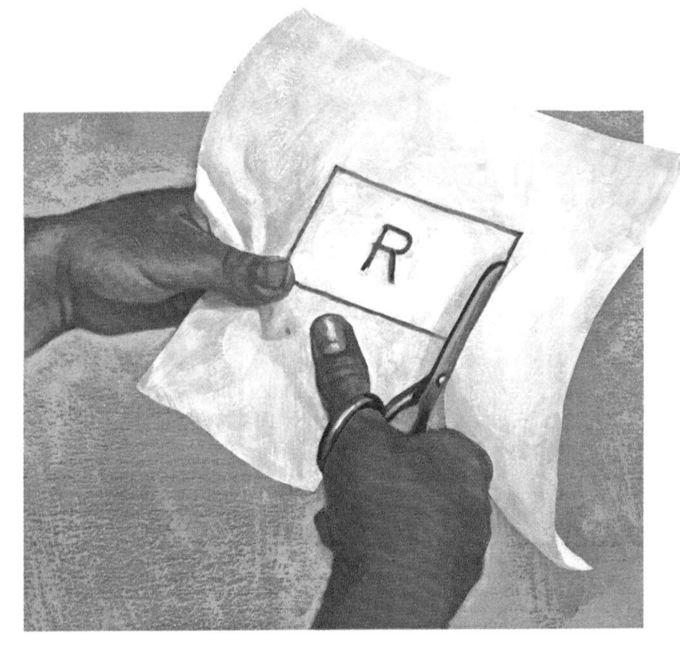

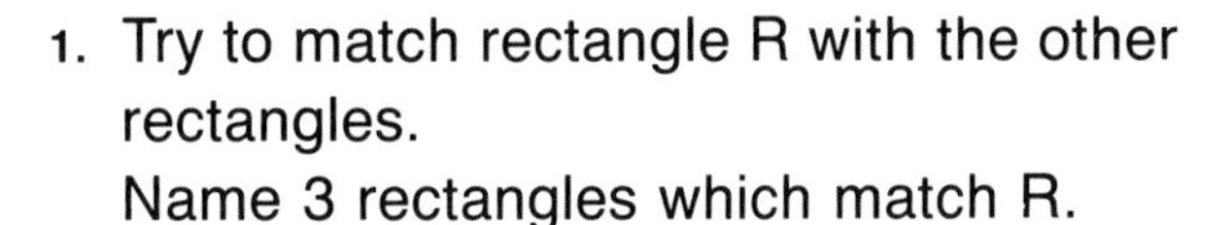

1. Try to match rectangle R with the other rectangles.
   Name 3 rectangles which match R.

2. Which of the rectangles are longer than R?

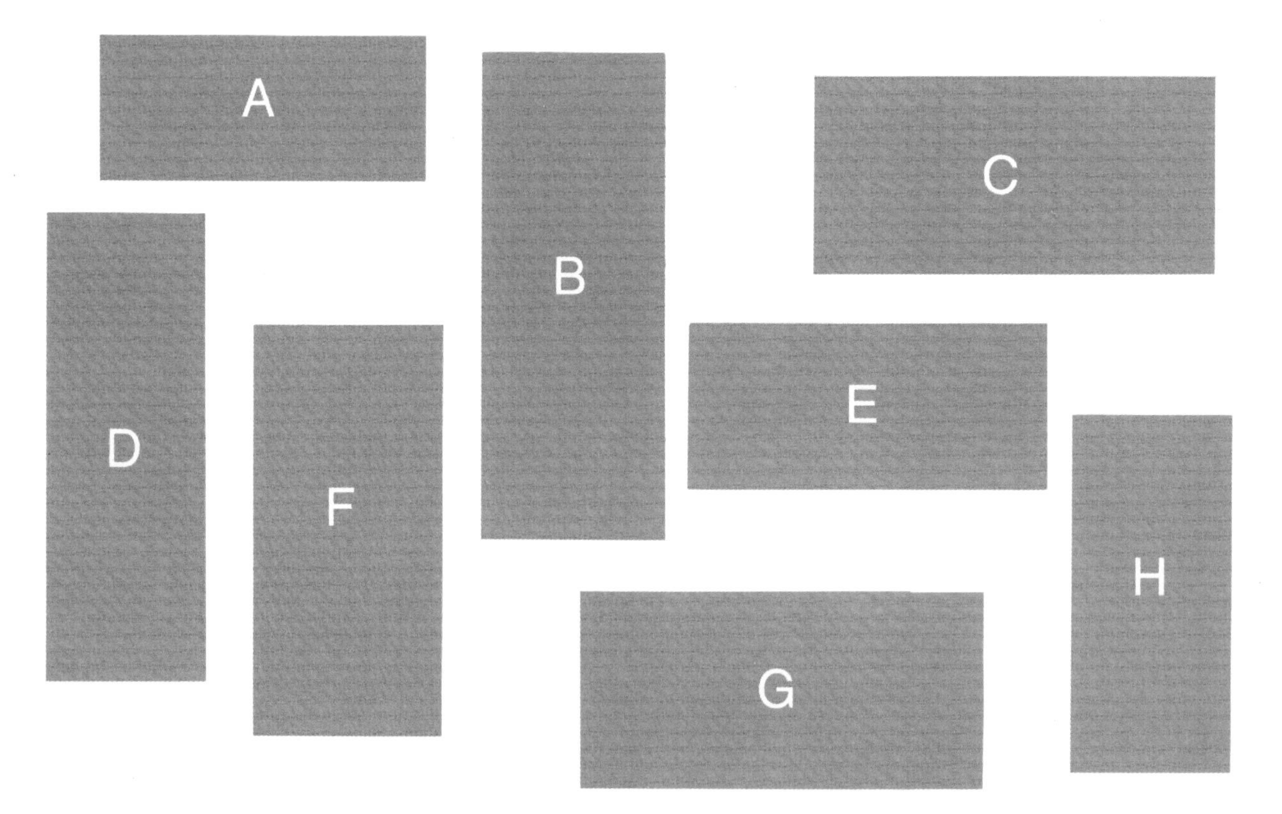

Rectangles

1. Write the length and width of each rectangle in a table.

Rectangle	Length	Width
A	3 cm	2 cm
B	▒ cm	▒ cm

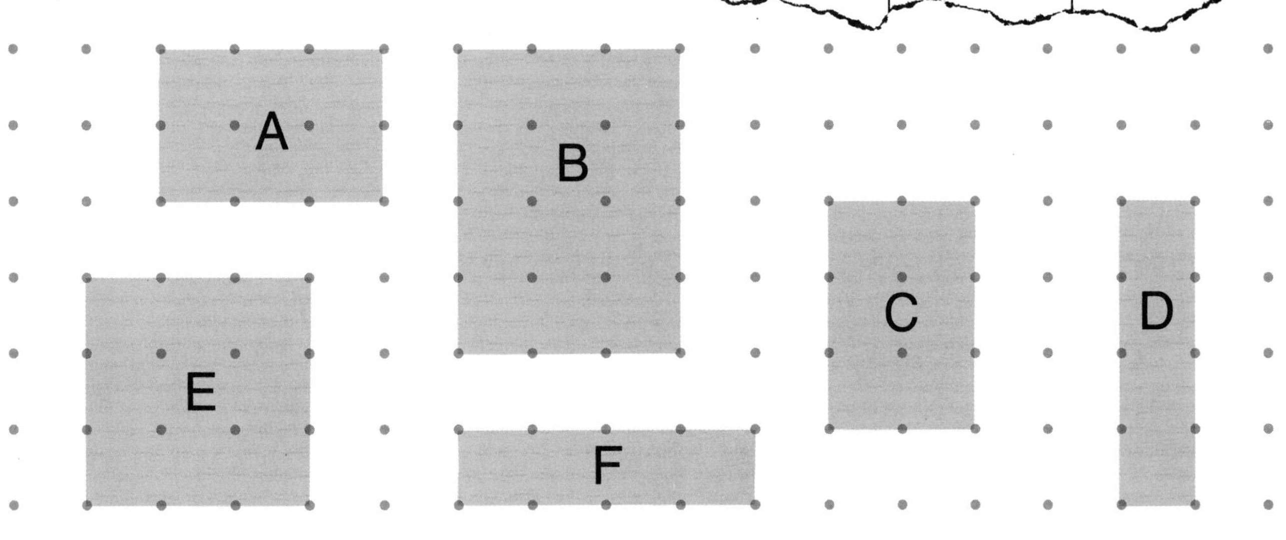

2. Rectangles with the same length and width match.
   Which of the rectangles above match?

3. Name the rectangles below which match X.

4. Is there a rectangle which matches Y?

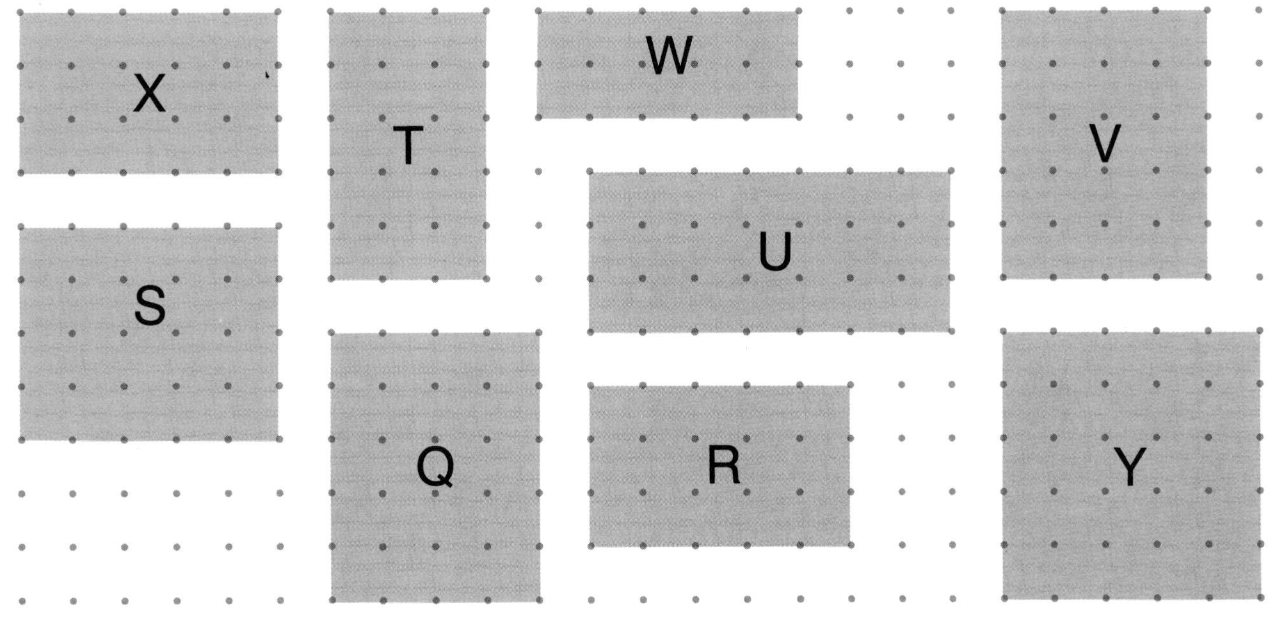

Matching Triangles

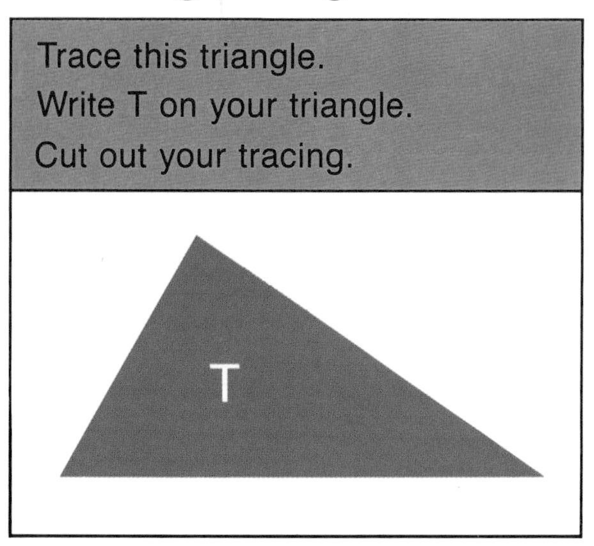

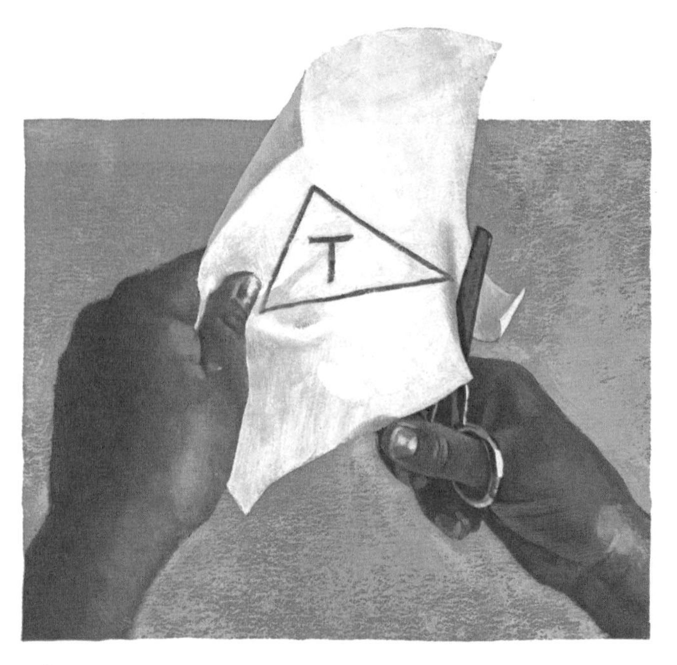

1. Try to match triangle T with the other triangles.
   Write the letters of the triangles which match T.

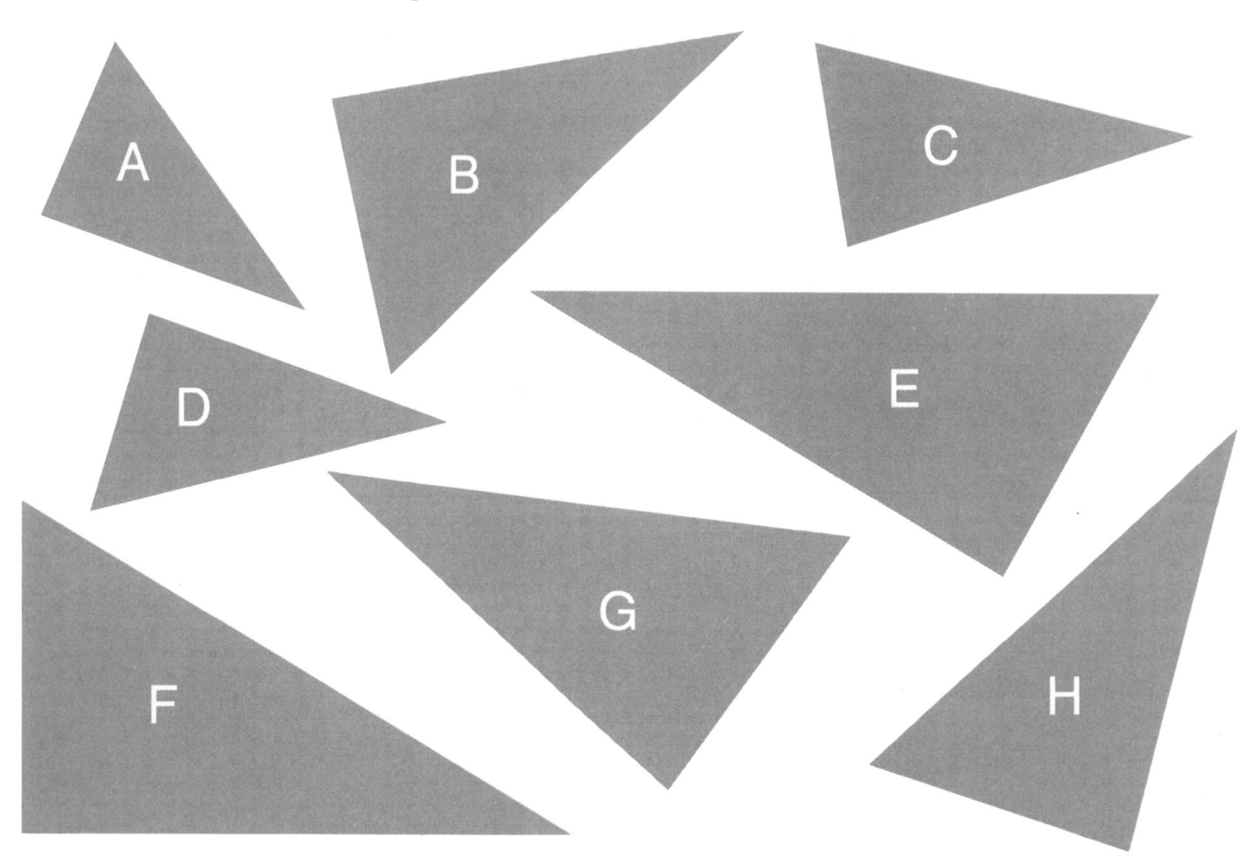

EXTENSIONS

Triangle Puzzles

Trace and cut out another triangle like triangle T.
Which figures below can be made using the 2 triangles?

6. What other figures can you make using the 2 triangles?

PROBLEM SOLVING

Constructing a Model

Sherlock has discovered the shoeprint of a criminal.

Which of the criminals made this shoeprint?

To find out, trace and cut out the shoeprint.

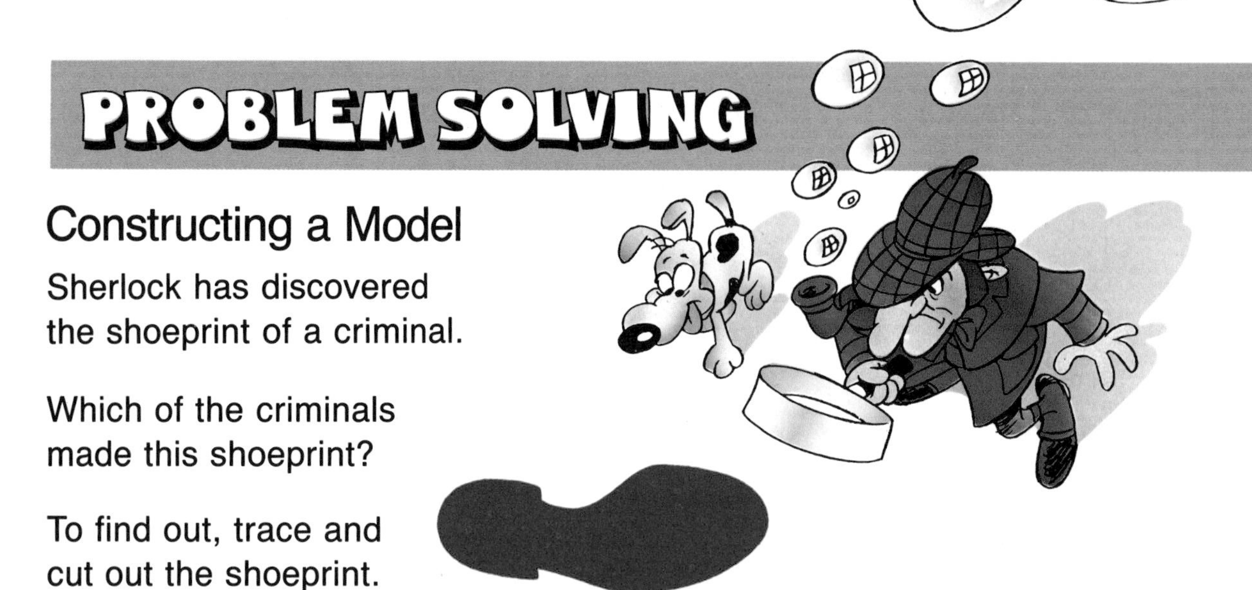

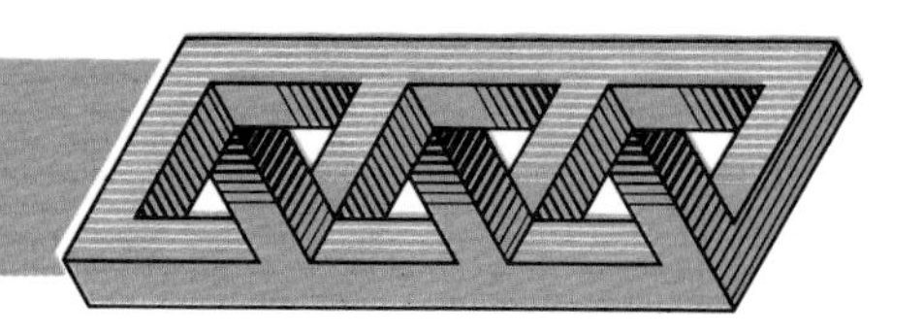

Identifying and Extending Patterns

Draw the next 2 figures in each pattern.

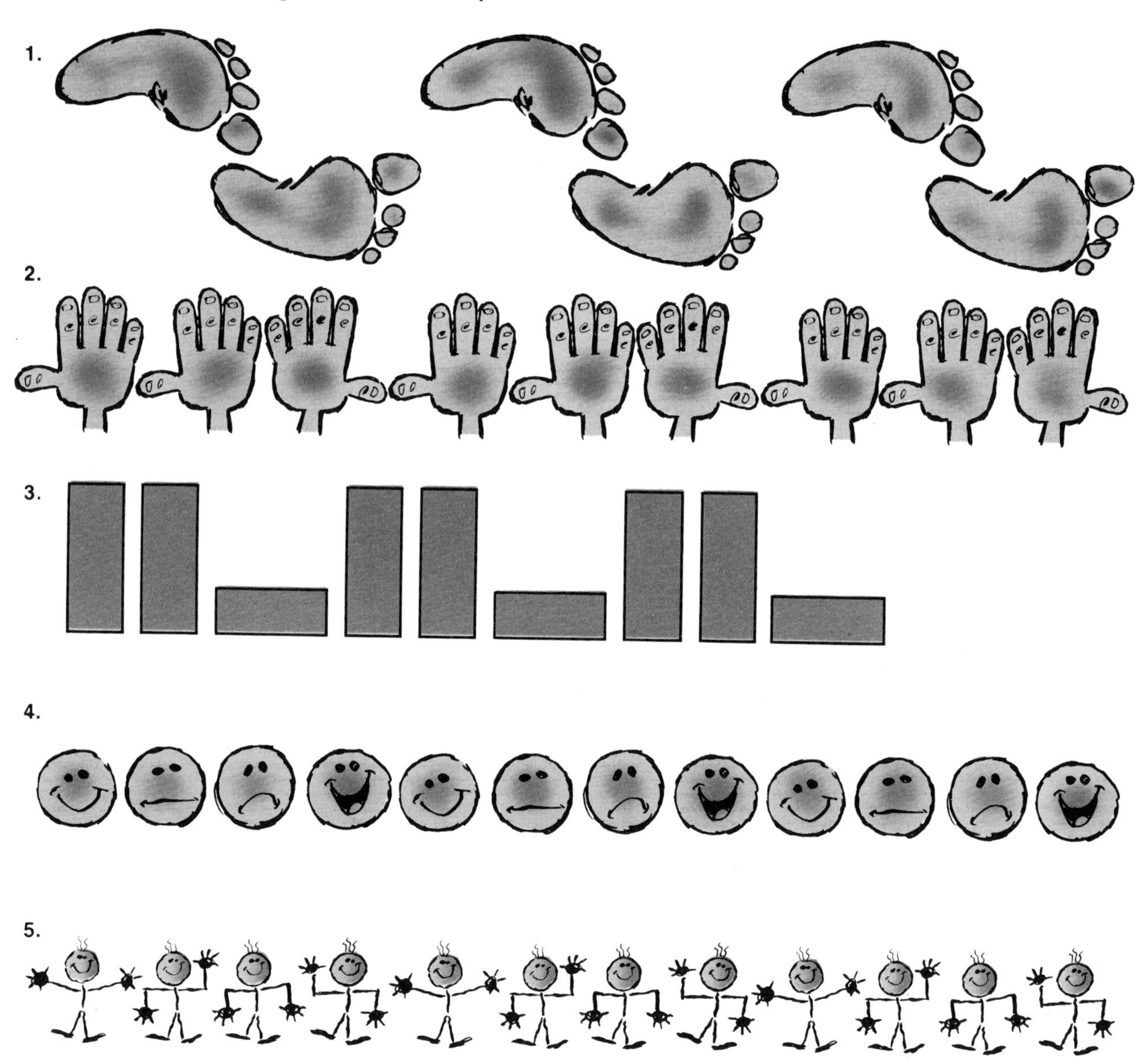

Symmetry

Many pictures are symmetrical.

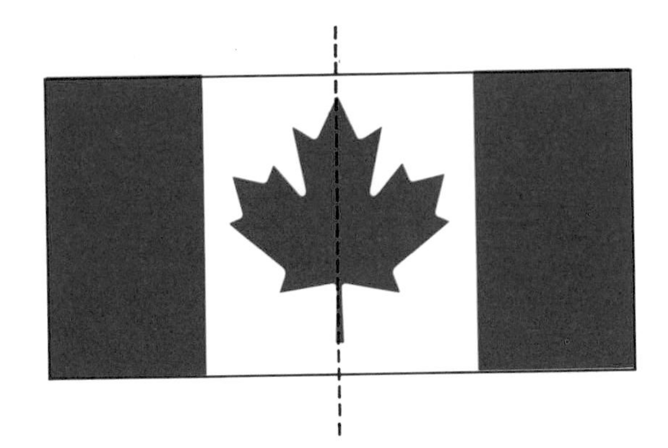

Which of the following pictures are symmetrical?

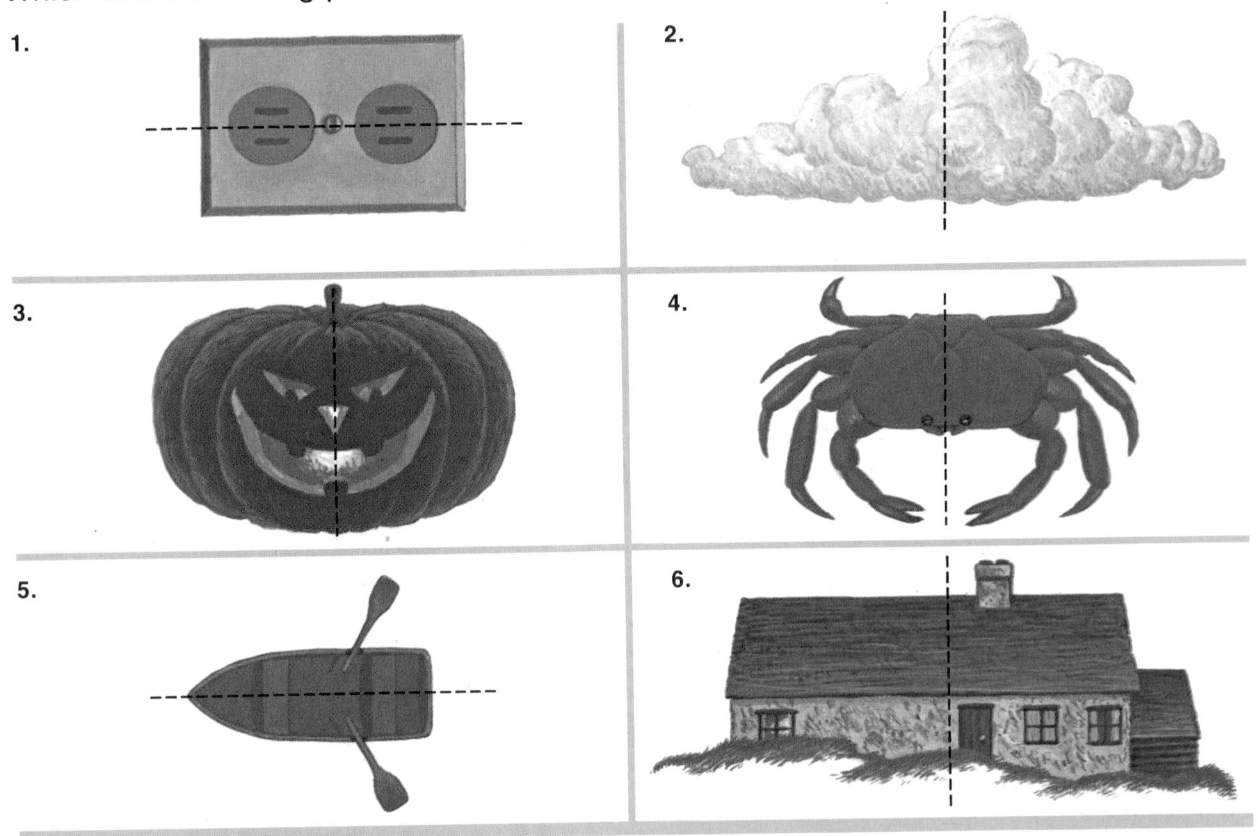

7. Find an object in your classroom that is symmetrical.

JUST FOR FUN

1. Fold a piece of paper.

2. Draw a figure as shown.

3. Cut out your figure.

4. Unfold. Are both parts the same?

Repeat the steps for these figures.

The line along the fold is called a line of symmetry. Why?

CHECK-UP

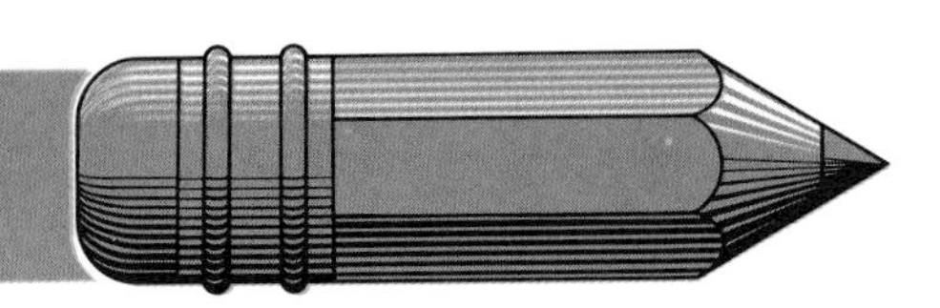

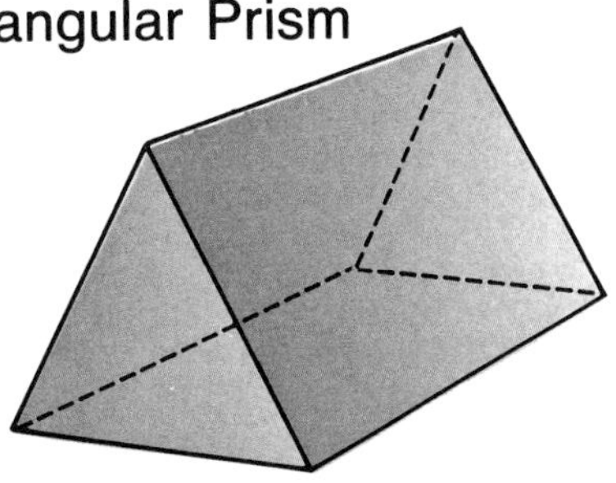

Cube

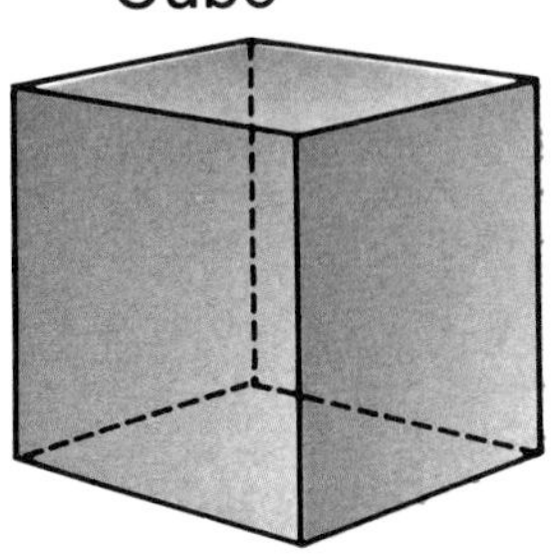

Cylinder

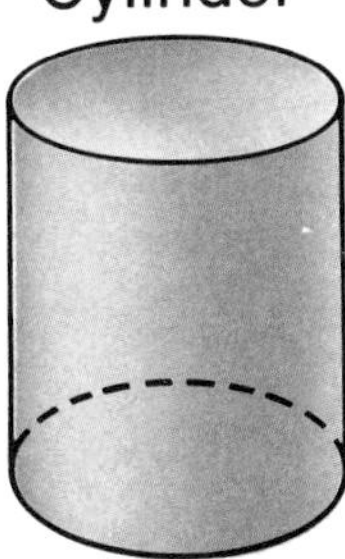

1. Each of the following drawings was made by tracing a face of one of the solids above. Name the solid.

A.

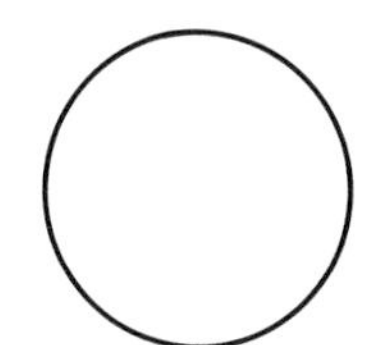

B.

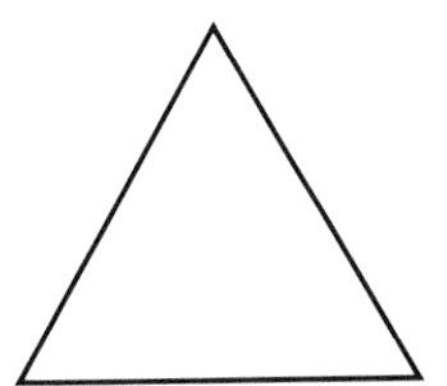

C.

2. Name the shape of each drawing in question 1.

---

3. Write the length and width of each rectangle.

A.

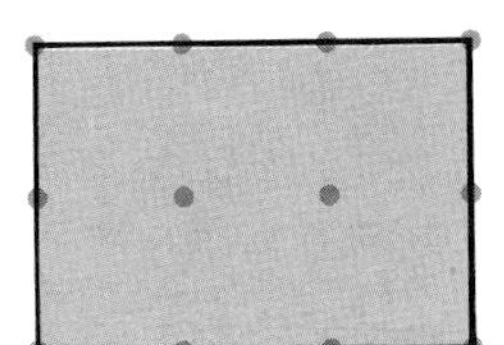

B.

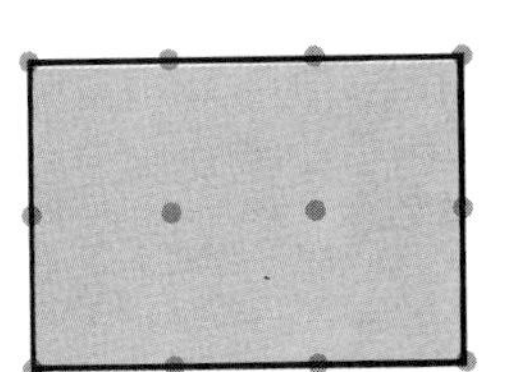

C.

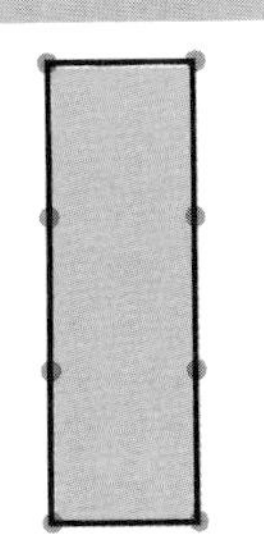

4. Which rectangle in question 3 matches rectangle A?

Which of the following pictures are symmetrical?

5.

6.

7.

CUMULATIVE REVIEW

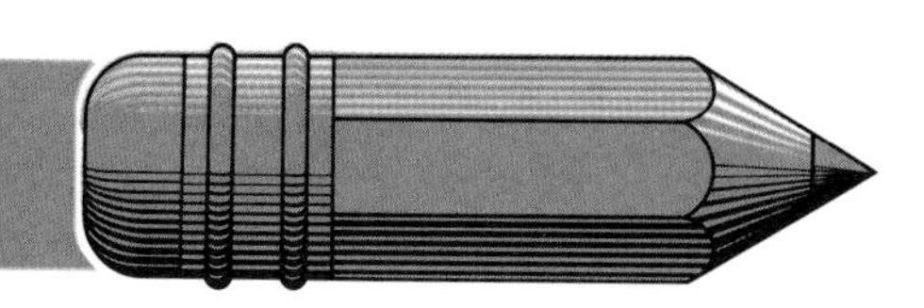

1. How many small cubes are shown?

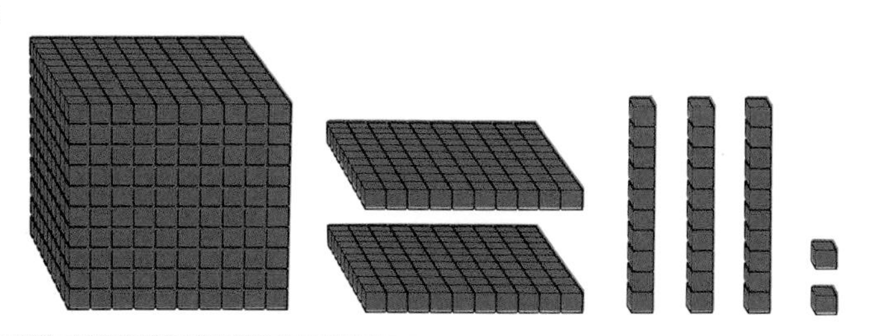

Write the meaning of each coloured digit.

2.  2879   3. 7043

Copy and complete.

4. 7582 = 7000 + ▒ + 80 + 2   5. 4018 = ▒ + 10 + 8

Write each numeral.

6. four thousand six hundred twenty-five   7. 3 thousands, 9 hundreds, 6 ones

Write the next 6 numbers in each pattern.

8. 1139, 1149, 1159,...   9. 7523, 7423, 7323,...

Use > (greater than) or < (less than) for each ▒ .

10. 6146 ▒ 6139   11. 1621 ▒ 1625   12. 4968 ▒ 4698

Add.

13. 283 + 614
14. 730 + 127
15. 208 + 47
16. 886 + 109
17. 647 + 181
18. 697 + 182
19. 148 + 367
20. 289 + 264
21. 596 + 37
22. 19 + 5
23. 62 + 25
24. 83 + 16

CUMULATIVE REVIEW

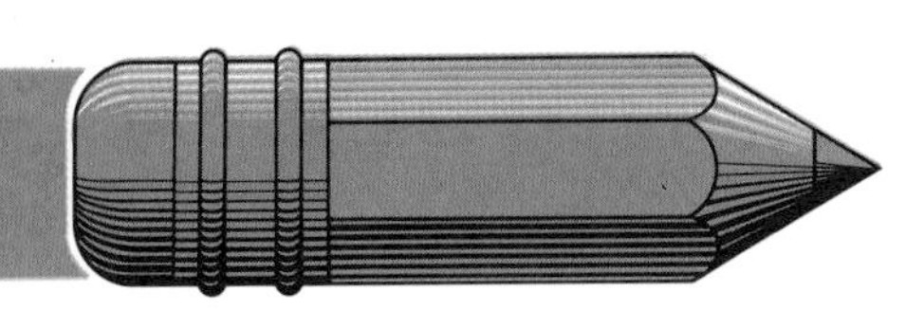

Subtract.

27. $\begin{array}{r} 596 \\ -372 \\ \hline \end{array}$
28. $\begin{array}{r} 824 \\ -713 \\ \hline \end{array}$
29. $\begin{array}{r} 942 \\ -824 \\ \hline \end{array}$
30. $\begin{array}{r} 533 \\ -216 \\ \hline \end{array}$
31. $\begin{array}{r} 623 \\ -281 \\ \hline \end{array}$
32. $\begin{array}{r} 614 \\ -283 \\ \hline \end{array}$
33. $\begin{array}{r} 342 \\ -148 \\ \hline \end{array}$
34. $\begin{array}{r} 856 \\ -378 \\ \hline \end{array}$
35. $\begin{array}{r} 765 \\ -289 \\ \hline \end{array}$
36. $\begin{array}{r} 450 \\ -128 \\ \hline \end{array}$
37. $\begin{array}{r} 703 \\ -187 \\ \hline \end{array}$
38. $\begin{array}{r} 800 \\ -235 \\ \hline \end{array}$

Complete. Watch the signs.

39. $\begin{array}{r} 438 \\ +341 \\ \hline \end{array}$
40. $\begin{array}{r} 465 \\ -223 \\ \hline \end{array}$
41. $\begin{array}{r} 732 \\ -514 \\ \hline \end{array}$
42. $\begin{array}{r} 169 \\ +273 \\ \hline \end{array}$
43. $\begin{array}{r} 507 \\ -219 \\ \hline \end{array}$
44. $\begin{array}{r} 723 \\ +285 \\ \hline \end{array}$

45. Find the cost of this figure.

Triangles cost 2¢.
Rectangles cost 4¢.
Circles cost 5¢.

46. This drawing shows 2 faces of the same solid.
Which solid was used: A, B, or C?

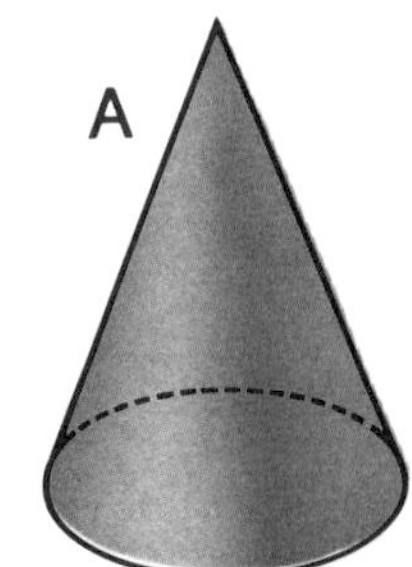

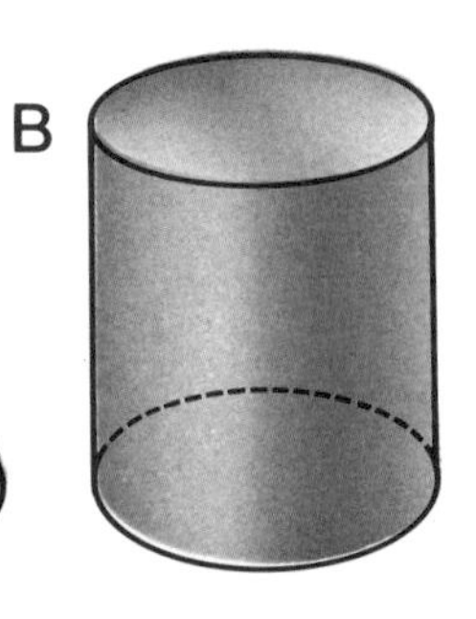

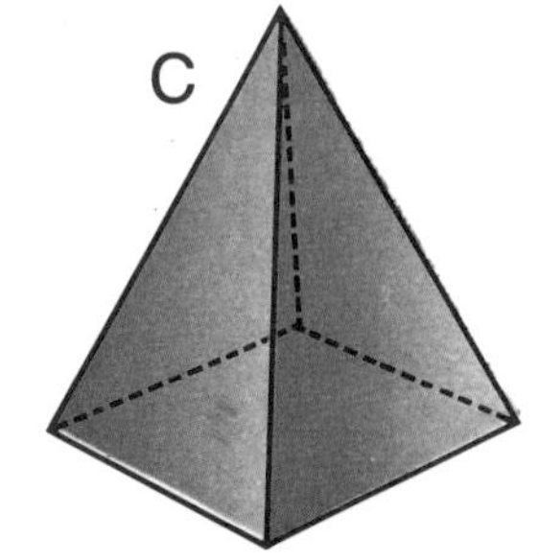

MULTIPLICATION FACTS TO 9 × 5

Biking

I ride my bike and the wheels go round
And carry me swiftly over the ground
With a rush and a hushed little slapping sound
That the tires always make.

At the end of the block I turn and then
I'm ready to ride and ride again,
Faster than fast till the moment when
I suddenly put on the brake.

—*Margaret Hillert*

Repeated Addition

How many bicycle wheels are there in all?
We see:

We think: $2 + 2 + 2 = 6$
or 3 groups of 2 equals 6
There are 6 bicycle wheels.

Copy and complete each sentence.

1.

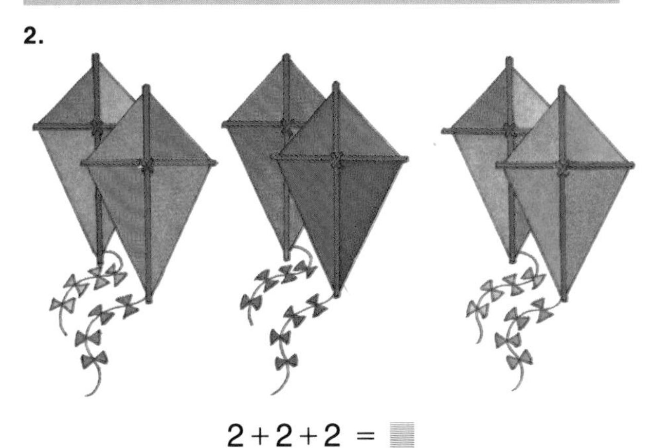

$3 + 3 + 3 + 3 = \square$
4 groups of 3 equals $\square$

2.

$2 + 2 + 2 = \square$
3 groups of 2 equals $\square$

3.

$5 + 5 = \square$
2 groups of 5 equals $\square$

Copy and complete each sentence.

1.

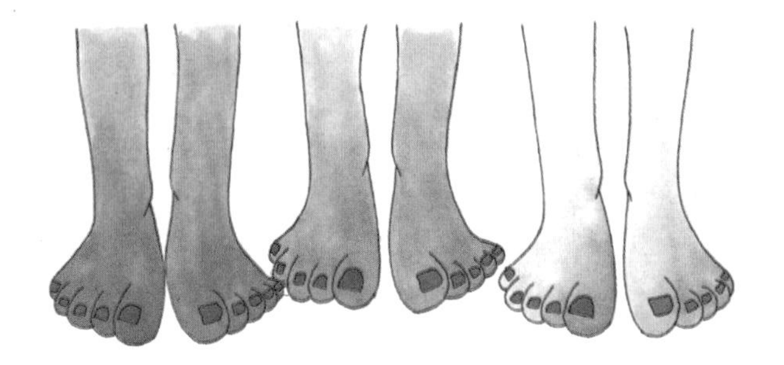

5 + 5 + 5 + 5 + 5 + 5 = ___

6 groups of 5 equals ___

2.

3 + 3 + 3 + 3 = ___

4 groups of 3 equals ___

3.

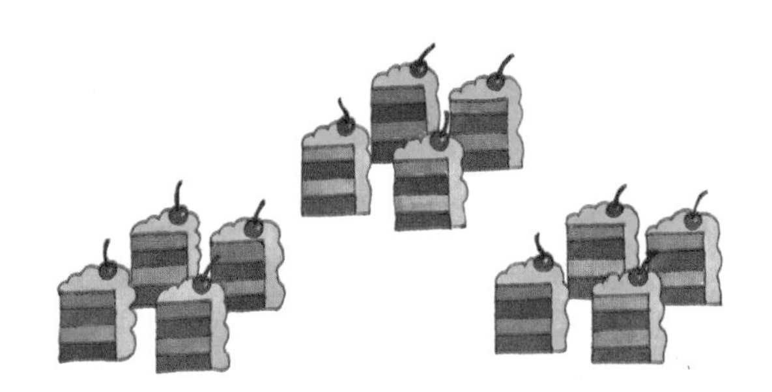

4 + 4 + 4 = ___

3 groups of 4 equals ___

4.

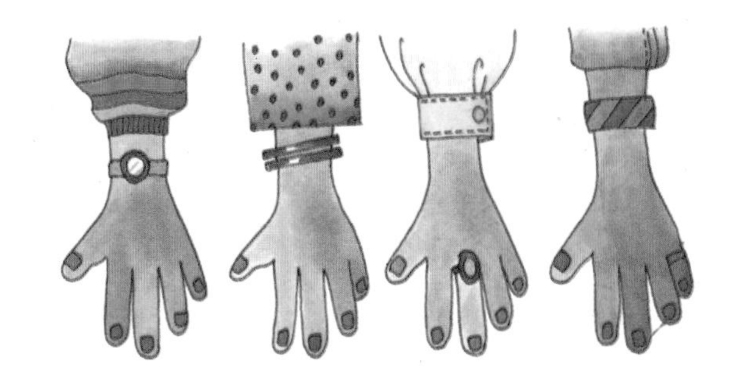

5 + 5 + 5 + 5 = ___

4 groups of 5 equals ___

5.

3 + 3 + 3 + 3 + 3 + 3 = ___

6 groups of 3 equals ___

6.

4 + 4 + 4 + 4 = ___

4 groups of 4 equals ___

Multiplication Sentences

We see:

We think:

$2 + 2 + 2 + 2 + 2 = 10$
or 5 groups of 2 equals 10.

We write: $5 \times 2 = 10$

Copy and complete each sentence.

1.  

2 groups of 3 equals ▒

$2 \times 3 =$ ▒

2. 

2 groups of 4 equals ▒

$2 \times 4 =$ ▒

3. 

4 groups of 3 equals ▒

$4 \times 3 =$ ▒

4. 

4 groups of 2 equals ▒

$4 \times 2 =$ ▒

5. 

3 groups of 4 equals ▒

$3 \times 4 =$ ▒

Copy and complete each sentence.

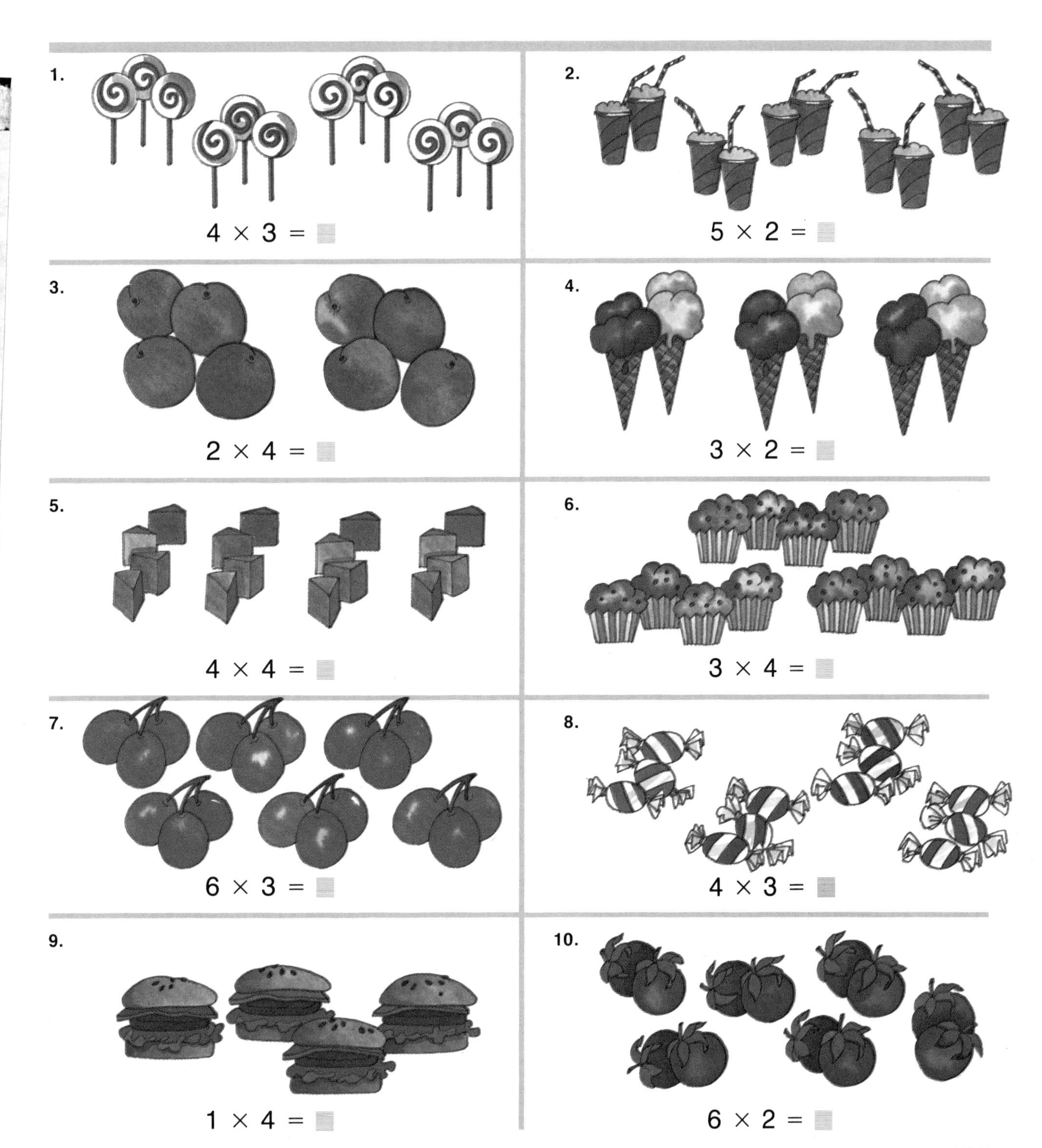

Writing Multiplication Sentences

How many chairs are shown?

We see:

We think: 4 groups of 4 equals 16.

We write: $4 \times 4 = 16$

There are 16 chairs.

Write a multiplication sentence to answer each question.

1. How many chairs are shown?

$3 \times \square = \square$

2. How many pictures do you see?

$2 \times \square = \square$

3. How many straws are there?

$4 \times \square = \square$

4. How many boxes are stacked?

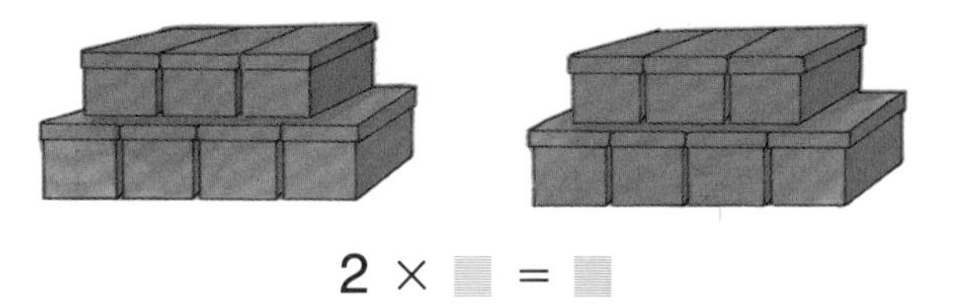

$2 \times \square = \square$

5. How many tins of juice are there altogether?

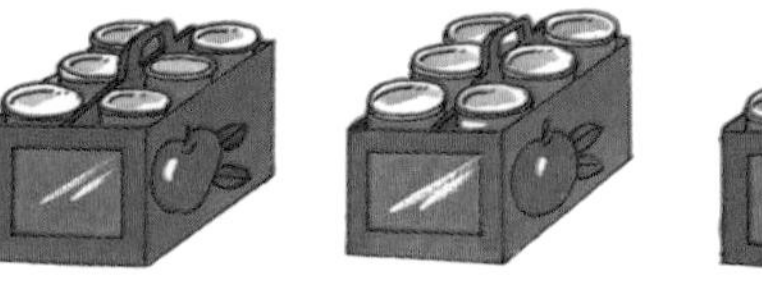

$3 \times \square = \square$

Write a multiplication sentence for each picture.

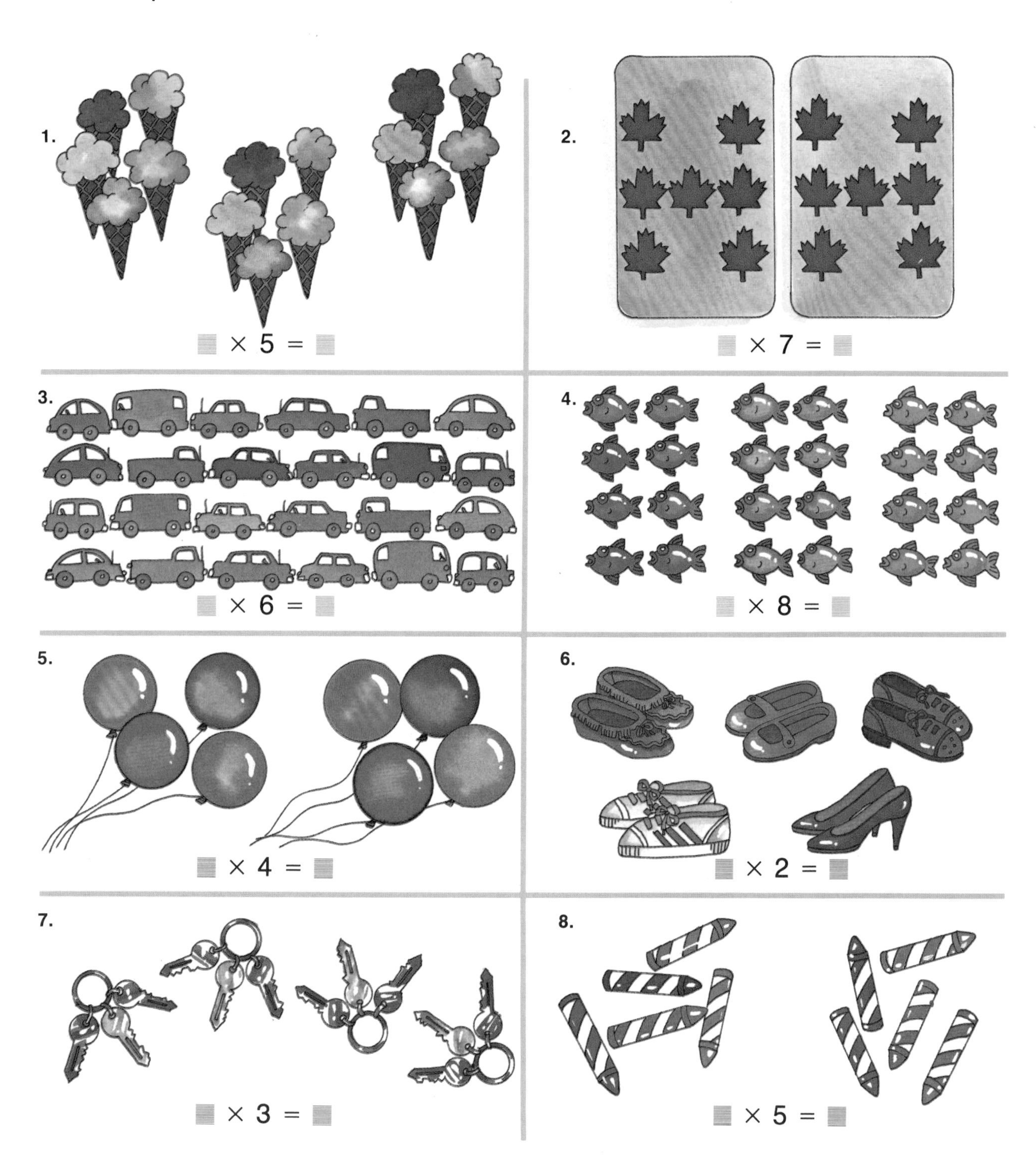

Interpreting Multiplication Pictures

How many faces are looking out?
We see:

We think: 3 groups of 4 equals 12.
We write: $3 \times 4 = 12$
There are 12 faces.

Write a multiplication sentence.

1. How many feet are there on all the cats?

2. How many ears are shown?

3. How many stamps were used?

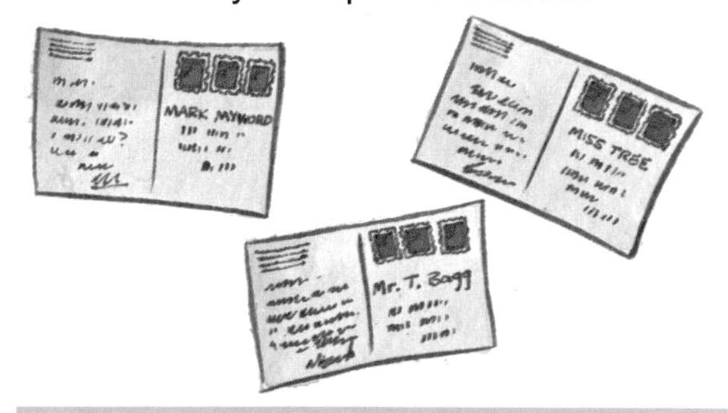

4. How many birds are in the nests?

Write a multiplication sentence for each.

1. How many cans are stacked?

2. How many arrows are there altogether?

3. How many wheels are there on both bicycles?

4. How many fish are in the bags?

Draw a picture for each multiplication sentence.

5. $3 \times 2 = 6$

6. $4 \times 5 = 20$

7. $2 \times 4 = 8$

Multiplication: Groups of 2

How many legs are shown?
We see:

We say: 4 times 2 equals 8.

We write: $4 \times 2 = 8$

There are 8 legs in all.

Write a multiplication sentence for each picture.

1. How many tusks do 3 elephants have?

2. How many eyes do 5 frogs have?

3. How many legs do these chickens have?

4. How many arms are there altogether on the monkeys?

5. How many ears do these bunnies have altogether?

Another Way to Write Multiplication Facts

How many eyes are shown altogether?
We see:

We say: 5 times 2 equals 10.

We write: $5 \times 2 = 10$

or $\begin{array}{r} 2 \\ \times\ 5 \\ \hline 10 \end{array}$

There are 10 eyes in all.

Multiply.

1. $6 \times 2 = \square$
2. $3 \times 2 = \square$
3. $5 \times 2 = \square$
4. $8 \times 2 = \square$
5. $4 \times 2 = \square$
6. $9 \times 2 = \square$
7. $\begin{array}{r} 2 \\ \times 1 \\ \hline \end{array}$
8. $\begin{array}{r} 2 \\ \times 7 \\ \hline \end{array}$
9. $\begin{array}{r} 2 \\ \times 2 \\ \hline \end{array}$
10. These 3 little kittens have lost all their mittens. How many mittens did they lose altogether?

PROBLEM SOLVING

A cube has 2 dots on each of its faces.
You can hold the cube any way you want.

1. How many dots are there in all?
2. What is the greatest number of dots you can see at one time?
3. What is the least number of dots you can see at one time?

Multiplication: Groups of 3

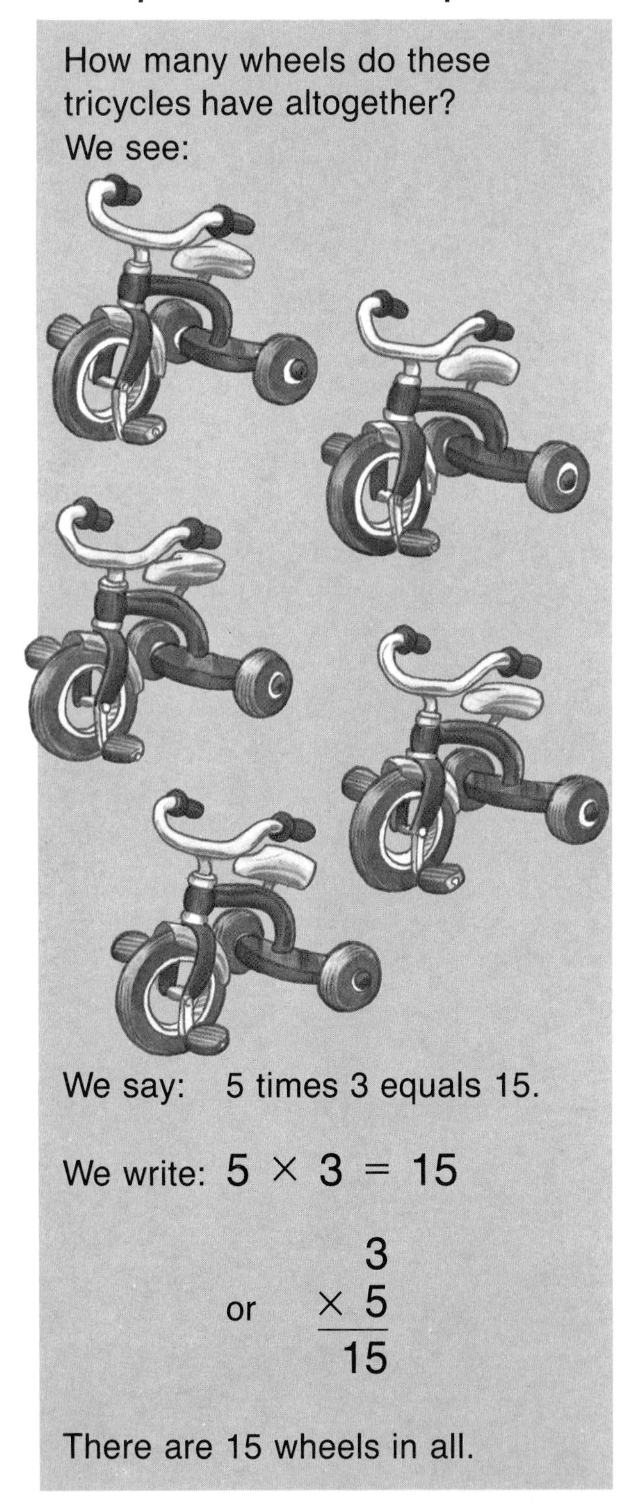

How many wheels do these tricycles have altogether?

We see:

We say: 5 times 3 equals 15.

We write: $5 \times 3 = 15$

or $\begin{array}{r} 3 \\ \times\ 5 \\ \hline 15 \end{array}$

There are 15 wheels in all.

Write a multiplication sentence for each picture.

1. How many wheels are shown in all?

2. How many legs are shown altogether?

3. How many maple leaves are there on the cards?

4. How many eggs are there in the bowls altogether?

5. How many bananas are there in all the bunches?

Multiply.

1. $4 \times 3 =$
2. $2 \times 3 =$
3. $3 \times 3 =$
4. $6 \times 3 =$

5. $7 \times 3 =$
6. $5 \times 3 =$
7. $9 \times 3 =$
8. $8 \times 3 =$

9. $\begin{array}{r} 3 \\ \times 4 \\ \hline \end{array}$
10. $\begin{array}{r} 3 \\ \times 6 \\ \hline \end{array}$
11. $\begin{array}{r} 3 \\ \times 5 \\ \hline \end{array}$
12. $\begin{array}{r} 3 \\ \times 8 \\ \hline \end{array}$
13. $\begin{array}{r} 3 \\ \times 7 \\ \hline \end{array}$
14. $\begin{array}{r} 3 \\ \times 9 \\ \hline \end{array}$

15. Ken Kangaroo hops 3 spaces every time.
Write the numbers he lands on.

0 1 2 3 4 5 6 7 8 9 10 11 12 13 14 15 16 17 18 19 20 21 22 23 24 25 26 27 28 29 30

16. There are 8 tricycles.
How many wheels are there in all?

Copy and complete.

17.

×	2
8	
4	
9	
7	

18.

×	3
7	
9	
6	
8	

19. Kim has 5 bunches of bananas.
There are 3 bananas in each bunch.
Kim also has 4 apples.
How many bananas does Kim have?

PROBLEM SOLVING

There are 15 tricycle wheels needed.
How many tricycles are there?

Multiplication: Groups of 4

How many feet do you see here?
We see:

We say: 5 times 4 equals 20.

We write: $5 \times 4 = 20$

or $\begin{array}{r} 4 \\ \times\ 5 \\ \hline 20 \end{array}$

There are 20 feet in all.

Write a multiplication sentence for each picture.

1. How many feet are there altogether?

2. How many chairs are shown?

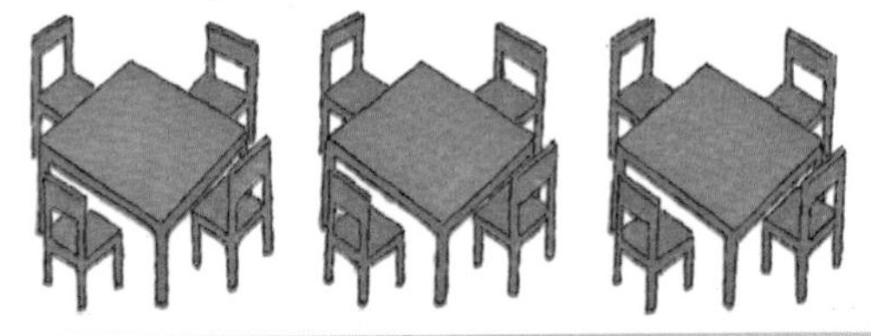

3. How many maple leaves are there?

4. How many oranges are there in the boxes altogether?

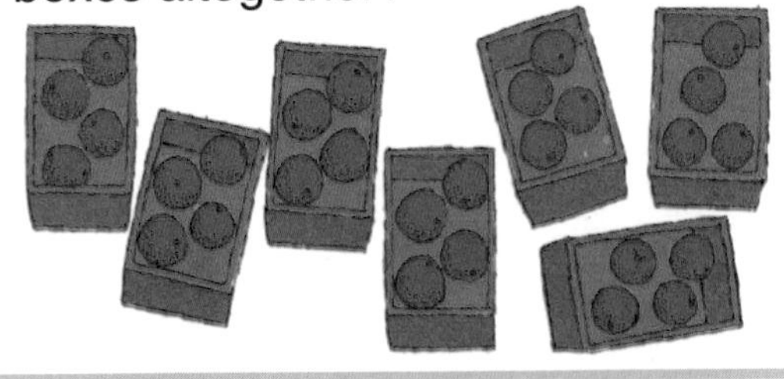

5. How many wheels are there in all?

Copy and complete.

1. $3 \times 4 =$ ▒
2. $2 \times 4 =$ ▒
3. $5 \times 4 =$ ▒
4. $6 \times 4 =$ ▒
5. $8 \times 4 =$ ▒
6. $9 \times 4 =$ ▒
7. $4 \times 4 =$ ▒
8. $7 \times 4 =$ ▒

9. $\begin{array}{r} 4 \\ \times 2 \\ \hline \end{array}$
10. $\begin{array}{r} 4 \\ \times 4 \\ \hline \end{array}$
11. $\begin{array}{r} 4 \\ \times 6 \\ \hline \end{array}$
12. $\begin{array}{r} 4 \\ \times 5 \\ \hline \end{array}$
13. $\begin{array}{r} 4 \\ \times 8 \\ \hline \end{array}$
14. $\begin{array}{r} 4 \\ \times 7 \\ \hline \end{array}$

15. Kathy Kangaroo hops 4 spaces every time.
Write the numbers she lands on.

Copy and complete.

16.

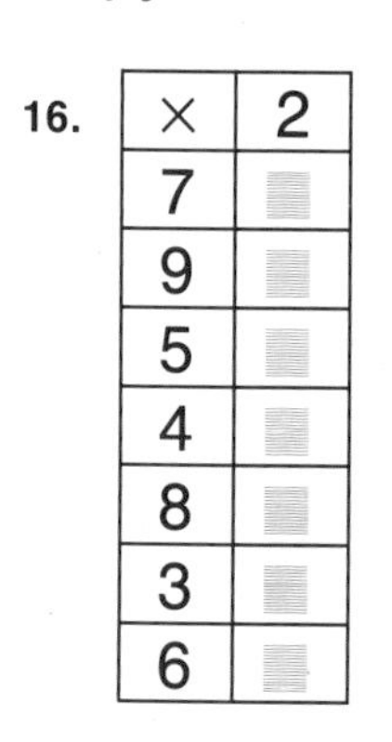

×	2
7	▒
9	▒
5	▒
4	▒
8	▒
3	▒
6	▒

17.

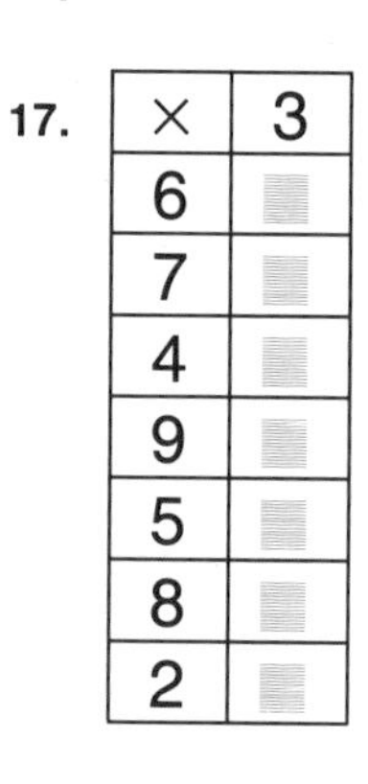

×	3
6	▒
7	▒
4	▒
9	▒
5	▒
8	▒
2	▒

18.

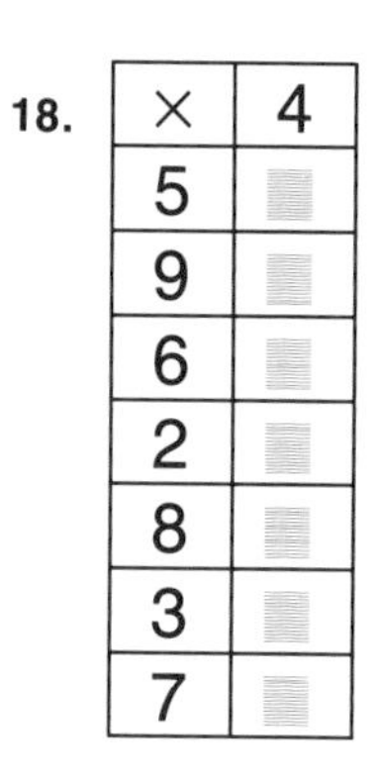

×	4
5	▒
9	▒
6	▒
2	▒
8	▒
3	▒
7	▒

19. A tractor has 4 tires.
A flat wagon has 6 tires.
How many tires are on 5 tractors?

PROBLEM SOLVING

There are 20 car wheels.
How many cars are there?

Multiplication: Groups of 5

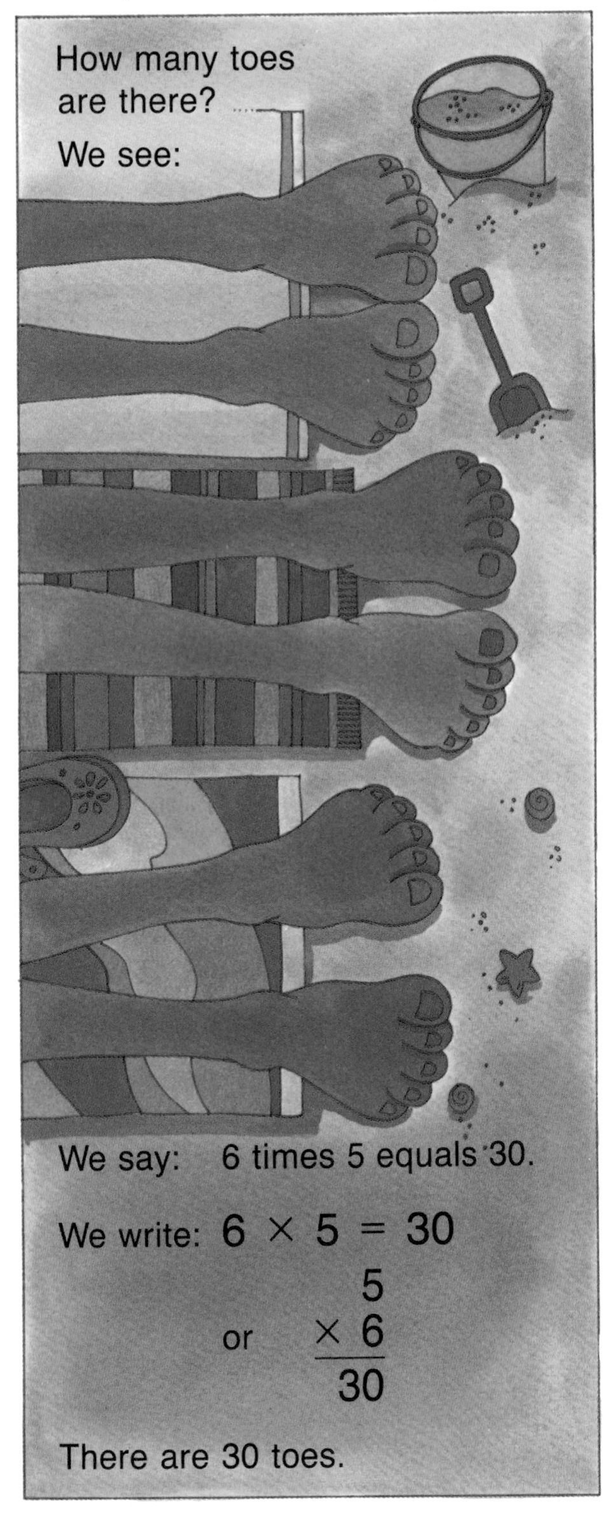

How many toes are there?

We see:

We say: 6 times 5 equals 30.

We write: $6 \times 5 = 30$

or $\begin{array}{r} 5 \\ \times\ 6 \\ \hline 30 \end{array}$

There are 30 toes.

Write a multiplication sentence for each picture.
Write each fact two ways.

1. How many fingers are shown? (Count the thumbs.)

2. How many faces are there altogether?

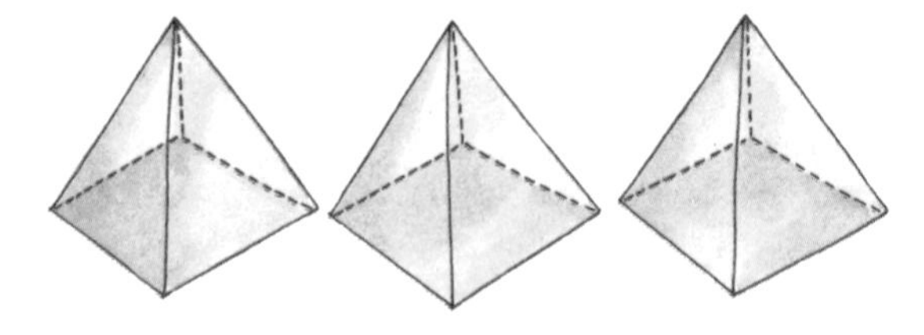

3. How many meatballs are on the plates?

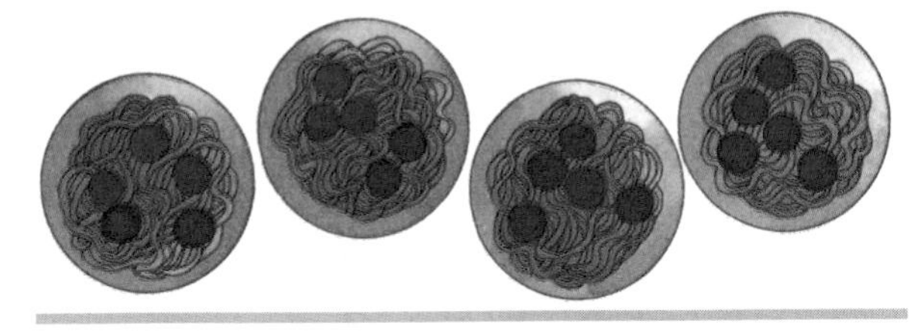

4. How many leaves do you see?

5. How many petals are on these flowers?

Copy and complete.

1. $4 \times 5 =$ ▒ 2. $3 \times 5 =$ ▒ 3. $7 \times 5 =$ ▒ 4. $2 \times 5 =$ ▒

5. $6 \times 5 =$ ▒ 6. $1 \times 5 =$ ▒ 7. $9 \times 5 =$ ▒ 8. $8 \times 5 =$ ▒

9. $\begin{array}{r} 5 \\ \times 6 \\ \hline \end{array}$ 10. $\begin{array}{r} 5 \\ \times 2 \\ \hline \end{array}$ 11. $\begin{array}{r} 5 \\ \times 7 \\ \hline \end{array}$ 12. $\begin{array}{r} 5 \\ \times 1 \\ \hline \end{array}$ 13. $\begin{array}{r} 5 \\ \times 8 \\ \hline \end{array}$ 14. $\begin{array}{r} 5 \\ \times 5 \\ \hline \end{array}$

15. How much money is shown?

16. A quarter is worth 5 nickels.
A dollar is worth 4 quarters.
How many nickels is a dollar worth?

17. How much would it cost for 9 of these stamps?

18. Each 卌 stands for 5.
What number is shown by this tally?

卌 卌 卌 卌 卌 卌 卌 卌

PROBLEM SOLVING

It takes the big hand 5 minutes to move from one number to the next.

How many minutes past 2 o'clock does the clock show?

Multiplication: Groups of 1

How many tails do the monkeys have altogether?

We see:

We say: 7 times 1 equals 7.

We write: $7 \times 1 = 7$

or $\begin{array}{r} 1 \\ \times\ 7 \\ \hline 7 \end{array}$

There are 7 tails in all.

Write a multiplication fact for each picture.

1. How many birds are in these cages?

2. How many fish are there in these bowls?

3. How many eggs are there altogether?

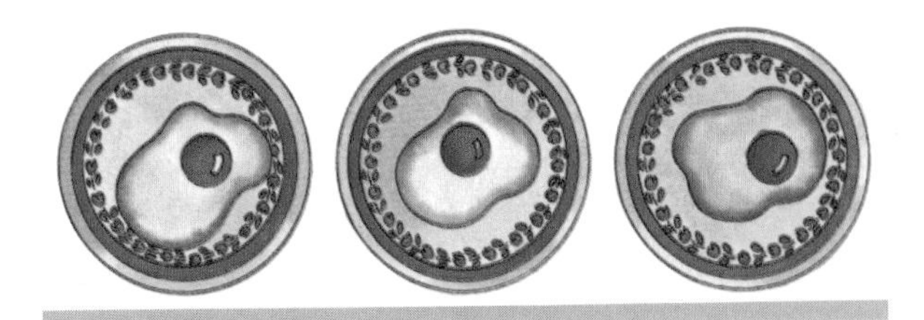

Multiply.

4. $\begin{array}{r} 1 \\ \times 3 \\ \hline \end{array}$ 5. $\begin{array}{r} 1 \\ \times 5 \\ \hline \end{array}$ 6. $\begin{array}{r} 1 \\ \times 2 \\ \hline \end{array}$

7. $\begin{array}{r} 1 \\ \times 6 \\ \hline \end{array}$ 8. $\begin{array}{r} 1 \\ \times 1 \\ \hline \end{array}$ 9. $\begin{array}{r} 1 \\ \times 4 \\ \hline \end{array}$

10. $\begin{array}{r} 1 \\ \times 9 \\ \hline \end{array}$ 11. $\begin{array}{r} 1 \\ \times 7 \\ \hline \end{array}$ 12. $\begin{array}{r} 1 \\ \times 8 \\ \hline \end{array}$

Multiplication Tables

21 is in the green square because

$$7 \times 3 = 21$$

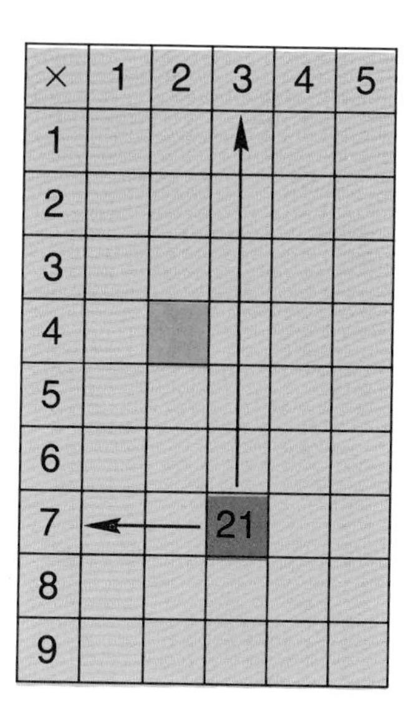

×	1	2	3	4	5
1					
2					
3					
4					
5					
6					
7			21		
8					
9					

What number would be in the pink square?
Why?

Make and complete this table.
Use squared paper.

Make and complete these tables.

1.

×	1	3	2	4	5
3					
1					
2					
4					

2.

×	2	3	5	4	1
2					
3					
5					
7					

3.

×	5	3	4	2	1
5					
2					
4					
6					

4.

×	4	1	5	2	3
3					
7					
9					
8					

5.

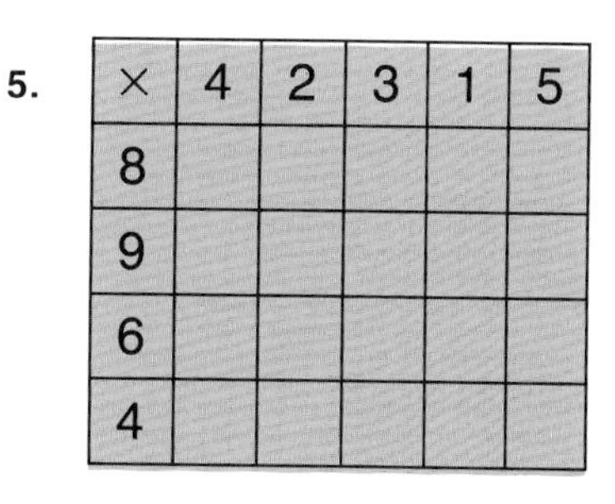

×	4	2	3	1	5
8					
9					
6					
4					

6.

×	3	5	1	2	4
7					
5					
3					
9					

Multiplying by Rows and Columns

The stamps are arranged in rows and columns.
We can write 2 sentences for this.

 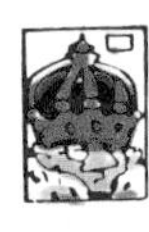
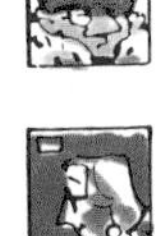 

Victoria writes:

$3 \times 5 = 15$

Owen writes:

$5 \times 3 = 15$

Write 2 multiplication sentences to match each picture.

1.

2.

3.

4.

5.

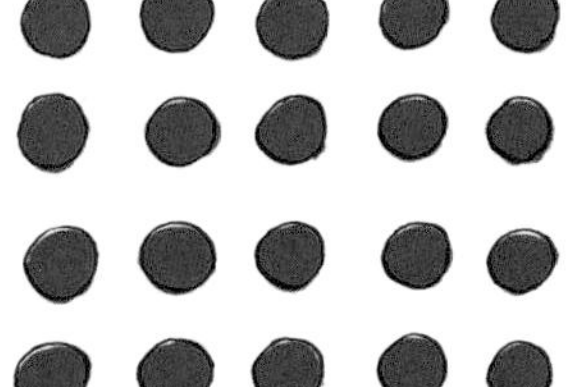

6.

Write 2 multiplication sentences to match each picture.

1.

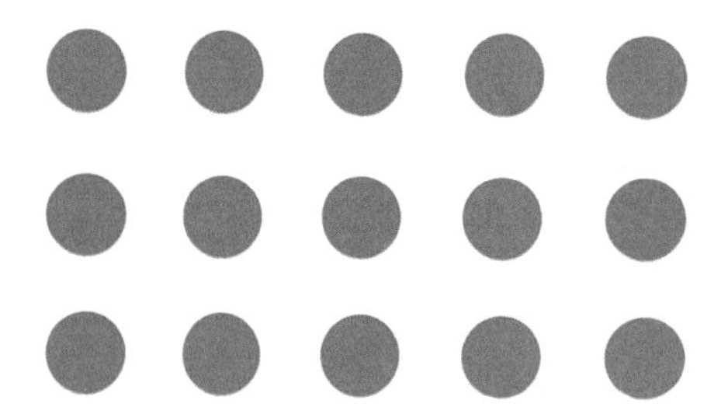

2.

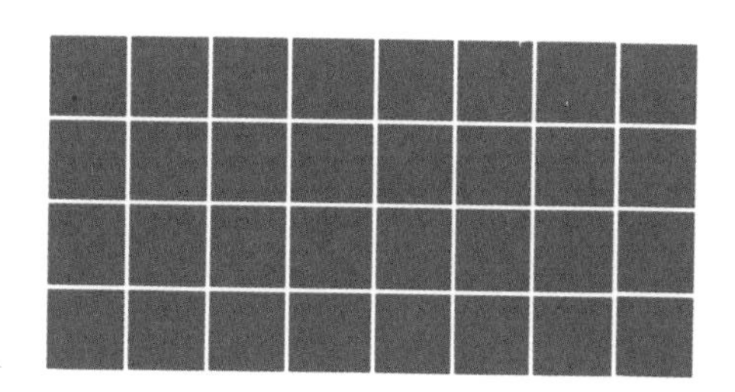

3. Nancy planted 6 rows of trees.
There were 4 trees in each row.
How many trees were planted in all?

4. Nick lined up 4 rows of toy soldiers.
There were 8 soldiers in each row.
How many toy soldiers did he have?

5. Ellen bought 6 envelopes and a sheet of stamps.
There were 5 rows of 7 stamps on the sheet.
How many stamps did she buy?

Continue the counting patterns.

6. 3, 6, 9, ▒, ▒, ▒, ▒, ▒, 27

7. 5, 10, 15, ▒, ▒, ▒, ▒, ▒, 45

8. 4, 8, 12, ▒, ▒, ▒, ▒, ▒, 36

Write a multiplication sentence to show how many tiles there are.

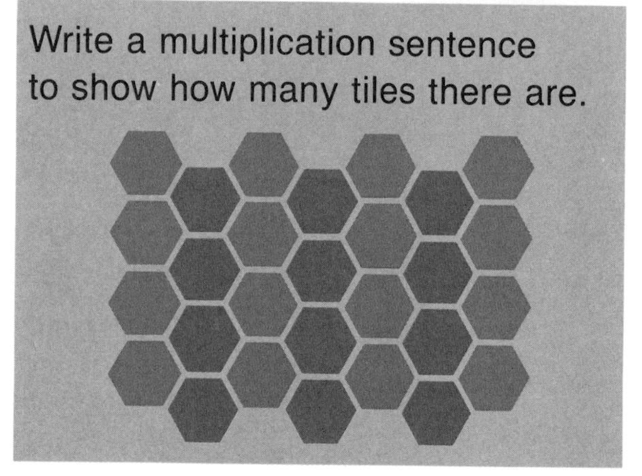

Multiplication: Groups of 0

How many fish do you see?

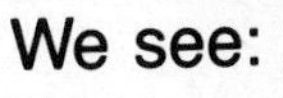

We see:

We say: 3 times 0 equals 0.

We write: $3 \times 0 = 0$ or $\begin{array}{r} 0 \\ \times 3 \\ \hline 0 \end{array}$

Each bowl has 0 fish.
There are 0 fish in all.

Multiply.

1. $\begin{array}{r} 0 \\ \times 5 \\ \hline \end{array}$
2. $\begin{array}{r} 0 \\ \times 3 \\ \hline \end{array}$
3. $\begin{array}{r} 0 \\ \times 8 \\ \hline \end{array}$
4. $\begin{array}{r} 0 \\ \times 1 \\ \hline \end{array}$
5. $\begin{array}{r} 0 \\ \times 7 \\ \hline \end{array}$
6. $\begin{array}{r} 0 \\ \times 9 \\ \hline \end{array}$
7. $\begin{array}{r} 0 \\ \times 6 \\ \hline \end{array}$
8. $\begin{array}{r} 0 \\ \times 4 \\ \hline \end{array}$
9. $\begin{array}{r} 0 \\ \times 2 \\ \hline \end{array}$

10. Copy and complete this table. Use squared paper.

×	0	1	2	3	4	5
0			0			
1				3		
2	0				8	
3		3		9		
4			8			
5						25

11. What happens when you multiply a number by 0? Write a rule.

ESTIMATING

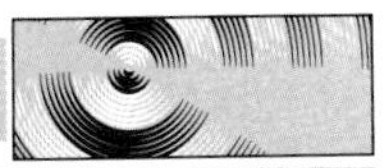

Estimate: How many muffins would fit in the pan?

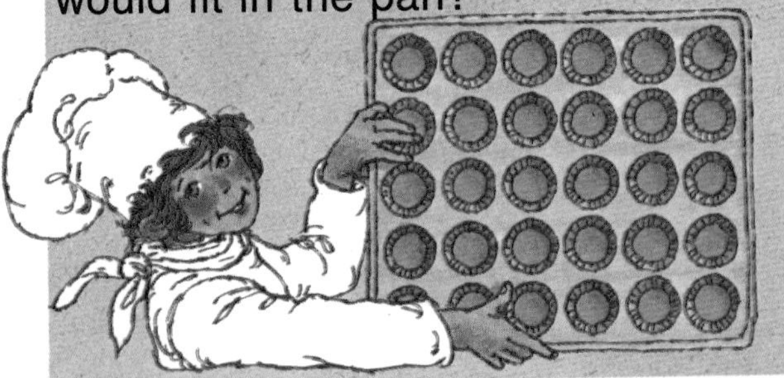

BITS AND BYTES

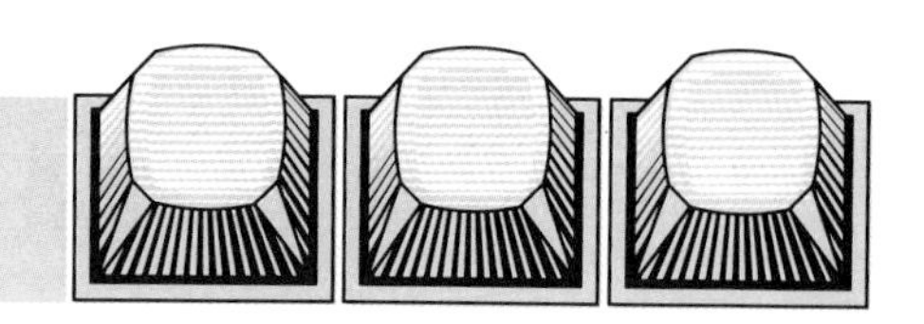

How many books are there in all?

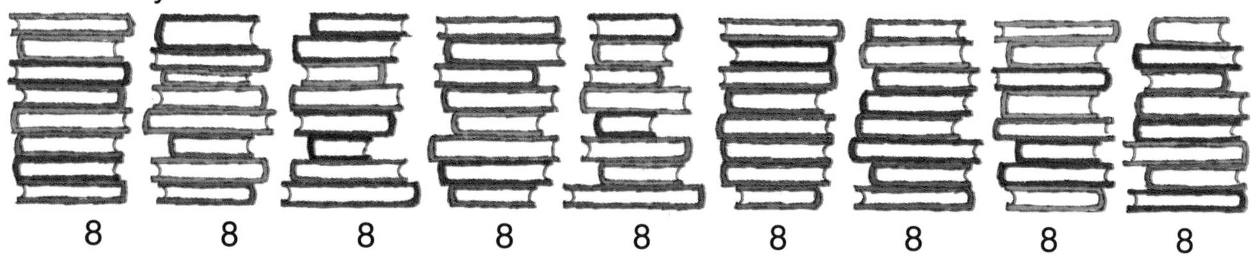

Kathy adds on her calculator.

Debbie multiplies on her calculator.

What answers do Kathy and Debbie get?
Use your calculator.
Why do they get the same answer?

Write each addition as a multiplication.
Use your calculator to find each answer.

1. $9 + 9 + 9 + 9 + 9 + 9$

2. $7 + 7 + 7 + 7 + 7 + 7 + 7 + 7$

3. $\begin{array}{r} 8 \\ 8 \\ 8 \\ 8 \\ 8 \\ +8 \\ \hline \end{array}$

4. $\begin{array}{r} 12 \\ 12 \\ 12 \\ +12 \\ \hline \end{array}$

5. $\begin{array}{r} 23 \\ 23 \\ 23 \\ +23 \\ \hline \end{array}$

6. $\begin{array}{r} 85 \\ 85 \\ 85 \\ 85 \\ +85 \\ \hline \end{array}$

7. $\begin{array}{r} 127 \\ 127 \\ 127 \\ +127 \\ \hline \end{array}$

JUST FOR FUN

Copy and complete.

×2	8	0	3	6	4	5	2	9	7	1
	16	0	6	12	8	10	4	18	14	2

1.

×3	8	4	1	7	0	6	2	5	9	3
			3				6	15		

2.

×5	8	5	1	7	3	0	4	9	6	2
	40		5							

3.

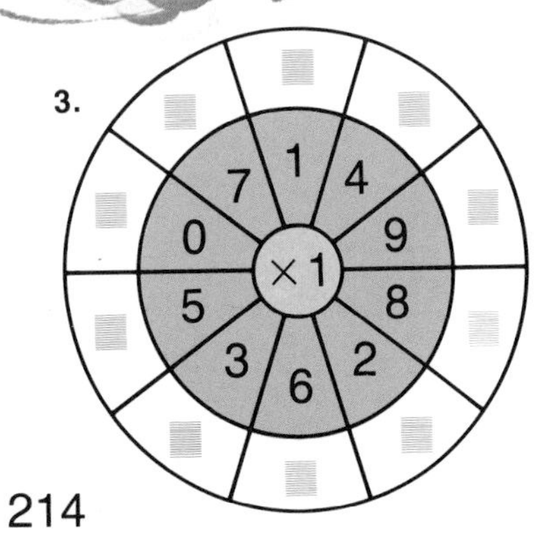

4.

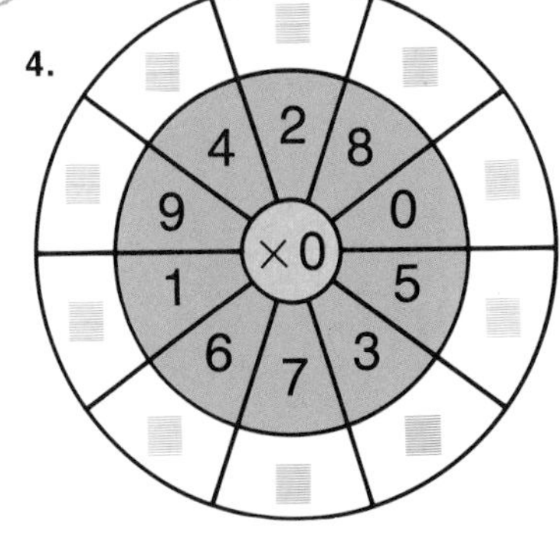

5.

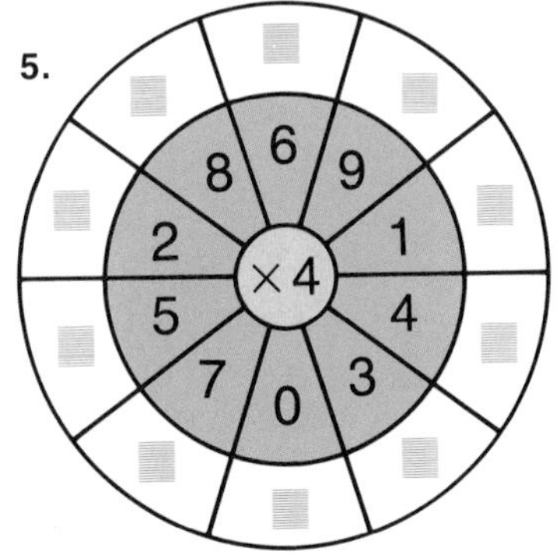

PROBLEM SOLVING

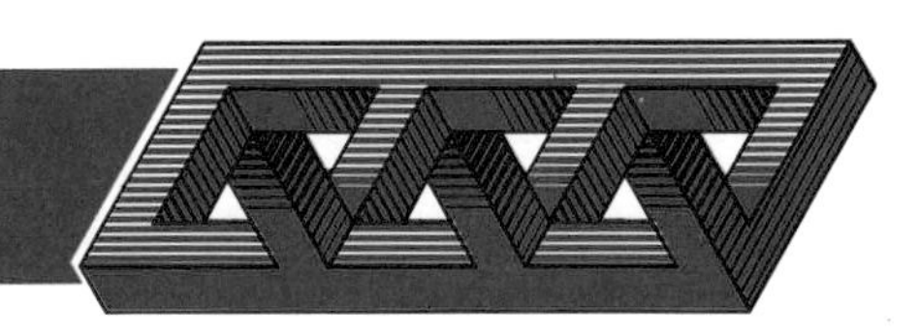

Constructing a Table

The clown has 2 pair of pants and 3 shirts.

Black Blue Yellow Red Green

He can make 6 different outfits.

Outfits with Black Pants

Outfits with Blue Pants

1. Copy and complete the table.

The Clown's Outfits

Pants	Shirt
Black	Yellow
Black	Red
Black	
Blue	Yellow
Blue	
	Green

2. Trudy's mother bought her a pair of blue jeans, a pair of brown slacks and a pair of white shorts. She also bought her a yellow blouse and a green shirt. Make a table to show how many different outfits she can wear.

3. Trudy's father also bought her a pink turtleneck sweater. Make a table to show how many outfits she now has.

PROBLEM SOLVING

Obtaining Information from a Picture

Betsy's Bug Collection

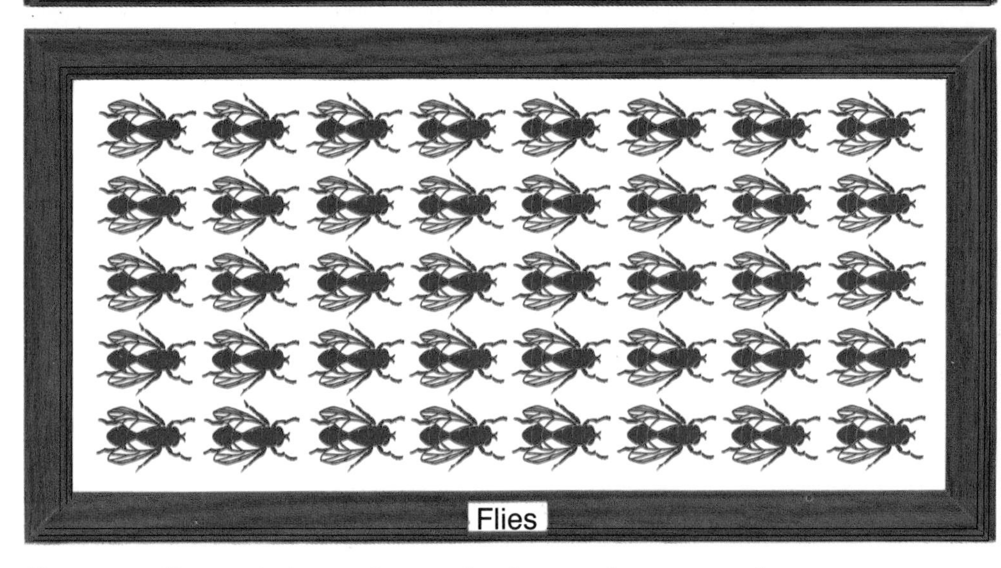

Betsy collected these bugs for her science project.

1. Write a multiplication sentence which shows how many grasshoppers there are.

2. Copy and complete.

Bugs	How Many?
Grasshoppers	
Flies	
Beetles	

3. How many more grasshoppers than flies are shown?

4. Write a multiplication sentence to show how many spiders there are in the first 3 rows.

5. Write a multiplication sentence to show how many spiders there are in the last 3 rows.

6. How many spiders are there in the bug collection?

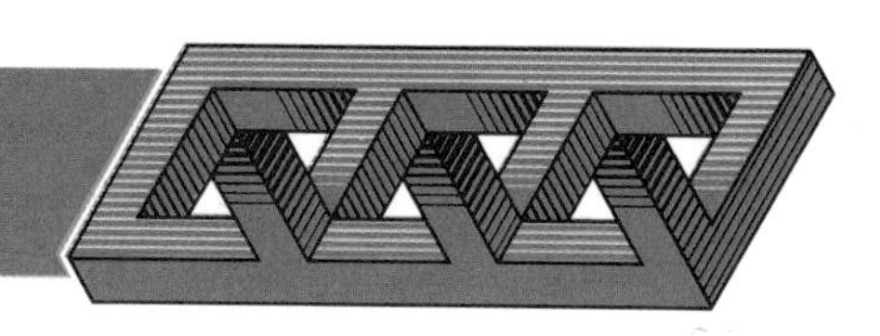

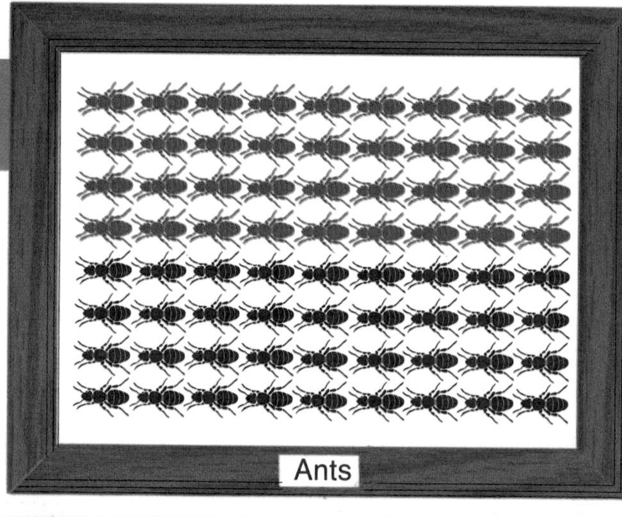

7. How many more flies are there than spiders?

8. Write a multiplication sentence to show how many ants are in the first 4 rows.

9. How many ants are in the last 4 rows?

10. How many ants are in the bug collection?

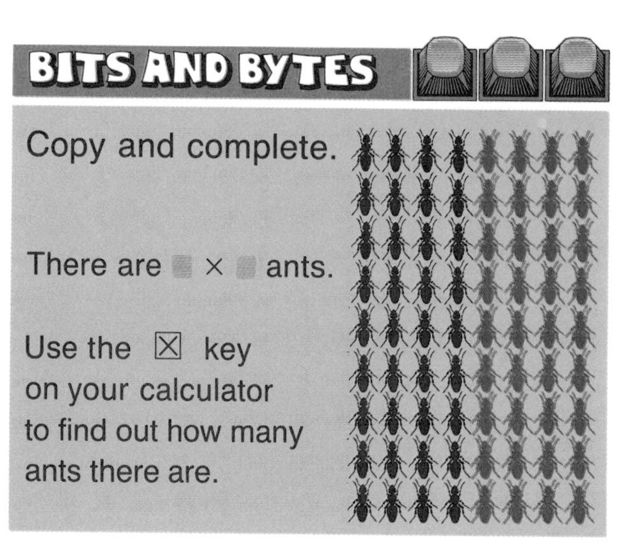

BITS AND BYTES

Copy and complete.

There are ▒ × ▒ ants.

Use the ⊠ key on your calculator to find out how many ants there are.

Story Problems

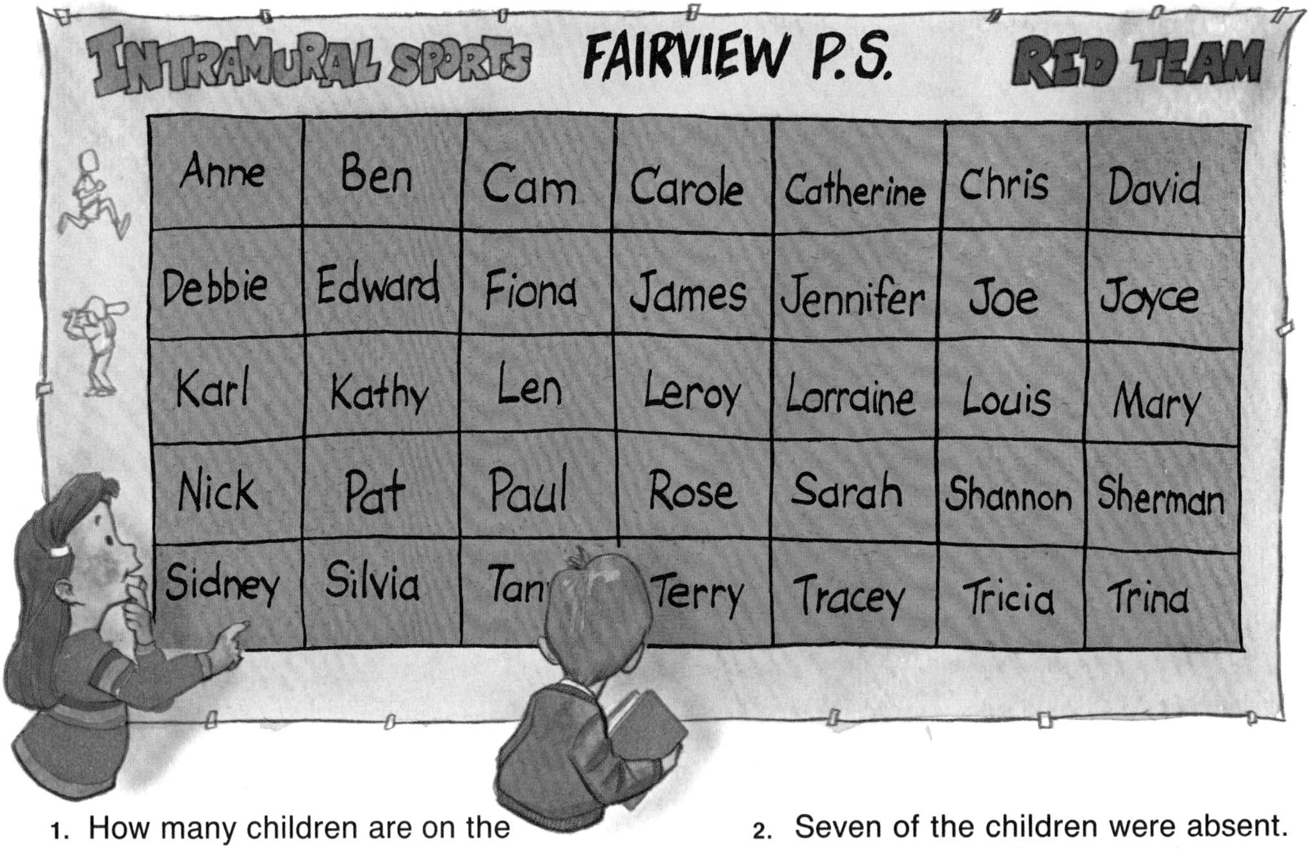

INTRAMURAL SPORTS FAIRVIEW P.S. RED TEAM

Anne	Ben	Cam	Carole	Catherine	Chris	David
Debbie	Edward	Fiona	James	Jennifer	Joe	Joyce
Karl	Kathy	Len	Leroy	Lorraine	Louis	Mary
Nick	Pat	Paul	Rose	Sarah	Shannon	Sherman
Sidney	Silvia	Tanya	Terry	Tracey	Tricia	Trina

1. How many children are on the Red Team?

2. Seven of the children were absent. How many children were present?

3. How many names are in the 3 bottom rows?

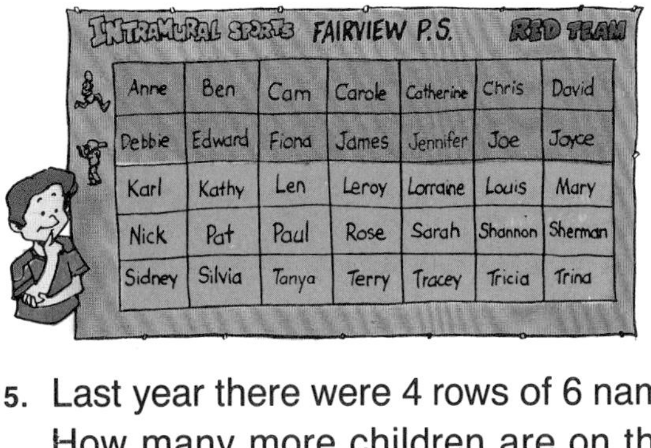

4. How many names are in the last 3 columns?

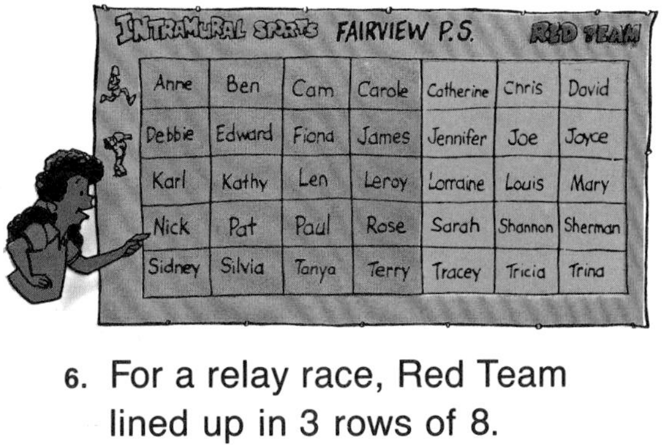

5. Last year there were 4 rows of 6 names. How many more children are on the Red Team this year?

6. For a relay race, Red Team lined up in 3 rows of 8. How many Red Team children were not in the relay race?

EXTENSIONS

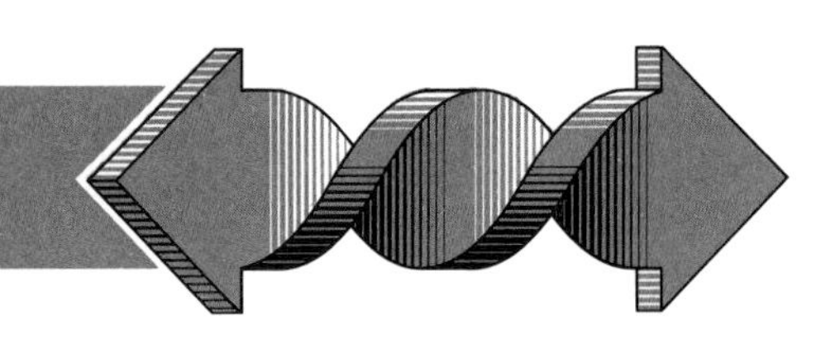

Number Patterns

1. Use a 100-chart.

A. Colour in yellow the numbers that have groups of 2.

B. Circle in blue the numbers that have groups of 3.

C. Continue each pattern to the end of the chart.

D. Which numbers are both coloured yellow and circled in blue?

1	2	3	4	5	6	7	8	9	10
11	12	13	14	15	16	17	18	19	20
21	22	23	24	25	26	27	28	29	30
31	32	33	34	35	36	37	38	39	40
41	42	43	44	45	46	47	48	49	50
51	52	53	54	55	56	57	58	59	60
61	62	63	64	65	66	67	68	69	70
71	72	73	74	75	76	77	78	79	80
81	82	83	84	85	86	87	88	89	90
91	92	93	94	95	96	97	98	99	100

2. Use squared paper to make charts like these.

A. Colour in yellow the numbers that have groups of 2.

B. Circle in blue the numbers that have groups of 3.

C. How do these patterns compare with the 100-chart patterns?

1	2	3	4	5	6	7
8	9	10	11	12	13	14
15	16	17	18	19	20	21
22	23	24	25	26	27	28
29	30	31	32	33	34	35
36	37	38	39	40	41	42
43	44	45	46	47	48	49
50	51	52	53	54	55	56
57	58	59	60	61	62	63
64	65	66	67	68	69	70

1	2	3	4	5	6
7	8	9	10	11	12
13	14	15	16	17	18
19	20	21	22	23	24
25	26	27	28	29	30
31	32	33	34	35	36
37	38	39	40	41	42
43	44	45	46	47	48
49	50	51	52	53	54
55	56	57	58	59	60

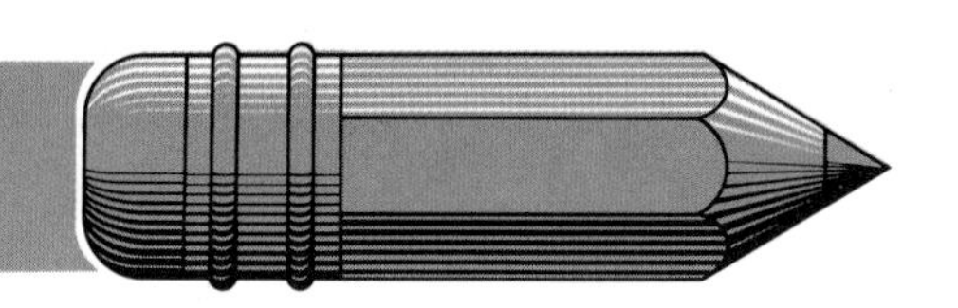

CHECK-UP

Copy and complete each sentence.

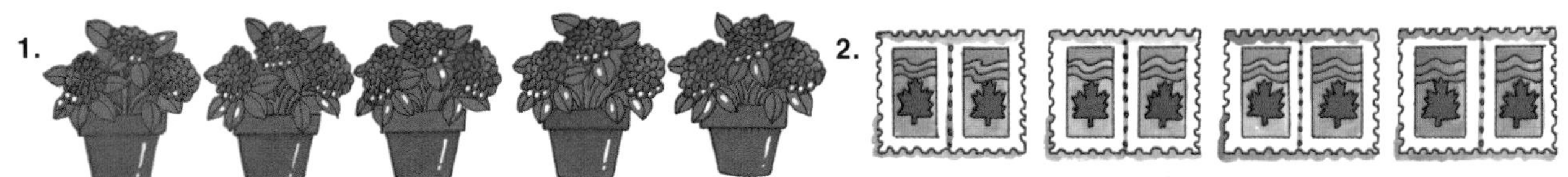

1. $3 + 3 + 3 + 3 + 3 = ▒$
   5 groups of 3 = ▒

2. 4 groups of 2 = ▒
   $4 \times 2 = ▒$

Write a multiplication sentence for each picture.

3.   

   $▒ \times 4 = ▒$

4.   

Multiply.

5. $6 \times 2 = ▒$
6. $9 \times 0 = ▒$
7. $6 \times 3 = ▒$

8. $\begin{array}{r} 3 \\ \times 9 \\ \hline \end{array}$
9. $\begin{array}{r} 1 \\ \times 8 \\ \hline \end{array}$
10. $\begin{array}{r} 5 \\ \times 7 \\ \hline \end{array}$
11. $\begin{array}{r} 2 \\ \times 3 \\ \hline \end{array}$
12. $\begin{array}{r} 4 \\ \times 5 \\ \hline \end{array}$
13. $\begin{array}{r} 3 \\ \times 8 \\ \hline \end{array}$

Write 2 multiplication sentences for each picture.

14. 

15. 

16. Copy and complete the table.

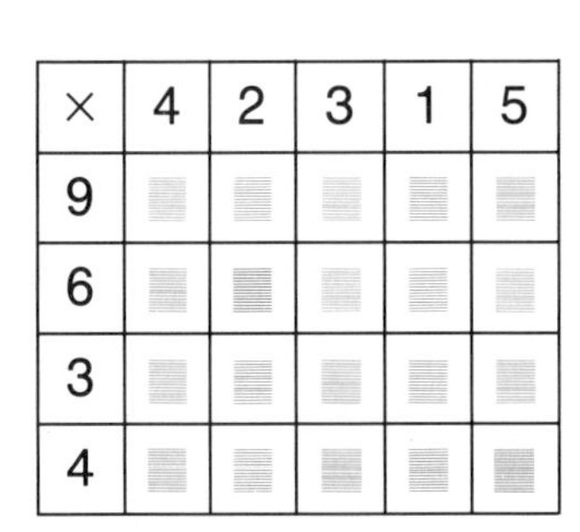

×	4	2	3	1	5
9	▒	▒	▒	▒	▒
6	▒	▒	▒	▒	▒
3	▒	▒	▒	▒	▒
4	▒	▒	▒	▒	▒

17. There are 4 players on each team.
    There are 7 teams.
    How many players are there altogether?

The Fishes of Kempenfelt Bay

Under the bubbles
Of Kempenfelt Bay,
The slippery fishes
Dawdle all day.

They park in the shallows
And wiggle and stray,
The slippery fishes
Of Kempenfelt Bay.

I ride on a bike.
I swing in the gym.
But I'd leave them behind
If I knew how to swim

With the slippery fishes
That dawdle all day,
Under the bubbles
Of Kempenfelt Bay.

—Dennis Lee

Sharing

Wendy shared 12 baseball cards equally among 4 children.

First, she gave each child a card.
She counted as she gave them.

She still had cards left, so she gave each child another card.

She still had cards left, so she gave each child another card.

How many cards did each child get?

12 cards shared among 4 children is 3 cards each.

Sometimes we say **divide** instead of **share**.
Divide 6 balloons among 3 children.

We see:

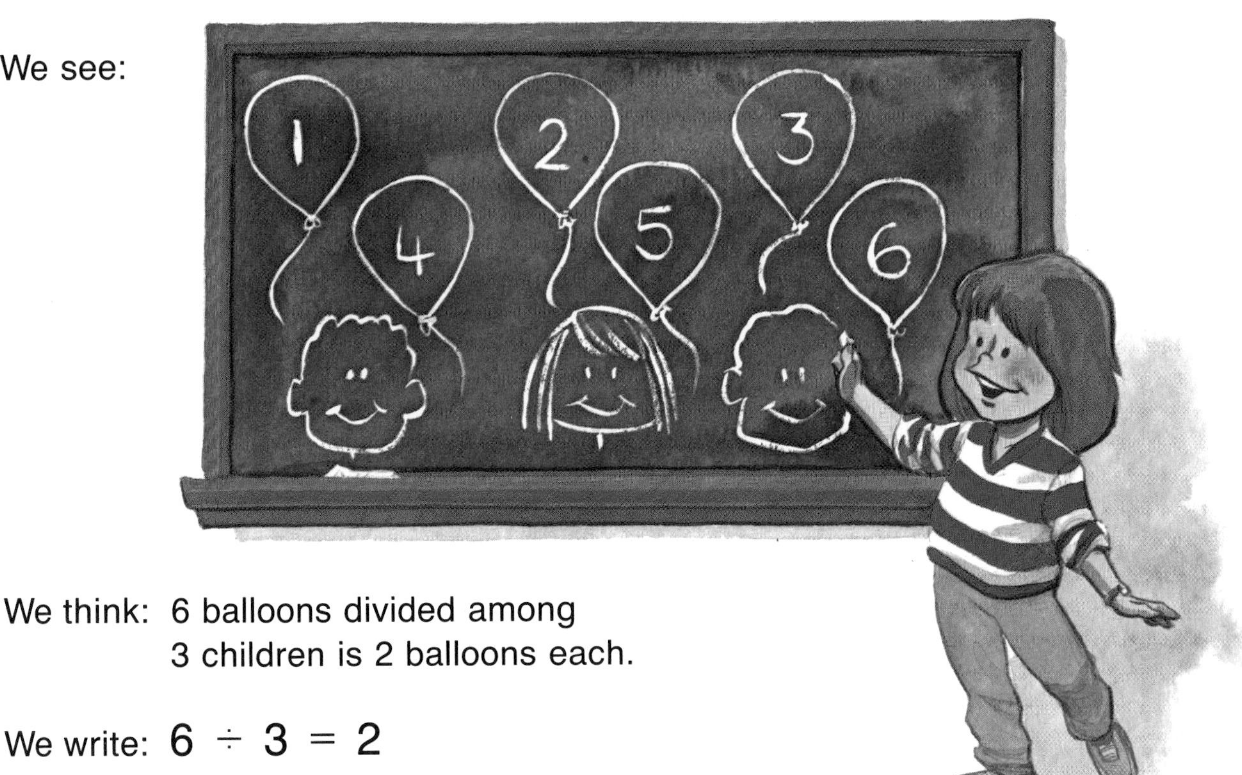

We think: 6 balloons divided among 3 children is 2 balloons each.

We write: $6 \div 3 = 2$

Draw a picture to show the dividing.
Copy and complete each sentence.

1. Divide 9 cards among 3 children.

   $9 \div 3 =$ ▒

2. Divide 8 balloons among 2 children.

   $8 \div 2 =$ ▒

3. Divide 12 fish among 3 bowls.

   $12 \div 3 =$ ▒

ESTIMATING

Estimate: About how many acorns are there for each squirrel?

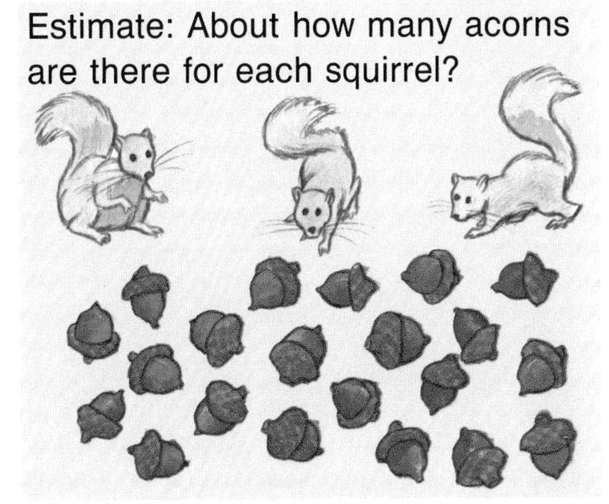

Equal Groups

There are 20 muffins.
There are 5 muffins on each plate.
How many plates are there?

We see:

We think: 20 divided into equal groups of 5 is 4 groups.

We write: $20 \div 5 = 4$

There are 4 plates.

Copy and complete each sentence.

1.

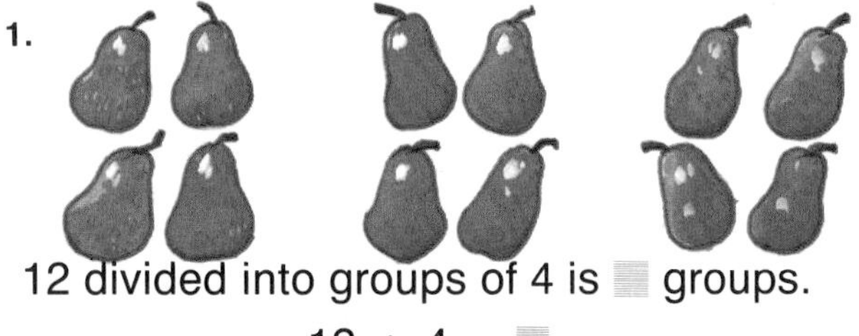

12 divided into groups of 4 is ▒ groups.

$12 \div 4 = $ ▒

2.

6 divided into groups of 2 is ▒ groups.

$6 \div 2 = $ ▒

3.

16 divided into groups of 4 is ▒ groups.

$16 \div 4 = $ ▒

4.

9 divided into groups of 3 is ▒ groups.

$9 \div 3 = $ ▒

5.

6 divided into groups of 3 is ▒ groups.

$6 \div 3 = $ ▒

6.

10 divided into groups of 2 is ▒ groups.

$10 \div 2 = $ ▒

There are 20 fish.
We put 4 fish in each bowl.
How many bowls do we need?

We see:

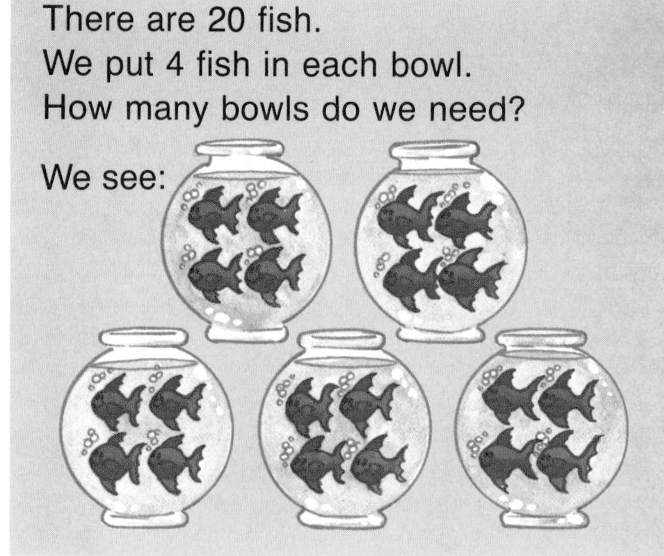

We think: 20 divided into equal groups of 4 is 5 groups.

We write: $20 \div 4 = 5$

We need 5 bowls.

Copy and complete each sentence.

1.     

___ divided into groups of 2 is ___ groups.

$\_\_\_ \div 2 = \_\_\_$

2. 

___ divided into groups of 3 is ___ groups.

$\_\_\_ \div 3 = \_\_\_$

3.   

___ divided into groups of 4 is ___ groups.

$\_\_\_ \div 4 = \_\_\_$

4. 

___ divided into groups of 3 is ___ groups.

$\_\_\_ \div 3 = \_\_\_$

5.   

___ divided into groups of 4 is ___ groups.

$\_\_\_ \div 4 = \_\_\_$

6. 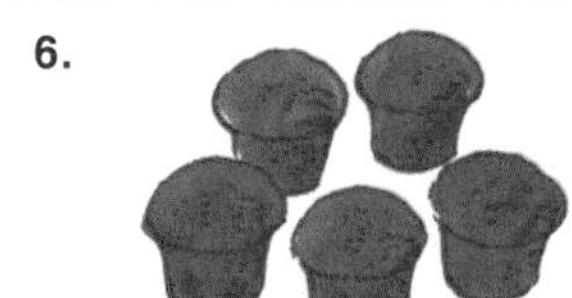 

___ divided into groups of 5 is ___ groups.

$\_\_\_ \div 5 = \_\_\_$

Interpreting Division Pictures

How many blocks of stamps are shown?

We see:

We think: There are 12 stamps.
There are 4 stamps in each block.
12 divided into groups of 4 is 3 groups.

We write: $12 \div 4 = 3$

There are 3 blocks of stamps.

Fill in the missing numbers to answer each question.

1. How many teams are shown?

▒ children

▒ on each team

▒ ÷ ▒ = ▒

2. How many cartons of juice are there?

▒ bottles

▒ in each carton

▒ ÷ ▒ = ▒

Fill in the missing numbers to answer each question.

1. How many groups of balloons are shown?

____ balloons

____ in each group

____ ÷ ____ = ____

2. How many sets of paints are there?

____ tubes

____ in each set

____ ÷ ____ = ____

3. How many bunches of bananas are there?

____ bananas

____ in each bunch

____ ÷ ____ = ____

4. How many pairs of shoes are in the boxes?

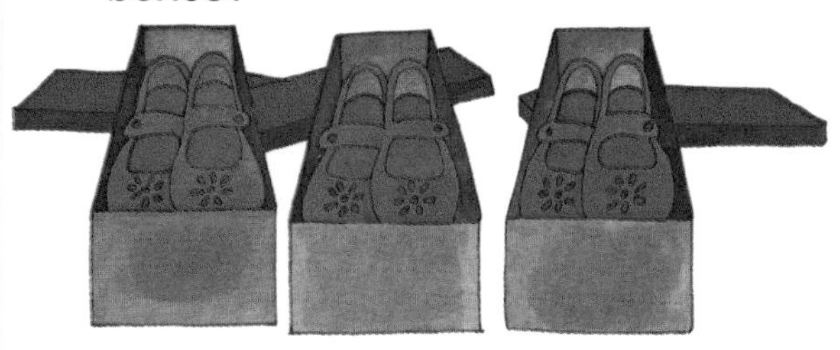

____ shoes

____ in each pair

____ ÷ ____ = ____

5. How many boxes of pears are there?

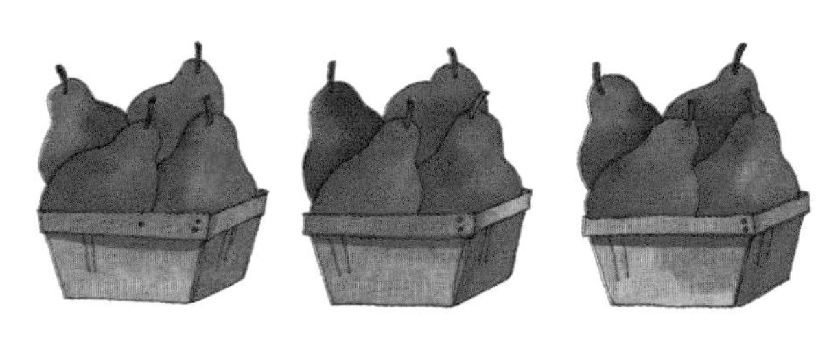

____ pears

____ in each box

____ ÷ ____ = ____

6. How many groups of brushes are shown?

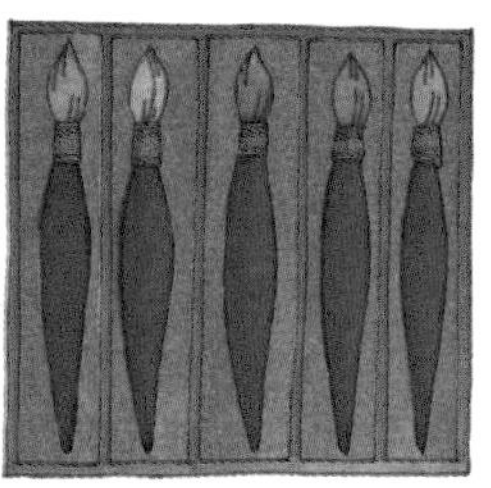

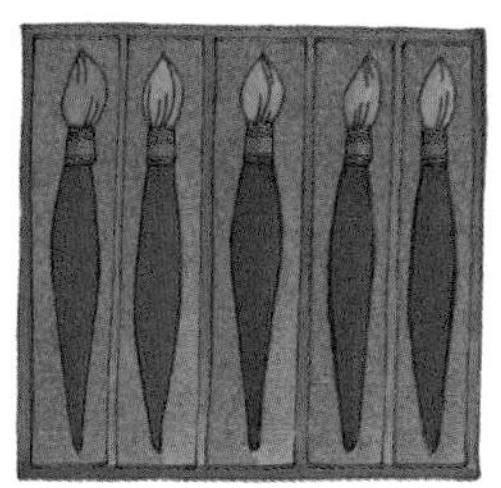

____ brushes

____ in each group

____ ÷ ____ = ____

Division: Groups of 2

There are 10 shoes in all.
There are 2 shoes in each pair.
How many pairs of shoes are there?

We see:

We think: 10 divided into groups of 2 is 5 groups.

We write: $10 \div 2 = 5$

There are 5 pairs of shoes.

Copy and complete each sentence.

1.

$\square \div 2 = \square$

2.

$\square \div 2 = \square$

3.

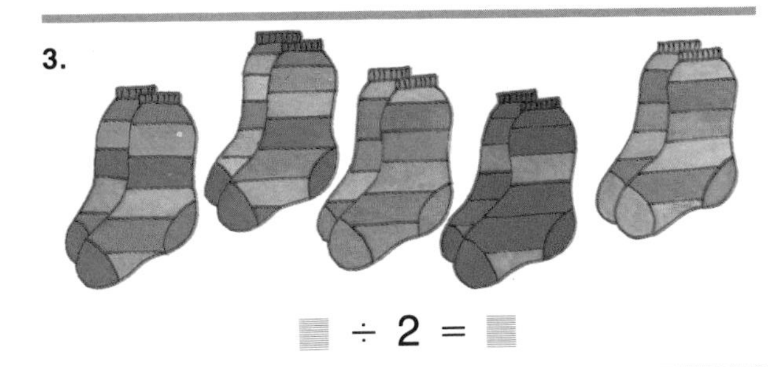

$\square \div 2 = \square$

4.

$\square \div 2 = \square$

5.

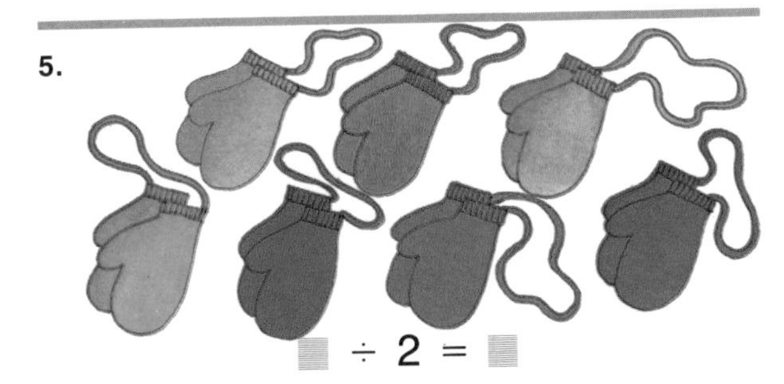

$\square \div 2 = \square$

Division: Groups of 3

There are 24 birds in all.
There are 3 birds in each group.
How many groups of birds are there?

We see:

We think: 24 divided into groups of 3 is 8 groups.

We write: $24 \div 3 = 8$

There are 8 groups of birds.

Copy and complete each sentence.

1. 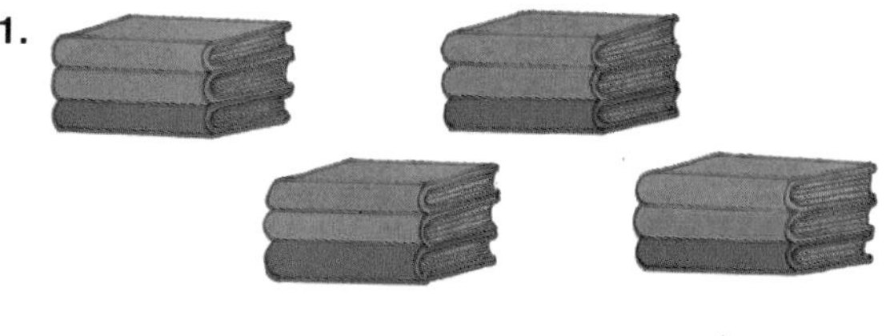

$\square \div 3 = \square$

2. 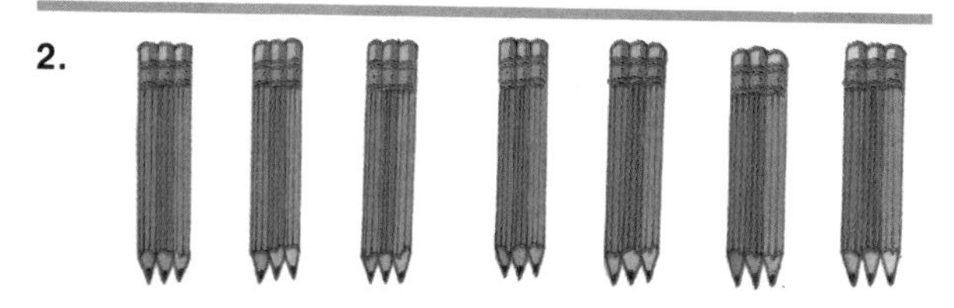

$\square \div 3 = \square$

3. 

$\square \div 3 = \square$

4. 

$\square \div 3 = \square$

5. 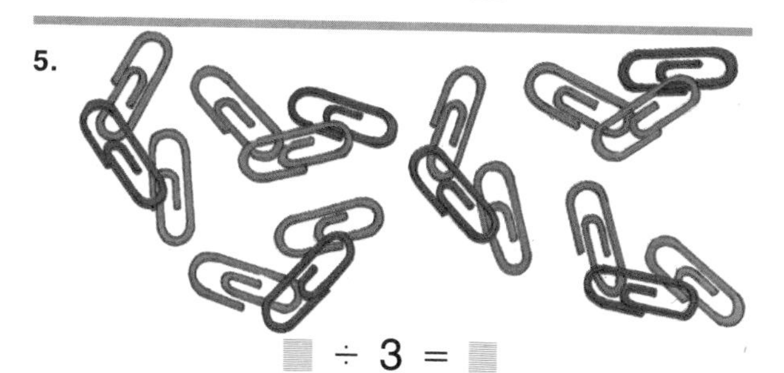

$\square \div 3 = \square$

Division: Groups of 4

There are 12 leaves in all.
There are 4 leaves on each plant.
How many plants are there?

We see:

We think: 12 divided into groups of 4 is 3 groups.

We write: $12 \div 4 = 3$

There are 3 plants.

Copy and complete each sentence.

1. 

$\square \div 4 = \square$

2. 

$\square \div 4 = \square$

3. 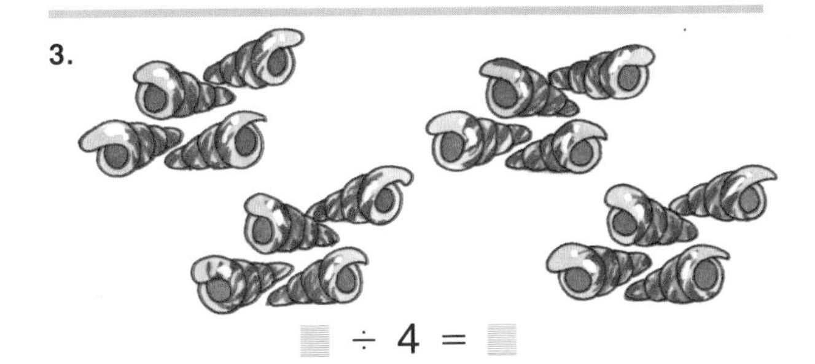

$\square \div 4 = \square$

4. 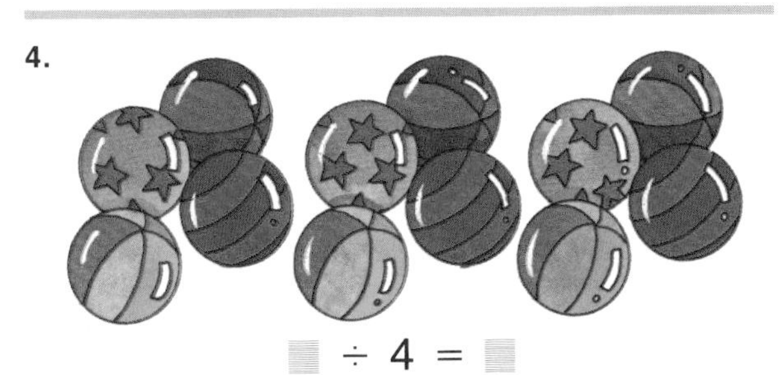

$\square \div 4 = \square$

5. 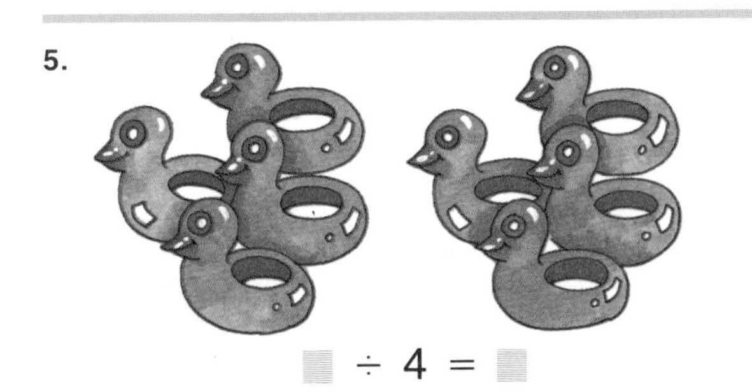

$\square \div 4 = \square$

Division: Groups of 5

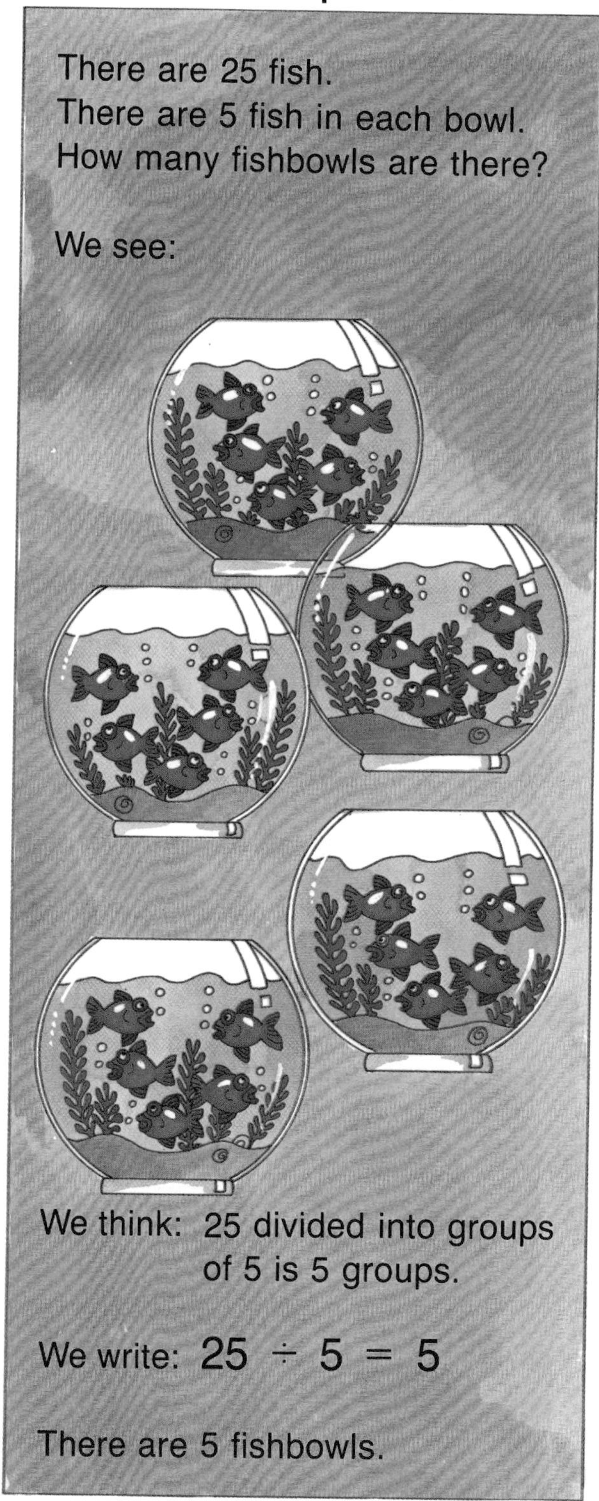

There are 25 fish.
There are 5 fish in each bowl.
How many fishbowls are there?

We see:

We think: 25 divided into groups of 5 is 5 groups.

We write: $25 \div 5 = 5$

There are 5 fishbowls.

Copy and complete each sentence.

1. 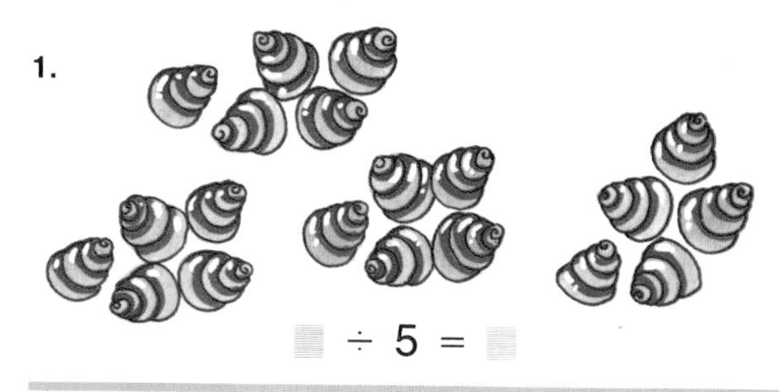

$\square \div 5 = \square$

2. 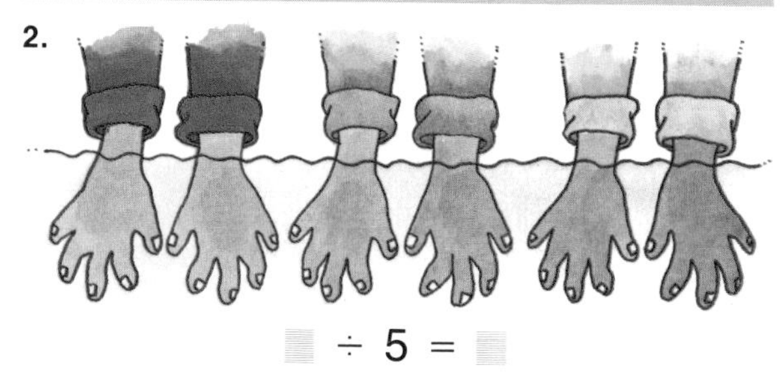

$\square \div 5 = \square$

3. 

$\square \div 5 = \square$

4. 

$\square \div 5 = \square$

5. 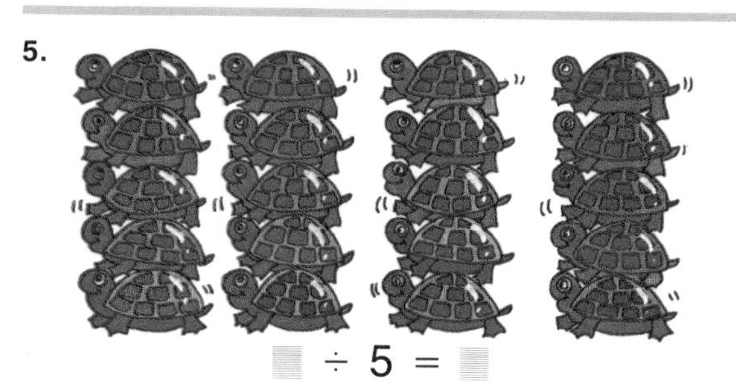

$\square \div 5 = \square$

Division: Groups of 1

There are 4 rubber ducks.
There is 1 duck in each tub.
How many tubs are there?

We see:

We think: 4 divided into groups of 1 is 4 groups.

We write: $4 \div 1 = 4$

There are 4 tubs.

Copy and complete each sentence.

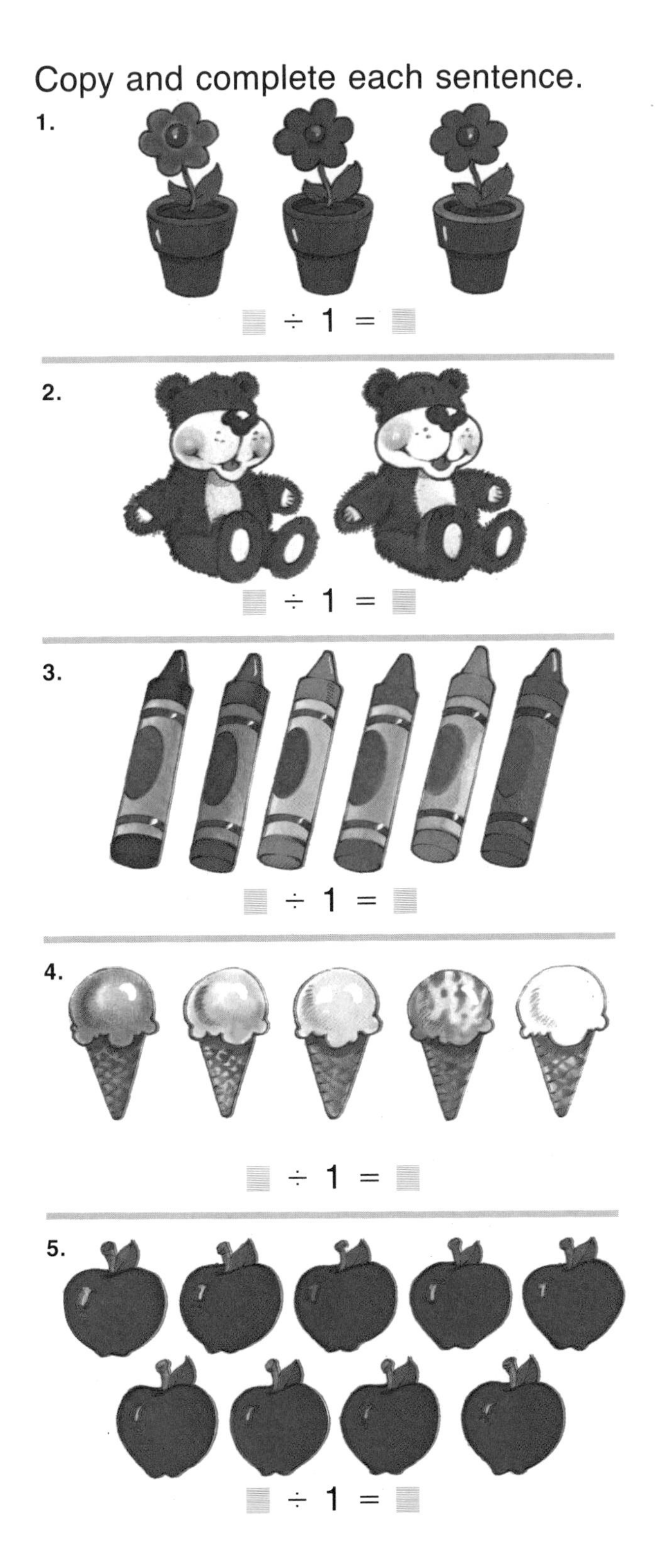

1. $\square \div 1 = \square$

2. $\square \div 1 = \square$

3. $\square \div 1 = \square$

4. $\square \div 1 = \square$

5. $\square \div 1 = \square$

Drawing Division Pictures

We see: $12 \div 3 = \square$

We think: There are 12 things in all.
There are 3 things in each group.
How many groups are there?

Draw a picture.

There are 4 groups.

$12 \div 3 = 4$

Draw a picture for each sentence.
Then complete the sentences.

1. $6 \div 3 = \square$
2. $20 \div 5 = \square$
3. $8 \div 4 = \square$
4. $5 \div 1 = \square$
5. $9 \div 3 = \square$
6. $10 \div 2 = \square$

Rows and Columns for Division

We can write 2 division sentences for each picture.

Bridget writes:

$15 \div 5 = 3$

Karl writes:

$15 \div 3 = 5$

Write 2 division sentences to match each picture.

1.

2.

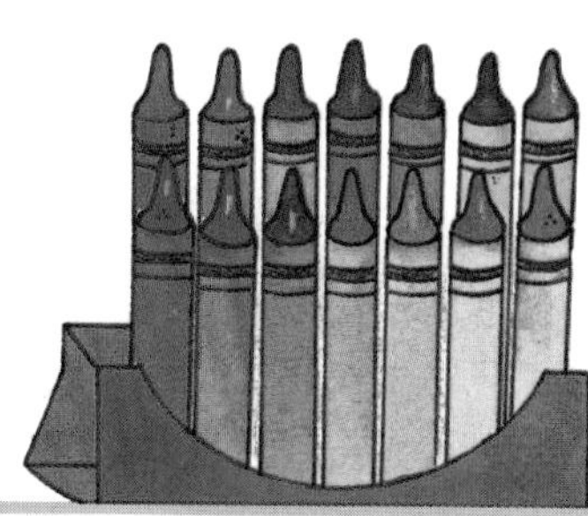

3.

4.

5.

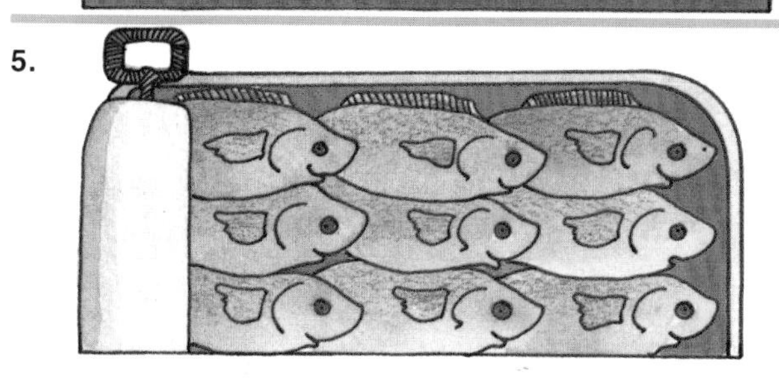

6.

Write 2 division sentences to match each picture.

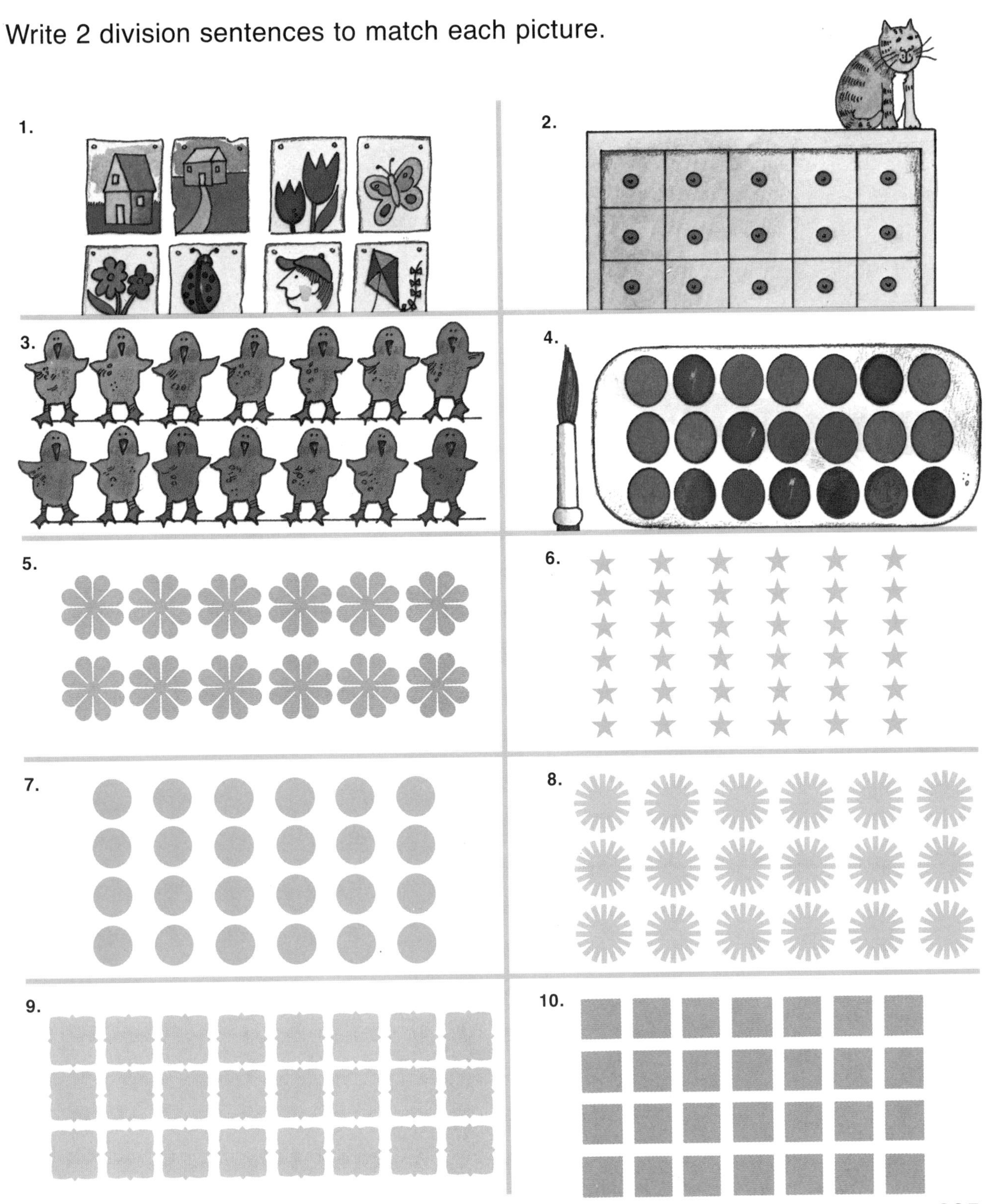

Fact Families

Two multiplication facts match this picture.

$$4 \times 5 = 20$$
$$5 \times 4 = 20$$

Two division facts match this picture.

$$20 \div 5 = 4$$
$$20 \div 4 = 5$$

These 4 facts are called a fact family.

Write a fact family for each picture.

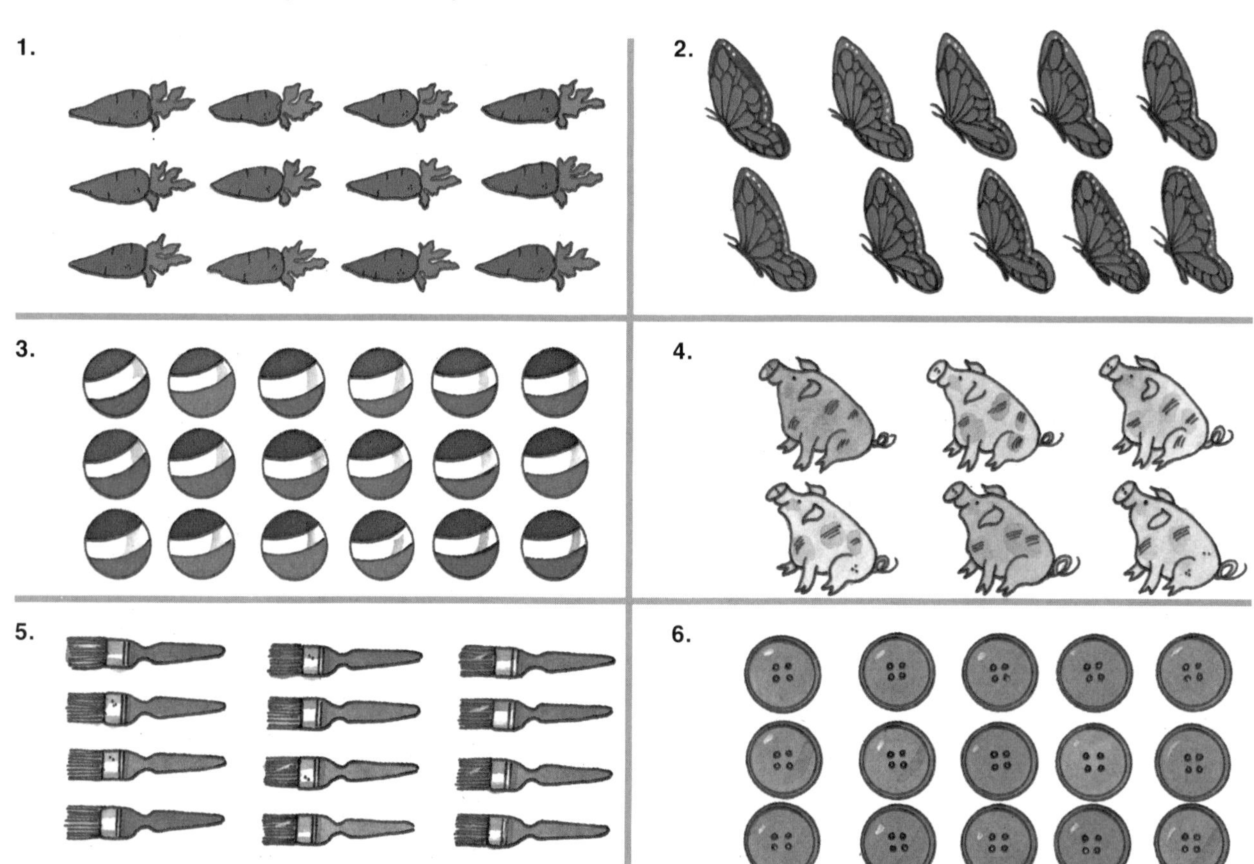

All the facts in a fact family use the same 3 numbers.

$4 \times 5 = 20$

$5 \times 4 = 20$

$20 \div 4 = 5$

$20 \div 5 = 4$

Write the fact family for each group of numbers.

Copy and complete.

4. $3 \times 4 = \square$
   $12 \div 4 = \square$

5. $6 \times 2 = \square$
   $12 \div 2 = \square$

6. $3 \times 5 = \square$
   $15 \div 5 = \square$

7. $4 \times 2 = \square$
   $8 \div 2 = \square$

8. $7 \times 2 = \square$
   $14 \div 2 = \square$

9. $6 \times 3 = \square$
   $18 \div 3 = \square$

10. $7 \times 3 = \square$
    $21 \div 3 = \square$

11. $9 \times 2 = \square$
    $18 \div 2 = \square$

12. $8 \times 3 = \square$
    $24 \div 3 = \square$

Using Fact Families to Divide

To solve: $12 \div 4 = \square$

We think: $\square \times 4 = 12$

We write: $3 \times 4 = 12$

so $12 \div 4 = 3$

Use fact families to solve these.

1. $6 \div 2 = \square$
2. $12 \div 3 = \square$
3. $8 \div 4 = \square$
4. $15 \div 3 = \square$
5. $9 \div 3 = \square$
6. $10 \div 2 = \square$
7. $12 \div 4 = \square$
8. $21 \div 3 = \square$
9. $6 \div 3 = \square$
10. $8 \div 2 = \square$
11. $20 \div 4 = \square$
12. $12 \div 2 = \square$
13. $10 \div 5 = \square$
14. $16 \div 4 = \square$
15. $16 \div 2 = \square$
16. $20 \div 5 = \square$
17. $32 \div 4 = \square$
18. $24 \div 4 = \square$
19. $30 \div 5 = \square$
20. $15 \div 5 = \square$
21. $6 \div 1 = \square$
22. $28 \div 4 = \square$
23. $14 \div 2 = \square$
24. $9 \div 1 = \square$
25. $25 \div 5 = \square$
26. $18 \div 3 = \square$
27. $27 \div 3 = \square$
28. $24 \div 3 = \square$
29. $35 \div 5 = \square$
30. $18 \div 2 = \square$

JUST FOR FUN

Copy and complete.

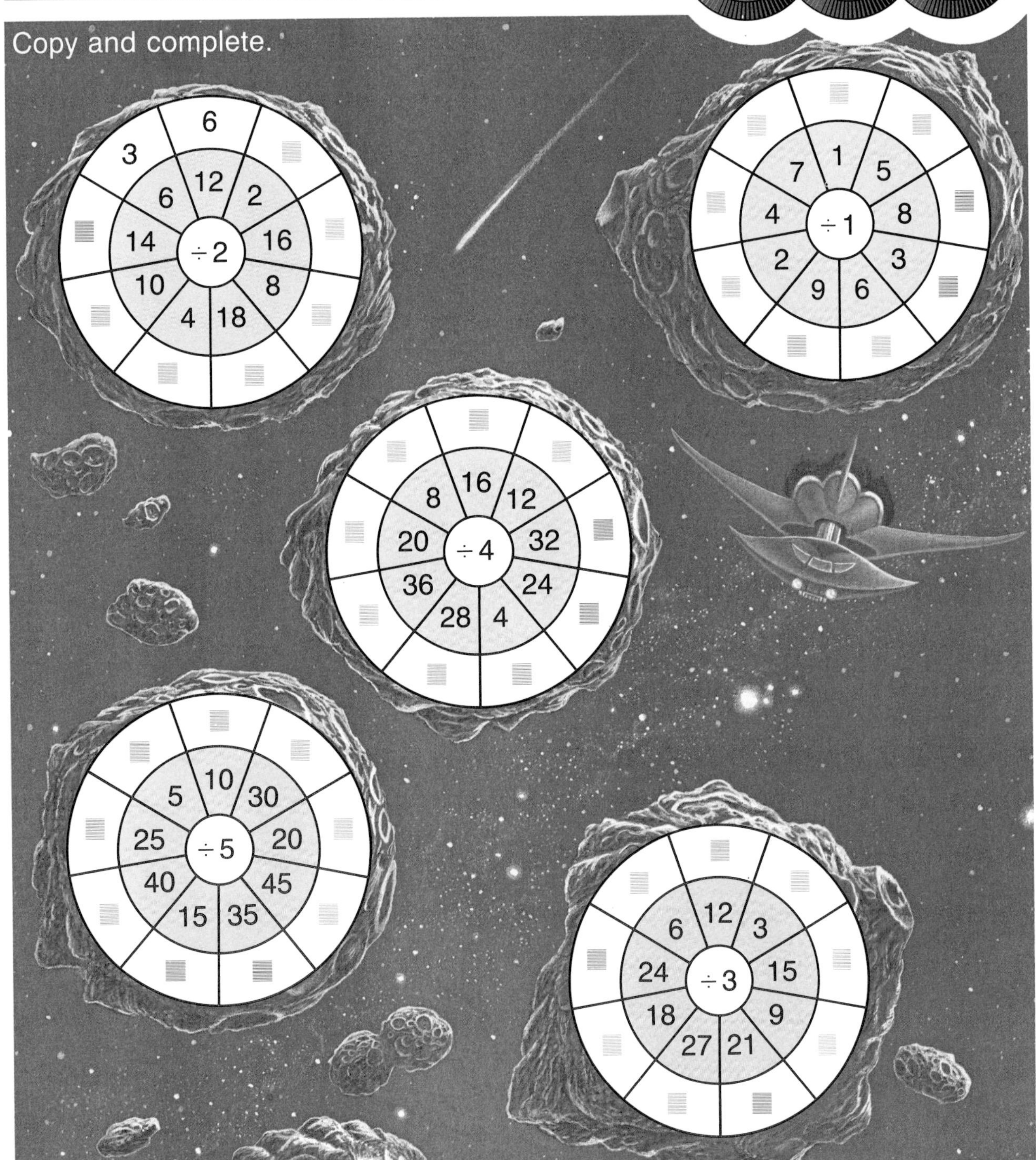

A New Way to Write Division

Copy and complete these tables.

1.

×	4
3	■

2.

×	2
6	■

3.

×	3
5	■

4.

×	4
7	■

5.

×	■
3	15

6.

×	■
4	20

7.

×	■
2	12

8.

×	■
5	25

Clara writes $\overset{■}{3\overline{)15}}$ instead of

×	■
3	15

$\overset{■}{3\overline{)15}}$ is a new way to write $15 \div 3 =$ ■

Write these division sentences another way.

9. $\overset{■}{3\overline{)12}}$

9. $12 \div 3 = 4$

10. $\overset{■}{2\overline{)8}}$

11. $\overset{■}{3\overline{)9}}$

12. $\overset{■}{4\overline{)20}}$

13. $\overset{■}{3\overline{)18}}$

14. $\overset{■}{5\overline{)15}}$

15. $\overset{■}{2\overline{)14}}$

Copy and complete.

16. $2\overline{)8}$

16. $\overset{4}{2\overline{)8}}$

17. $3\overline{)12}$

18. $4\overline{)8}$

19. $3\overline{)6}$

20. $4\overline{)24}$

21. $3\overline{)18}$

22. $4\overline{)12}$

23. $5\overline{)20}$

24. $4\overline{)16}$

25. $3\overline{)24}$

26. $4\overline{)20}$

The Jungle's Best Problem Solvers

Chimpanzees are the smartest animals in the jungle. They sometimes gather in groups to solve problems. This group might be working on a division question.

There are 18 chimpanzees in a family. How many groups of 3 are there?

We think: How many groups of 3 in 18?

We write: $3\overline{)18}$ with quotient 6

There are 6 groups of chimpanzees.

Divide.

1. $2\overline{)10}$
2. $3\overline{)15}$
3. $2\overline{)12}$
4. $4\overline{)16}$
5. $3\overline{)24}$
6. $2\overline{)16}$
7. $3\overline{)21}$
8. $5\overline{)30}$
9. $2\overline{)18}$
10. $4\overline{)24}$
11. $5\overline{)35}$
12. $3\overline{)27}$
13. How many groups of 4 in a family of 20 chimpanzees?
14. 28 bananas were eaten. Each chimpanzee ate 4. How many chimpanzees were there?

Applications

SUSAN'S BIRTHDAY PARTY

1. 7 friends came to Susan's party. Her brother and sister came too. How many children were at the party?

2. They went to the movies in 2 cars. How many children rode in each car?

3. The movie tickets cost $2 each. How much did they cost for all 10 children?

4. 24 people were waiting at the refreshment stand. There were 4 people in each line. How many lines were there?

5. A burger at Elmer's Foodhouse costs $2. How many burgers can be bought for $12?

6. Each chicken dinner costs $3.
   How many dinners can be bought for $24?

7. Elmer's Foodhouse sold tokens for computer games.
   It cost $1 for 3 tokens.
   How much did it cost for 12 tokens?

8. Susan's father bought 5 packages of balloons.
   There were 8 balloons in each package.
   How many balloons were there in all?

9. How many balloons were there for each child at Susan's party?

10. Write your own problem about the party.

PROBLEM SOLVING

Susan is 24 years younger than her mother.
How old is Susan's mother?

CHECK-UP

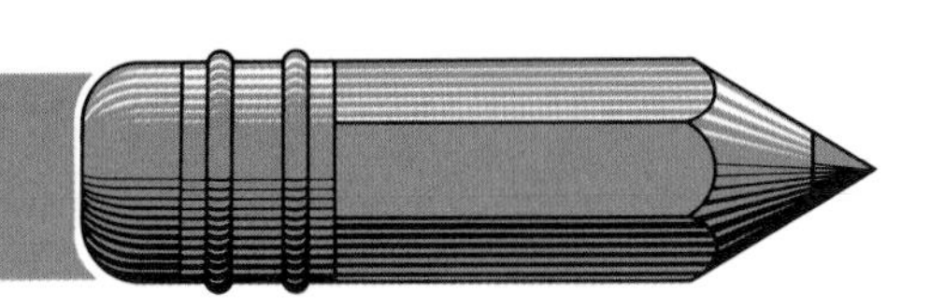

Draw a picture to show the sharing.
Copy and complete the sentence.

1. Share 6 oranges among 3 children

   $6 \div 3 = ▒$

2. Share 8 carrots between 2 rabbits.

   $8 \div 2 = ▒$

Copy and complete.

3. 

   12 divided into groups of
   4 is ▒ groups.
   $12 \div 4 = ▒$

4. 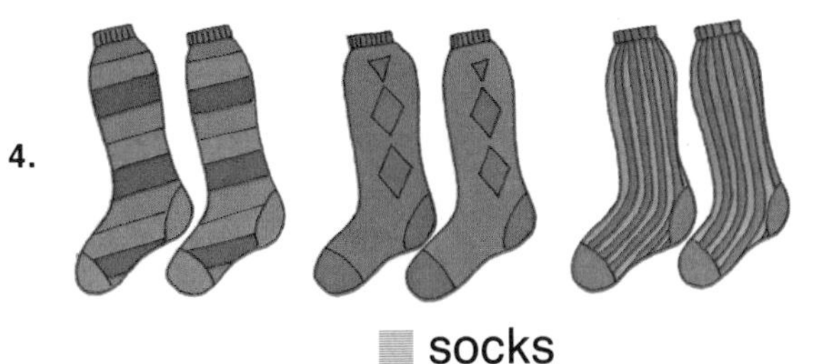

   ▒ socks
   ▒ in each pair
   $▒ \div ▒ = ▒$

Divide.

5. $20 \div 5 = ▒$
6. $15 \div 3 = ▒$
7. $18 \div 2 = ▒$
8. $4\overline{)24}$
9. $5\overline{)35}$
10. $3\overline{)15}$

Write 2 division sentences to match each picture.

11. 

12. 

13. There are 28 children.
    There are 7 children on each team.
    How many teams are there?

14. 32 flowers were planted.
    4 flowers were planted in each pot.
    How many pots were needed?

LENGTHS OF TIME

Time is peculiar
And hardly exact.
Though minutes are minutes,
You'll find for a fact
(As the older you get
And the bigger you grow)
That time can
Hurrylikethis
Or plod, plod, slow.

—Phyllis McGinley

Covering Surfaces

How many squares of each size
do you need to cover these figures?
Record your answers in a table.

Figure	Blue Squares	Green Squares
A		
B		

A.

B.

Area

Maria is covering a floor with tiles.
Each tile is 1 square unit.

 = 1 square unit

It takes 15 tiles to cover the floor.

The area of the floor is
15 square units.

Write the area of each floor.

1.

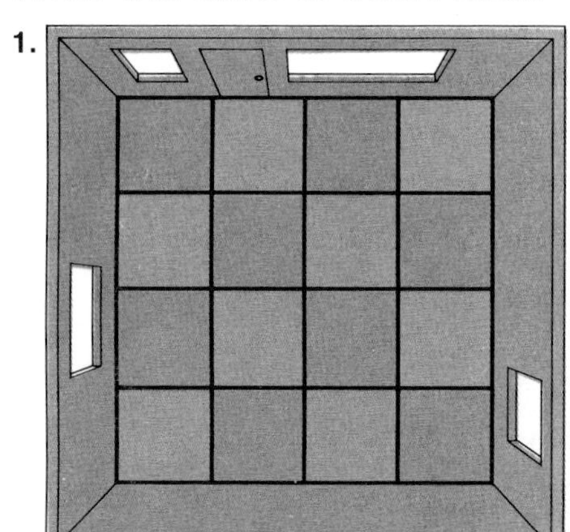

2.

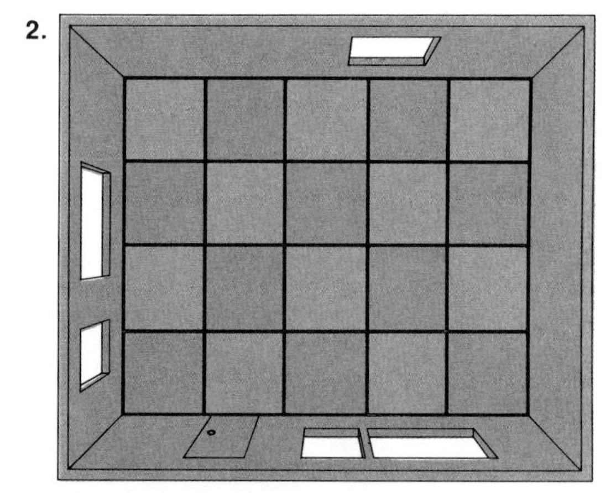

3.

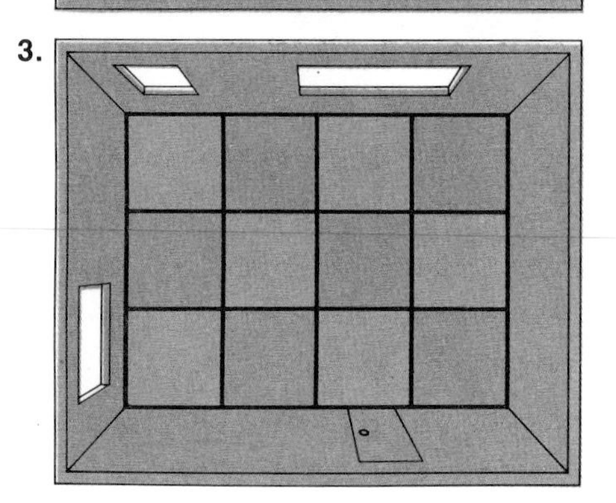

4.

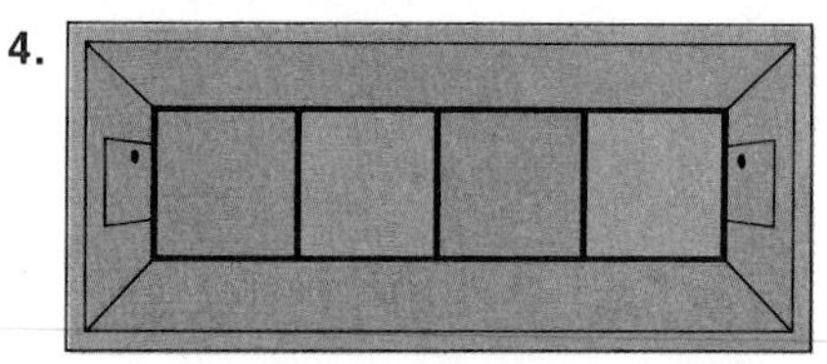

5.

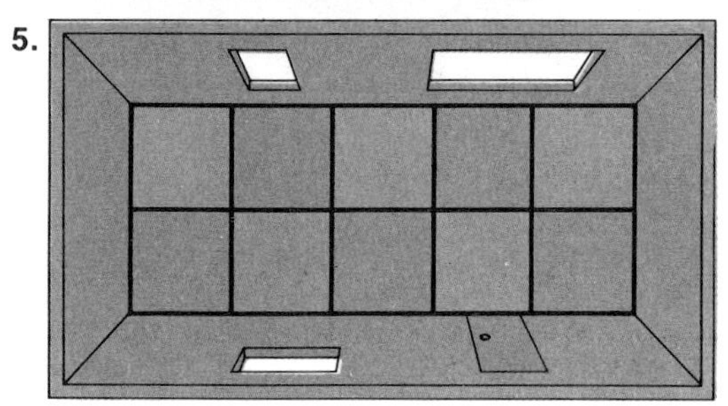

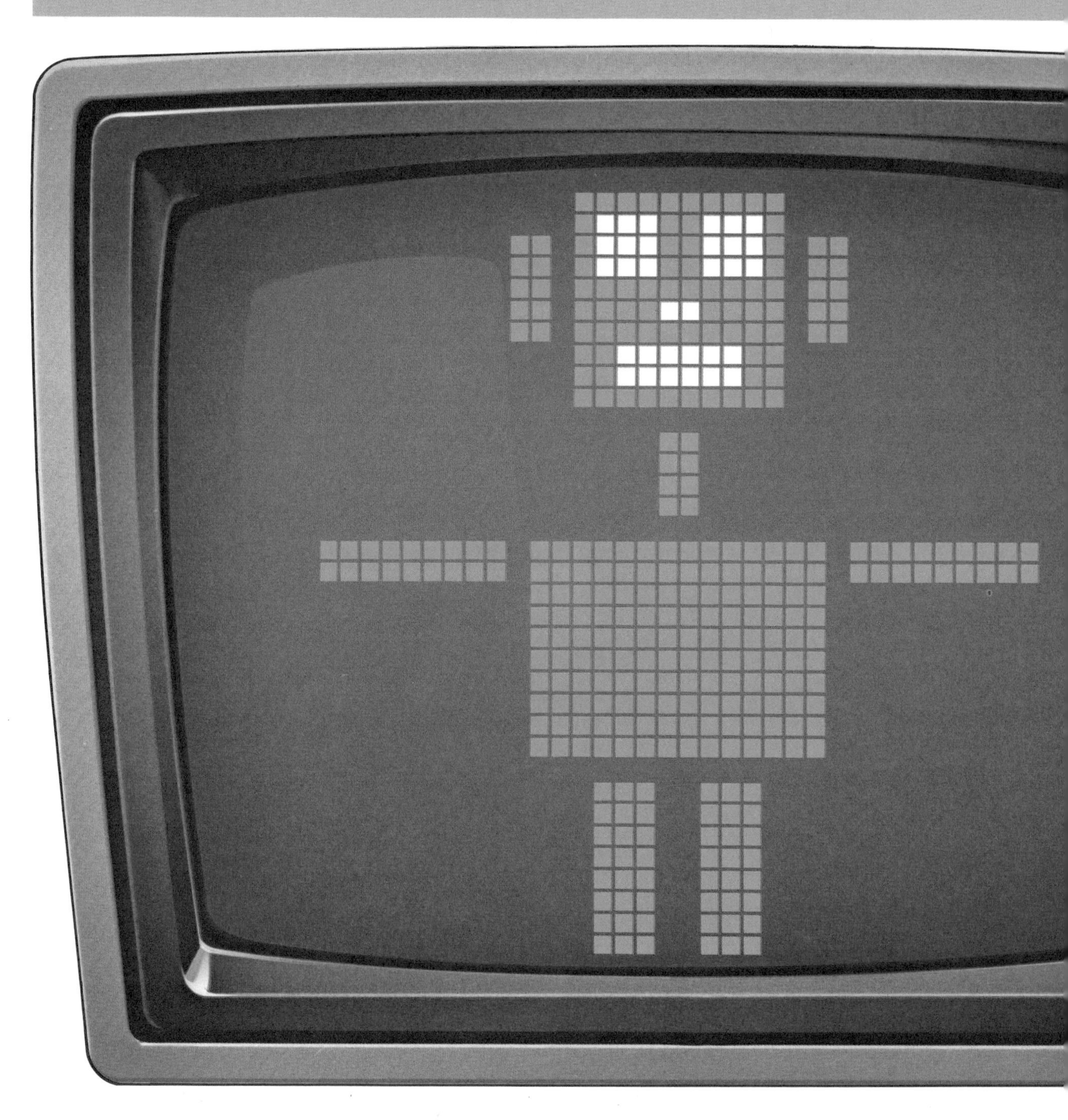

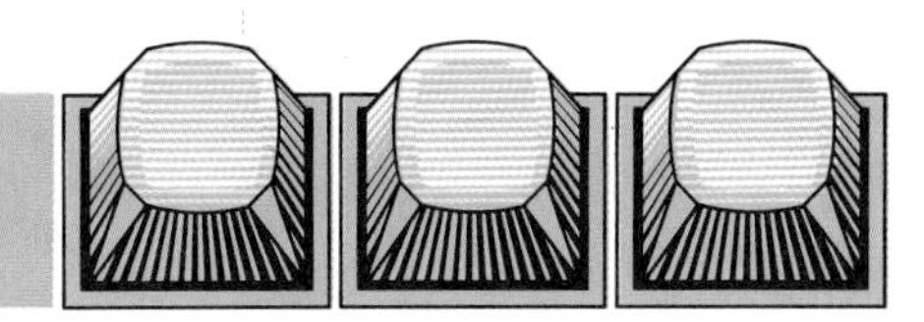

1. Write the area of Robbie's parts in a table like this.

Part	Area
Nose	2 square units
Eye	▒ square units
Ear	▒ square units
Mouth	▒ square units
Neck	▒ square units
Arm	▒ square units
Leg	▒ square units

2. Which of Robbie's parts has the smallest area?

3. How many square units are there in Robbie's 2 eyes?

4. How many square units are there in Robbie's 2 eyes, nose and mouth?

5. Which body part has greater area than the leg, but smaller area than the body?

6. Make your own robot on squared paper. What is the total area of your robot?

Area: Square Centimetres

This square has an area of one square centimetre.

What is the area of the pink region?

We say: eight square centimetres.

We write: $8\ cm^2$

Find the area of each region.

EXTENSIONS

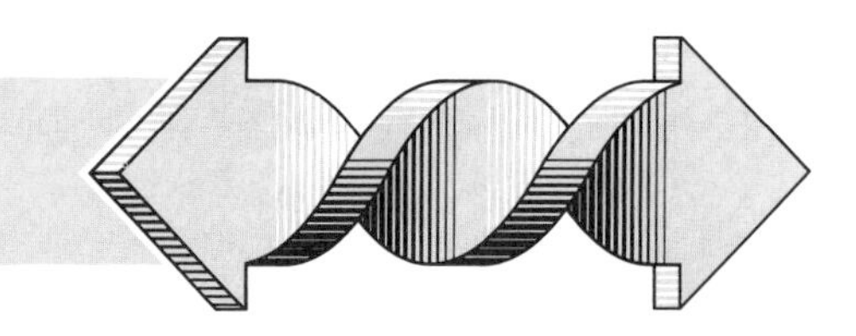

Calculating Area

1. Each stamp below is a rectangle.
   Find the length, width, and area of each.
   Write your answer in a table.

Stamp	Length	Width	Area
A	4 cm	3 cm	12 cm^2
B	cm	cm	cm^2

2. Study the table.
   How can we find the area if we know the length and width?

3. Use squared paper.
   Draw stamps with these areas.
   A. 16 cm^2 B. 12 cm^2 C. 20 cm^2

4. Draw a stamp of your own.
   Find its length, width and area.

Filling Boxes

Og found 2 boxes.
He packed each box with blocks.
All the blocks are the same size.

Then he emptied the blue box.
These are the blocks it held.

Then Og emptied the red box too.
These are the blocks it held.

Which box is bigger?

Layers of Blocks

Each wagon has one layer of blocks.

There are 6 blocks in this wagon.

How many blocks are there in each wagon?

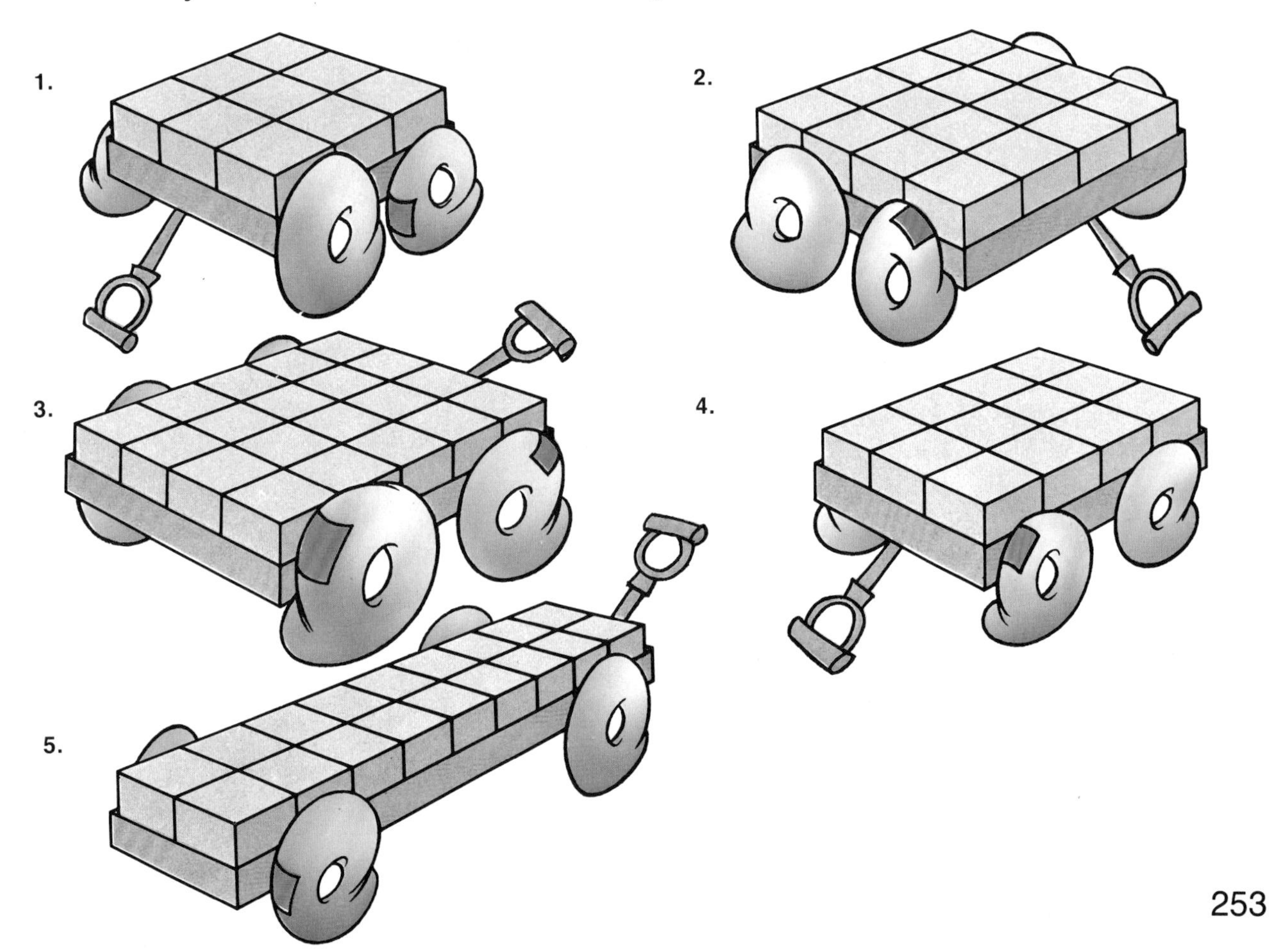

Counting Cubes

This is a cube. 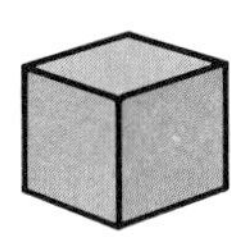

Megan used 8 cubes
to make this building.

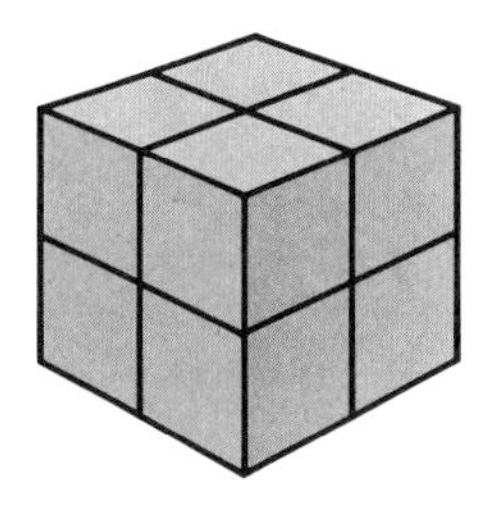

How many cubes are in each building?

One Layer	Two Layers	Three Layers
1.	2.	3.
4.	5.	6.

How many cubes are in each building?

One Layer	Two Layers	Three Layers
7. 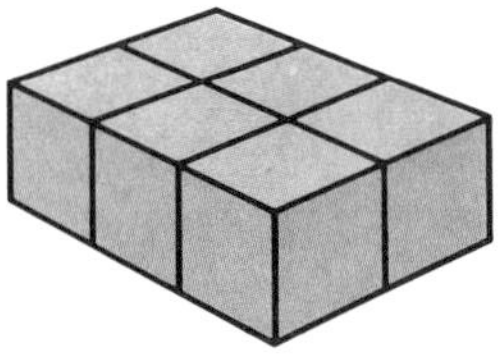	8. 	9. 
10. 	11. 	12. 
13. 	14. 	15. 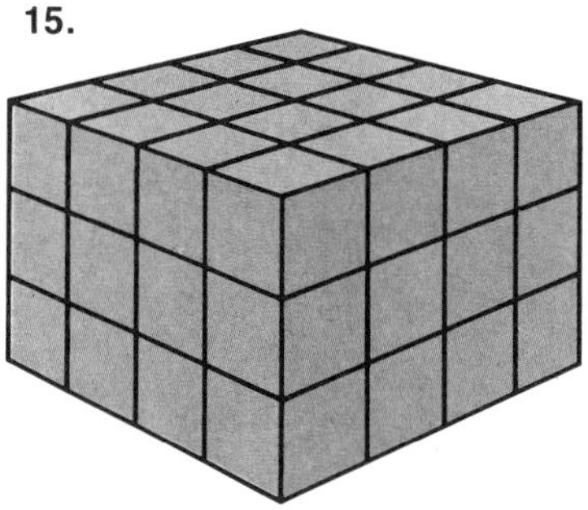

16. Make a building of your own. How many cubes did you use?

17. How many different buildings can you make using 12 cubes?

Irregular Buildings

Four friends made buildings from cubes.

This is Lisa's building.

This is Jamie's building.

This is Colin's building.

This is Sarah's building.

1. How many cubes are in each building?
2. List the buildings in order from smallest to largest.

How many cubes are in each of these buildings?

EXTENSIONS

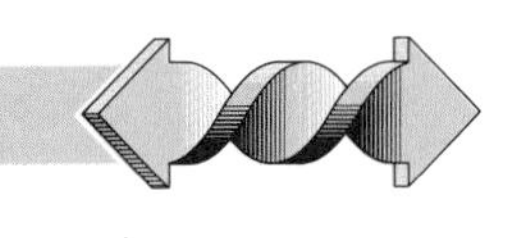

Use 24 cubes.
Construct this building.

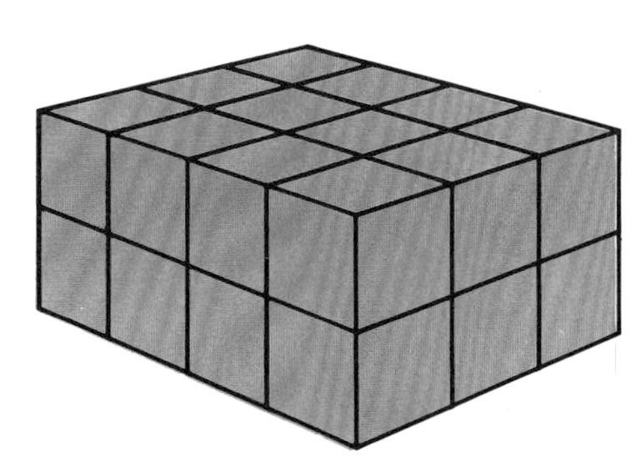

Make as many different buildings as possible using all 24 cubes each time.

Make a picture of each building you make.

Filling Containers

Leslie wanted to compare these 3 containers.

She used this pitcher to help her.

First she filled the pitcher with water. She emptied it into the first container. This is what happened.

She filled the pitcher again and emptied it into the second container. This is what happened.

She filled the pitcher again. She emptied it into the third container. This is what happened.

1. Which container holds more than the pitcher?
2. Which container holds less than the pitcher?
3. Which container holds the same amount as the pitcher?

Litres

This carton holds one litre.

We write: **1 L**

1 L of milk can fill about 4 glasses.

1. Write the letters of the containers which hold more than 1 L.

2. Which container holds the most?

ESTIMATING

Estimating Capacity in Litres

Get 3 large containers.

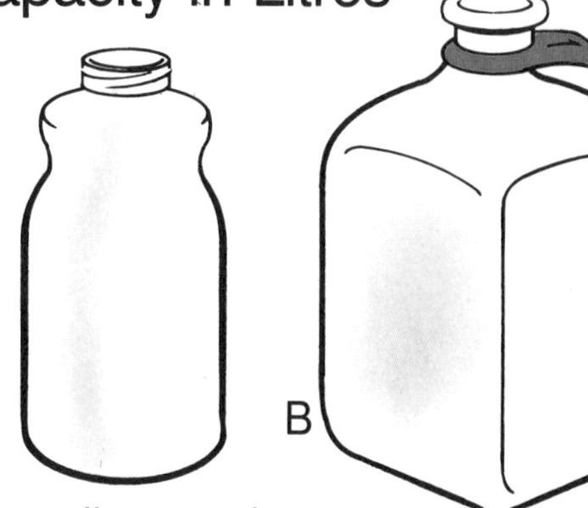

1. Estimate how many litres each container can hold.

Container	Estimate
A	▒ L
B	▒ L
C	▒ L

2. Fill one of your containers with water.
   Count how many times the container fills a 1 L container.

   Write the number of litres each container can hold.
   Write this in your table.

Container	Estimate	Measurement
A	▒ L	▒ L
B	▒ L	▒ L
C	▒ L	▒ L

Compare each measurement with your estimate.

3. List the containers from greatest capacity to least capacity.

PROBLEM SOLVING

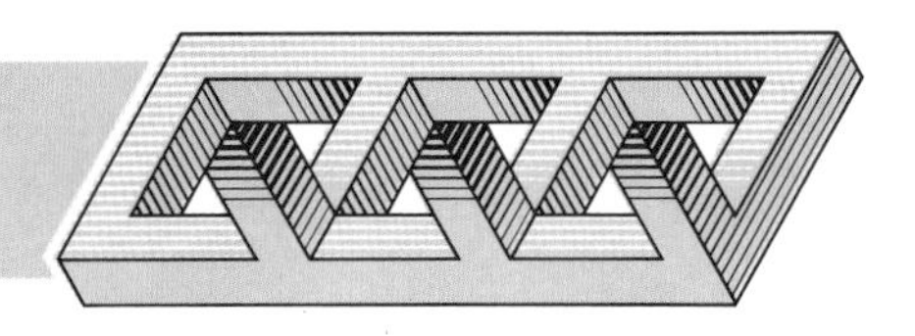

Multi-step Problems

The pitcher holds 1 L when full.

The milk jug holds 4 L when full.

1. The bucket can fill these pitchers. How much water can it hold?
2. How much liquid can 5 milk jugs hold?
3. Which will hold more:

5 jugs or 4 buckets?

4. Mr. Stevens bought 3 jugs of milk. How many glasses of milk can he pour?

Remember:

TELLING TIME

The little hand is the hour hand.
It tells how many hours past
12 o'clock.

The big hand is the minute hand.
It tells how many minutes past
the hour.

We see:

We say: The time is nine forty.

We write: **9:40**

Write the time shown on each clock.

1.

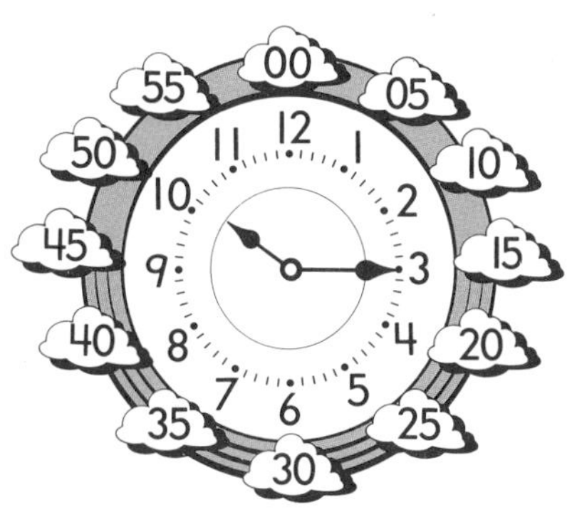

2.

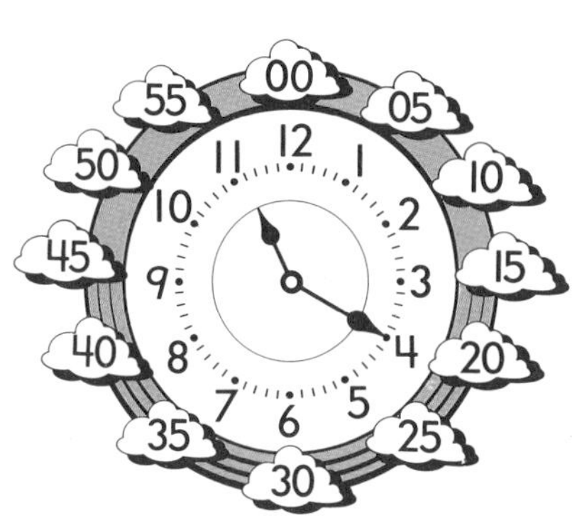

3.

4.

5.

6.

Earlier and Later

Bill was
10 minutes early.
He arrived at 3:20.

Janet arrived for the
meeting on time.
She arrived at 3:30.

Audrey came
20 minutes late.
She arrived at 3:50.

Write the time shown on each clock.
Then answer the question.

1. What time was it
   5 minutes earlier?

2. What time will it
   be 25 minutes later?

3. What time will it
   be in 10 minutes?

4. What time was it
   40 minutes ago?

5. In 20 minutes what
   time will it be?

6. What time was it
   15 minutes earlier?

Time to the Nearest Minute

With these clocks, we can
tell time to the nearest minute.
Both clocks show the same time.

We see:

or

We see:

We write: 7:12

We say: The time is seven twelve.

Each clock has a partner which
shows the same time.
Write the number and letter of
the matching clocks.

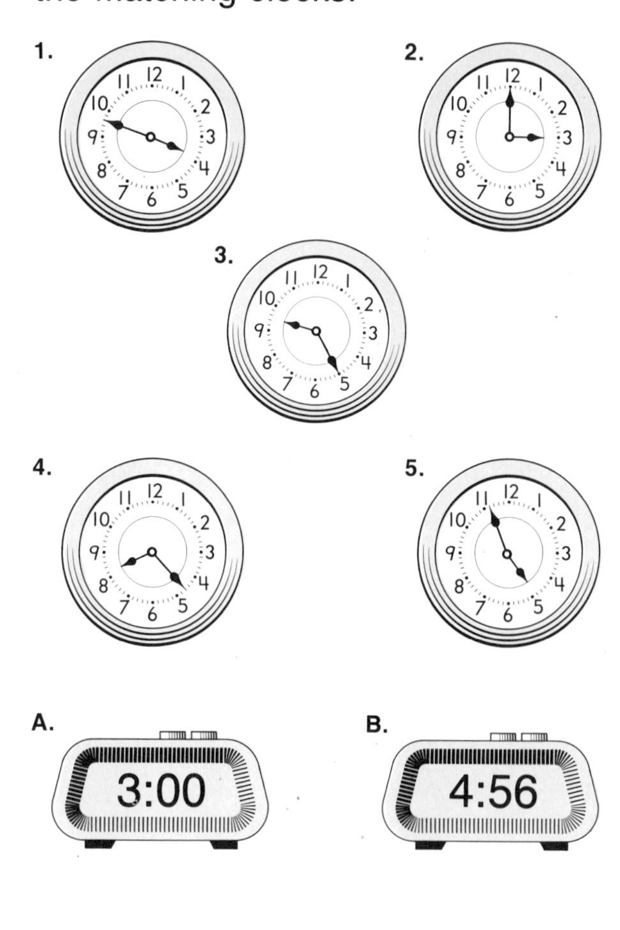

C.

D.

E.

Using Time

1. Kathy let the cat out at 7:25.
   It was outside for 45 minutes.
   When did it come back inside?

2. The volleyball game started at 3:48.
   It lasted for 20 minutes.
   What time did the game end?

3. It takes Helen 25 minutes
   to walk home.
   Supper was at 7:10.
   What time did she start
   to walk home?

4. David looked at the clock at 11:33.
   There had been a fire drill
   16 minutes earlier.
   When was the fire drill?

5. What is your favorite tv program?
   If it started at 7:30, what time
   would it end?

This calendar shows some of the things Julie will do at the summer day camp.

The date on which the camp starts is Friday, August 3.

AUGUST						
Sunday	Monday	Tuesday	Wednesday	Thursday	Friday	Saturday
			1	2	3 first day of camp	4
5	6 zoo visit	7	8	9 bicycle rodeo	10	11
12	13 first computer day	14	15 overnight hike	16	17 visit from parents	18
19	20	21 canoe trip	22	23	24 swim meet	25
26	27	28	29 wiener roast	30 second computer day	31 last day of camp	

Use the calendar
to answer these questions.

1. In which month will Julie be at day camp?

2. On what date will Julie go to visit the zoo?

3. On what day of the week does camp end? Write the date.

4. How many weeks will Julie spend at camp?

5. How many days will Julie spend at camp? (She does not go on Saturdays or Sundays.)

6. What event takes place 6 days after the bicycle rodeo?

7. How many days after the first computer day is the second computer day?

8. How many days after the overnight hike is the wiener roast?

9. Julie will visit her grandmother on the Sunday after camp ends. Write the date of her visit.

THE EARTH TAKES A TRIP

FEBRUARY

S	M	T	W	T	F	S
				1	2	3
4	5	6	7	8	9	10
11	12	13	14	15	16	17
18	19	20	21	22	23	24
25	26	27	28			

MARCH

S	M	T	W	T	F	S
				1	2	3
4	5	6	7	8	9	10
11	12	13	14	15	16	17
18	19	20	21	22	23	24
25	26	27	28	29	30	31

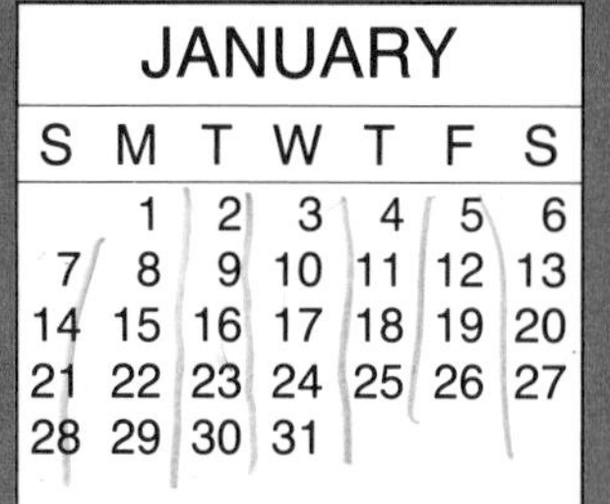

JANUARY

S	M	T	W	T	F	S
	1	2	3	4	5	6
7	8	9	10	11	12	13
14	15	16	17	18	19	20
21	22	23	24	25	26	27
28	29	30	31			

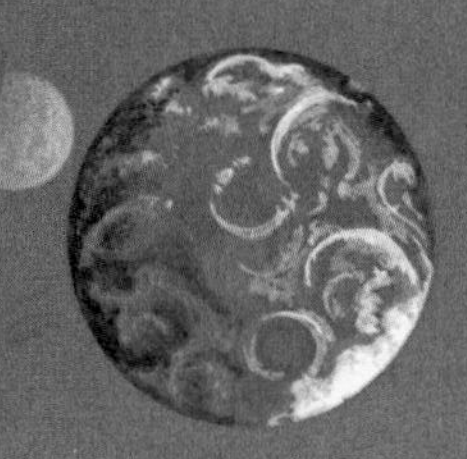

WINTER

It takes the earth 12 months to travel around the sun.
This period of time is called a year.
There are usually 365 days in 1 year.

DECEMBER

S	M	T	W	T	F	S
						1
2	3	4	5	6	7	8
9	10	11	12	13	14	15
16	17	18	19	20	21	22
23	24	25	26	27	28	29
30	31					

NOVEMBER

S	M	T	W	T	F	S
				1	2	3
4	5	6	7	8	9	10
11	12	13	14	15	16	17
18	19	20	21	22	23	24
25	26	27	28	29	30	

FA

OCTOBER

S	M	T	W	T	F	S
	1	2	3	4	5	6
7	8	9	10	11	12	13
14	15	16	17	18	19	20
21	22	23	24	25	26	27
28	29	30	31			

1. Write the names of all the months of the year in order.
2. Write the names of the months which have exactly 30 days.
3. Which month has the fewest days?
4. Does any month have more than 5 Sundays?
5. What is the 3rd month of the year?
6. Which month comes 6 months after July?
7. In which month is the first day of spring?

Applications

APRIL

S	M	T	W	T	F	S
1	2	3	4	5	6	7
8	9	10	11	12	13	14
15	16	17	18	19	20	21
22	23	24	25	26	27	28
29	30					

MAY

S	M	T	W	T	F	S
		1	2	3	4	5
6	7	8	9	10	11	12
13	14	15	16	17	18	19
20	21	22	23	24	25	26
27	28	29	30	31		

JUNE

S	M	T	W	T	F	S
					1	2
3	4	5	6	7	8	9
10	11	12	13	14	15	16
17	18	19	20	21	22	23
24	25	26	27	28	29	30

SPRING

SUMMER

JULY

S	M	T	W	T	F	S
1	2	3	4	5	6	7
8	9	10	11	12	13	14
15	16	17	18	19	20	21
22	23	24	25	26	27	28
29	30	31				

AUGUST

S	M	T	W	T	F	S
			1	2	3	4
5	6	7	8	9	10	11
12	13	14	15	16	17	18
19	20	21	22	23	24	25
26	27	28	29	30	31	

SEPTEMBER

S	M	T	W	T	F	S
						1
2	3	4	5	6	7	8
9	10	11	12	13	14	15
16	17	18	19	20	21	22
23	24	25	26	27	28	29
30						

8. In which month is the last day of fall?

9. How many months long is summer?

10. What is the date 3 weeks after Sunday, June 3?

11. What is the date one week before Thursday, August 2?

12. Use the calendar to find how many days from January 1 to March 17.

13. How many days are there from now until your next birthday?

Reading a Thermometer

Complete.

We use a thermometer to measure temperature.

We see:

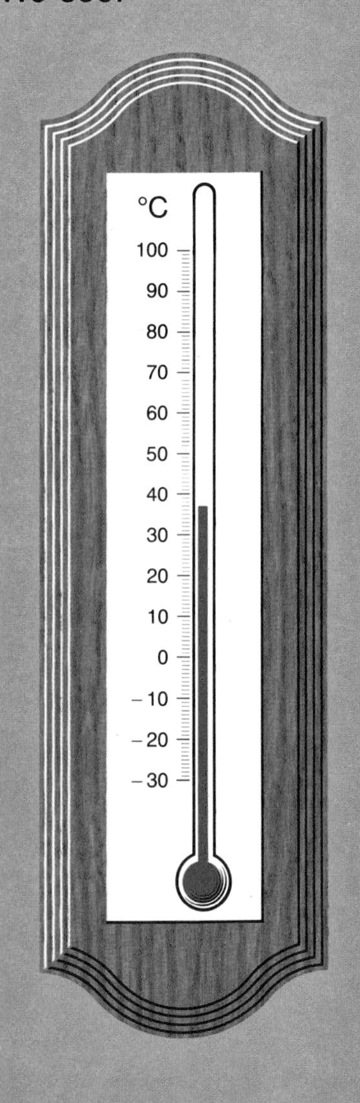

We say: The temperature is 37 degrees Celsius.

We write: 37°C

1. We see:

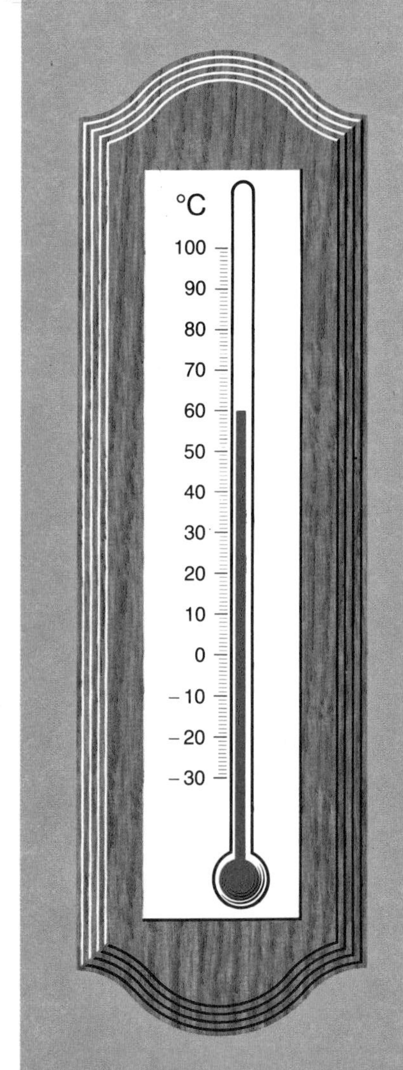

We say: The temperature is 60 degrees Celsius.

We write:

2. We see:

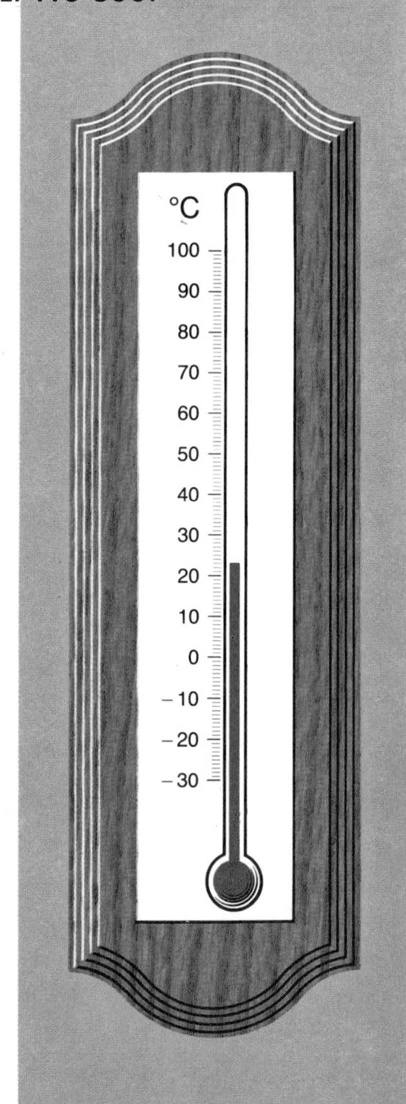

We say: The temperature is 23 degrees Celsius.

We write:

Sometimes we look closely at a thermometer to read the temperature.

Read the temperature shown on each thermometer to complete each sentence.

1. Water freezes at ▒ °C.

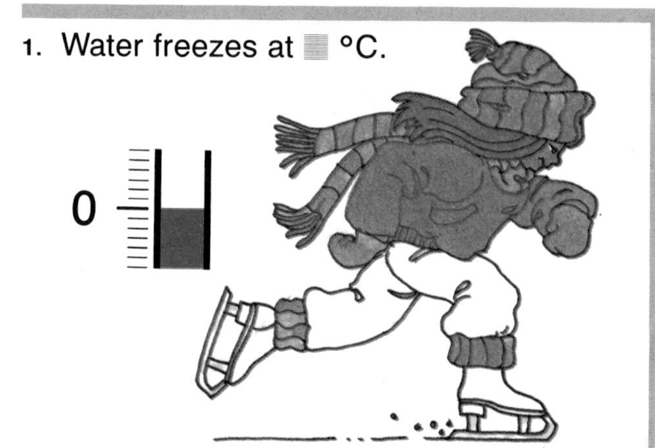

2. Room temperature is about ▒ °C.

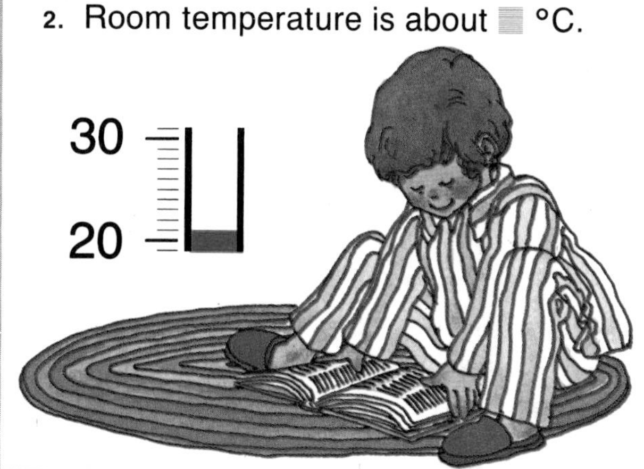

3. On a hot summer day it is about ▒ °C.

4. Normal body temperature is about ▒ °C.

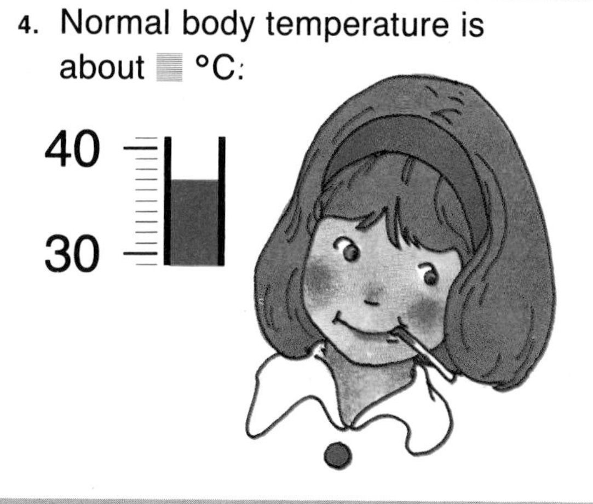

5. A hot drink is about ▒ °C.

6. Water boils at ▒ °C.

Heavier and Lighter

Who is heavier,
Mark or Melanie?

Name the heavier object
in each picture.

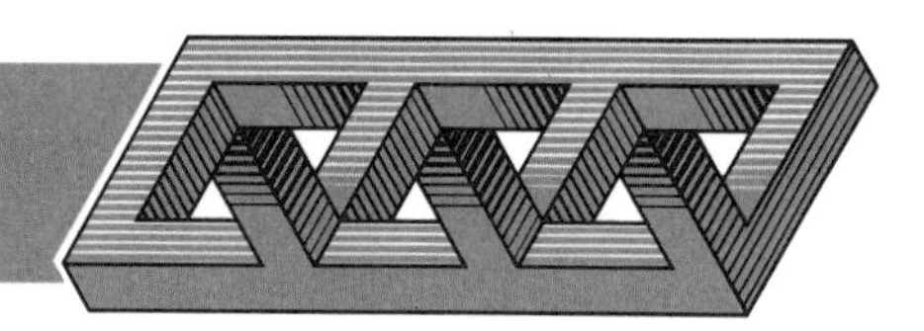

PROBLEM SOLVING

Obtaining Information from a Picture

Look at these pictures.
Then write the names of the heaviest and lightest persons.

The Kilogram

How heavy is a litre of milk?

A litre of milk has a mass of about one kilogram.

We say: one kilogram

We write: **1 kg**

Write the mass of each object in kilograms.

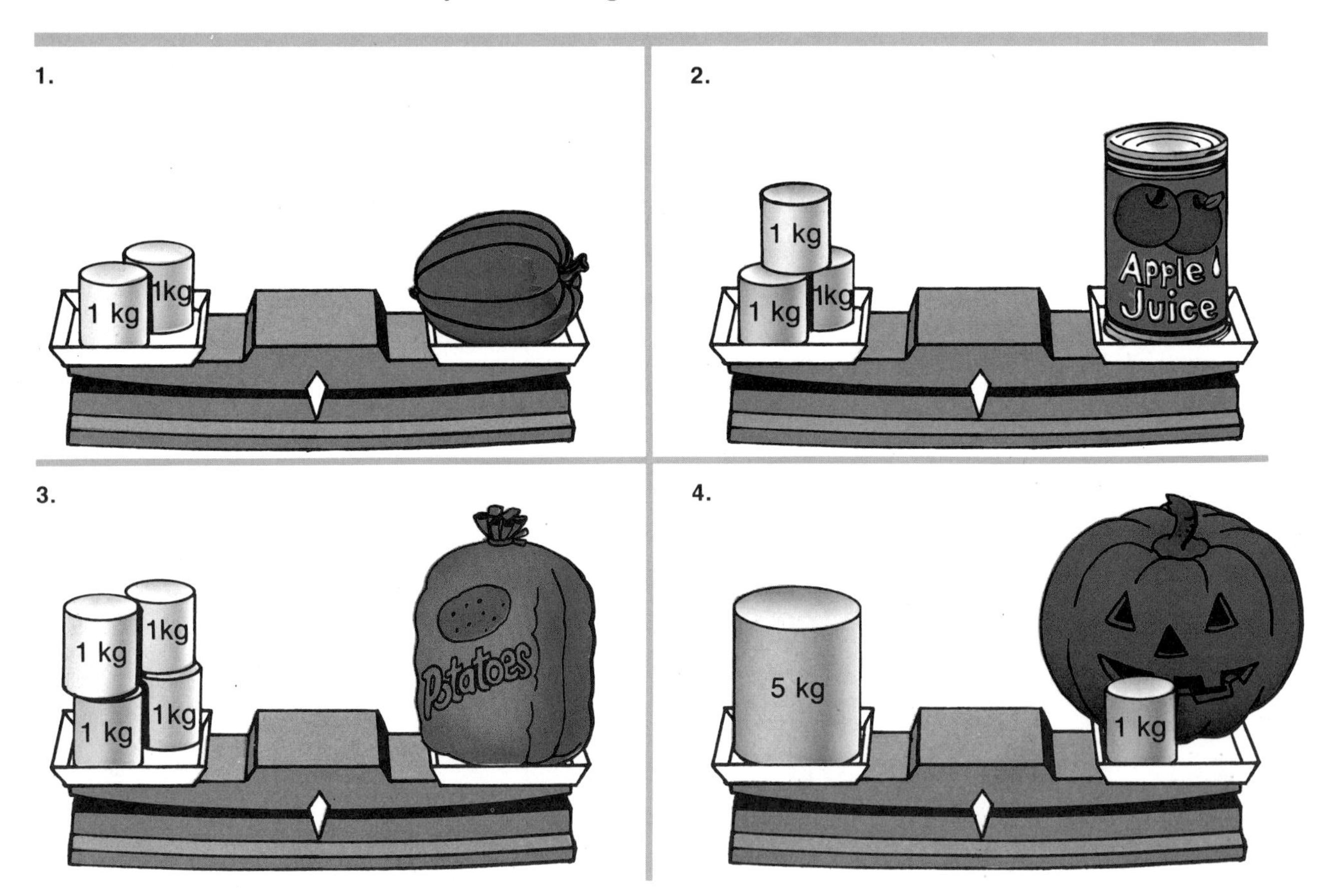

Living DRAGONS

Story Problems

There are no real dragons that breathe fire. The only real dragons today live on the tiny island of Komodo. These giant lizards are called Komodo dragons. They have a mass of about 100 kg. That is about 4 times as heavy as most grade 3 children.

There is a law against killing Komodo dragons. Why do you think this law was made?

This Komodo dragon is hungry. It will jump on the first animal that comes near. It can eat a goat in 10 minutes and a 40 kg buffalo in 15 minutes.

1. What is the mass of a Komodo dragon?
2. How many grade 3 children together have about the same mass as a Komodo dragon?
3. How much heavier is a Komodo dragon than a 40 kg buffalo?
4. How much longer does it take a Komodo dragon to eat a buffalo than a goat?
5. How many goats could a Komodo dragon eat in 20 minutes?

CHECK-UP

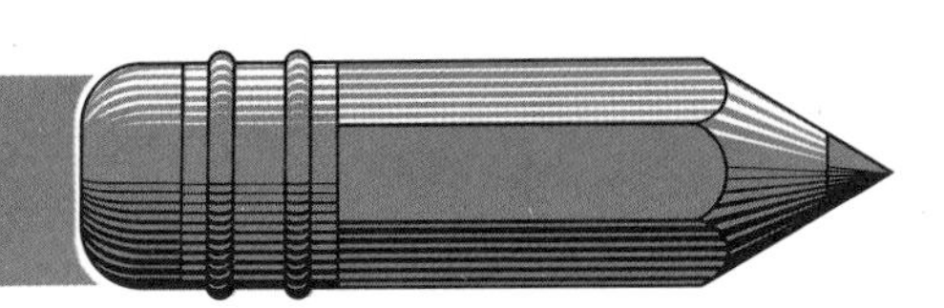

Each ☐ is 1 cm^2
Find the area of each region.

1.

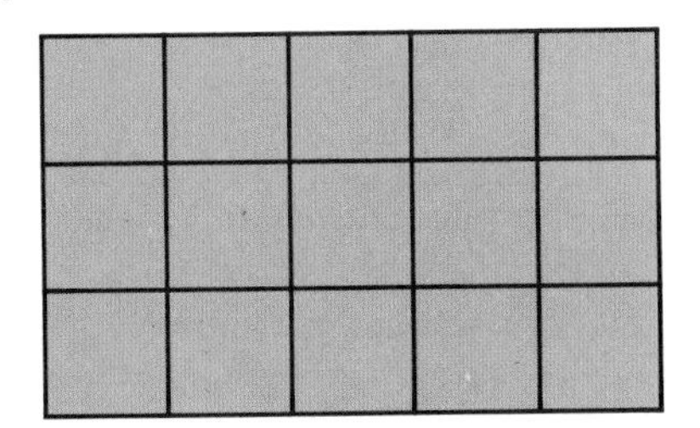

2.

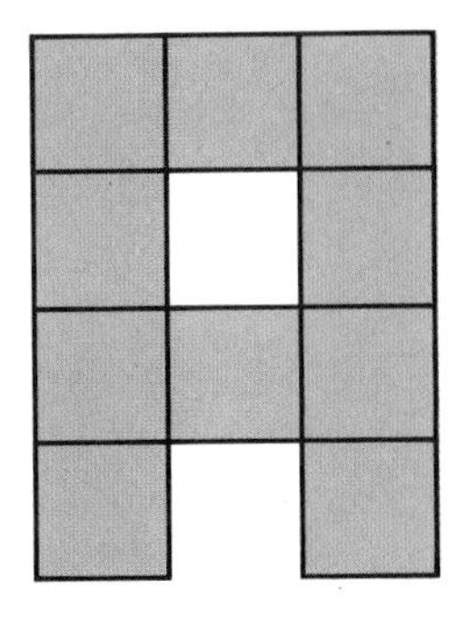

How many cubes are in each building?

3.

4.

5.

Write the time shown on each clock.

6.

7.

8.

Use the calendar.

9. Write the date of the first Monday in July.

10. Write the date of the third Wednesday in July.

JULY						
S	M	T	W	T	F	S
		1	2	3	4	5
6	7	8	9	10	11	12
13	14	15	16	17	18	19
20	21	22	23	24	25	26
27	28	29	30	31		

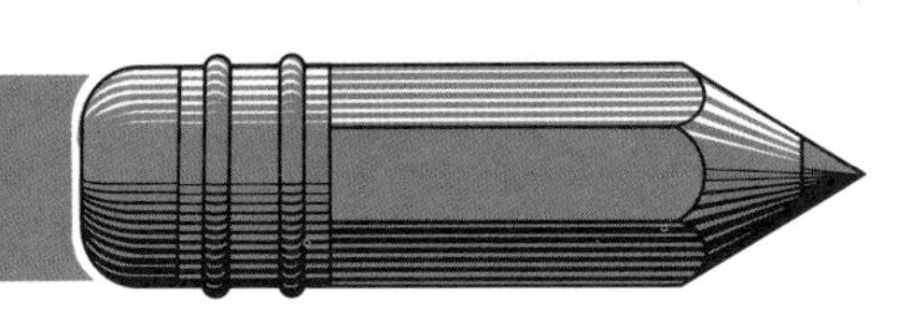

CUMULATIVE REVIEW

Copy and complete each sentence.

1.   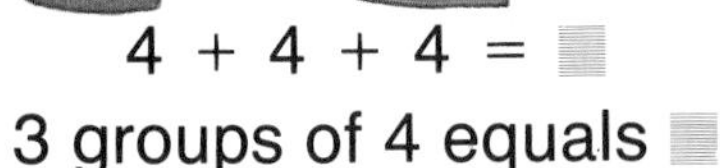

$4 + 4 + 4 = \square$

3 groups of 4 equals $\square$

2.  

2 groups of 3 equals $\square$

$2 \times 3 = \square$

Write a multiplication sentence to answer each question.

3. How many hats are shown?

$4 \times \square = \square$

4. How many stamps were used?

 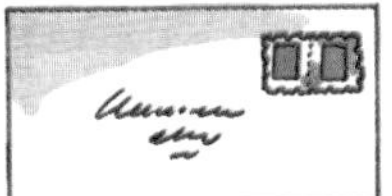 

Multiply.

5. 2×8	6. 5×3	7. 0×7	8. 1×4	9. 1×6	10. 5×7
11. 3×4	12. 2×1	13. 5×9	14. 2×6	15. 3×3	16. 0×8

Fill in the missing numbers.

17. 

8 divided into groups of 2

is $\square$ groups

$8 \div 2 = \square$

18. 

$\square$ books

$\square$ books in each pile

$\square \div \square = \square$

CUMULATIVE REVIEW

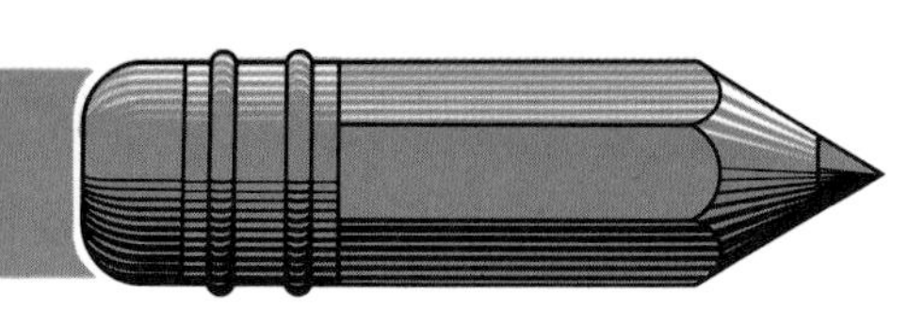

Divide.

19. $3\overline{)18}$ **20.** $5\overline{)15}$ **21.** $4\overline{)28}$ **22.** $2\overline{)14}$ **23.** $1\overline{)8}$ **24.** $2\overline{)12}$

25. $4\overline{)4}$ **26.** $2\overline{)18}$ **27.** $3\overline{)24}$ **28.** $1\overline{)6}$ **29.** $5\overline{)40}$ **30.** $3\overline{)15}$

Each ☐ is 1 cm^2
Find the area of each region.

31.

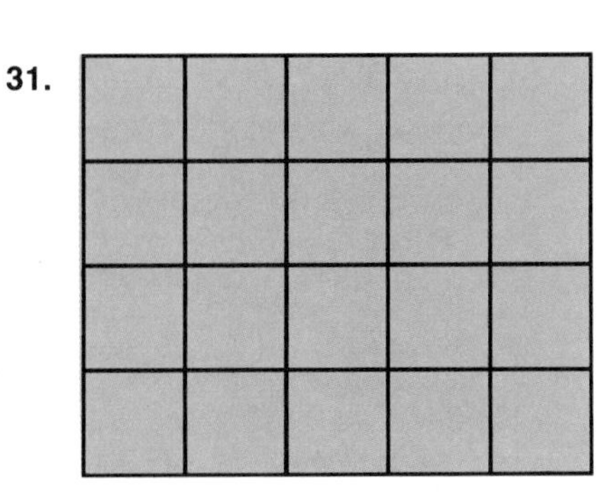

32.

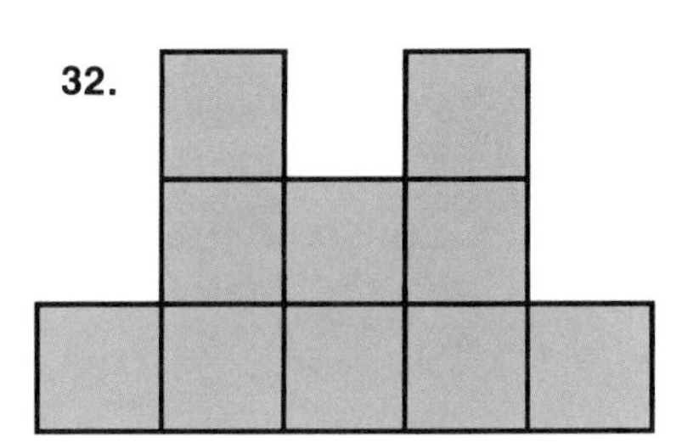

Write the time shown on each clock. Then answer the question.

33.

What time was it 5 minutes earlier?

34.

What time will it be in 10 minutes?

Use the calendar.

35. On which day of the week is Valentine's Day?

36. What is the date of the fifth Wednesday in February?

FEBRUARY						
S	M	T	W	T	F	S
			1	2	3	4
5	6	7	8	9	10	11
12	13	14	15	16	17	18
19	20	21	22	23	24	25
26	27	28	29			

Write the mass of each object in kilograms.

37.

38.

Sugaring Off

Drip, drip, drip
Sap from the maple tree.
The nights of frost and days of sun
Are here, and now the sap will run,
And sugaring off has just begun.

Run, run, run
Into the shiny pails.
The farmer's tapped the maple tree,
And maple sap is flowing free,
And brimming pails hang heavily.

Full, full, full
Then away to the sugar house
Where fires burn and leap and flare
While sap is boiling, bubbling clear,
And steaming fragrance fills the air.

Bubble, bubble, bubble
Then hurrah for the sugaring off!
The fire burns with a steady glow,
The paddle's stirring syrup-o,
And sugar's cooling on the snow.

—Helen Guiton

Using Related Multiplication Sentences

We see: $3 \times 6 = \square$

We think: We have not learned 3×6, so we think of the related multiplication sentence.

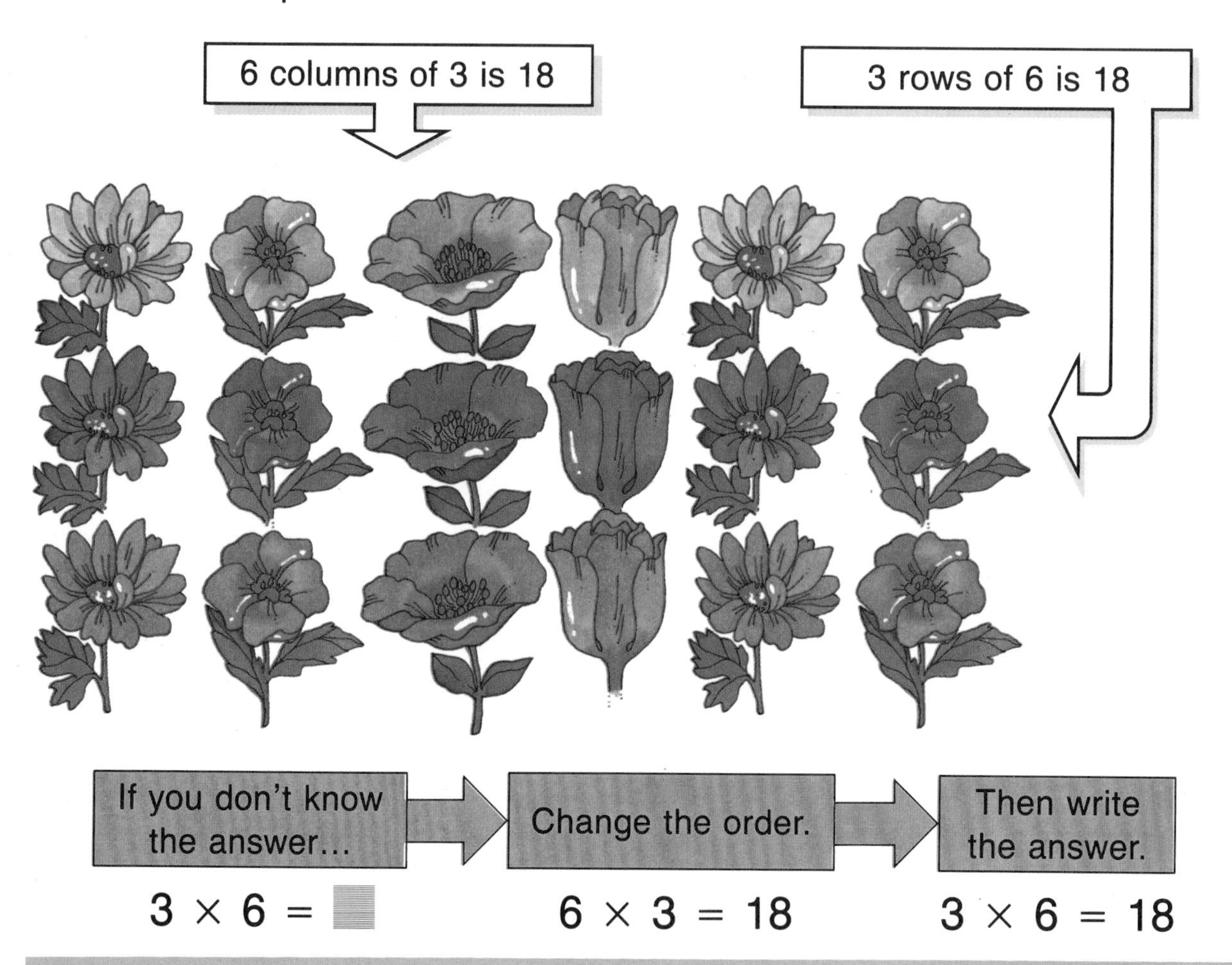

Write the related multiplication sentence.
Then complete the given sentence.

1. $2 \times 9 = \square$

1. $9 \times 2 = 18$
   $2 \times 9 = 18$

2. $3 \times 7 = \square$
3. $5 \times 6 = \square$
4. $5 \times 3 = \square$
5. $3 \times 8 = \square$
6. $2 \times 7 = \square$
7. $2 \times 5 = \square$
8. $4 \times 8 = \square$
9. $4 \times 6 = \square$
10. $5 \times 9 = \square$
11. $4 \times 7 = \square$
12. $5 \times 8 = \square$
13. $3 \times 9 = \square$
14. $2 \times 6 = \square$

Copy and complete each sentence.
Think of a related multiplication sentence if necessary.

1. $\begin{array}{r} 8 \\ \times 3 \\ \hline \end{array}$	2. $\begin{array}{r} 6 \\ \times 4 \\ \hline \end{array}$	3. $\begin{array}{r} 3 \\ \times 9 \\ \hline \end{array}$	4. $\begin{array}{r} 7 \\ \times 4 \\ \hline \end{array}$	5. $\begin{array}{r} 6 \\ \times 0 \\ \hline \end{array}$	6. $\begin{array}{r} 9 \\ \times 5 \\ \hline \end{array}$
7. $\begin{array}{r} 6 \\ \times 2 \\ \hline \end{array}$	8. $\begin{array}{r} 7 \\ \times 5 \\ \hline \end{array}$	9. $\begin{array}{r} 2 \\ \times 6 \\ \hline \end{array}$	10. $\begin{array}{r} 8 \\ \times 4 \\ \hline \end{array}$	11. $\begin{array}{r} 9 \\ \times 0 \\ \hline \end{array}$	12. $\begin{array}{r} 7 \\ \times 3 \\ \hline \end{array}$
13. $\begin{array}{r} 8 \\ \times 0 \\ \hline \end{array}$	14. $\begin{array}{r} 9 \\ \times 2 \\ \hline \end{array}$	15. $\begin{array}{r} 7 \\ \times 0 \\ \hline \end{array}$	16. $\begin{array}{r} 8 \\ \times 2 \\ \hline \end{array}$	17. $\begin{array}{r} 4 \\ \times 3 \\ \hline \end{array}$	18. $\begin{array}{r} 6 \\ \times 3 \\ \hline \end{array}$
19. $\begin{array}{r} 9 \\ \times 4 \\ \hline \end{array}$	20. $\begin{array}{r} 5 \\ \times 9 \\ \hline \end{array}$	21. $\begin{array}{r} 8 \\ \times 1 \\ \hline \end{array}$	22. $\begin{array}{r} 6 \\ \times 5 \\ \hline \end{array}$	23. $\begin{array}{r} 1 \\ \times 8 \\ \hline \end{array}$	24. $\begin{array}{r} 7 \\ \times 2 \\ \hline \end{array}$
25. $\begin{array}{r} 6 \\ \times 1 \\ \hline \end{array}$	26. $\begin{array}{r} 9 \\ \times 3 \\ \hline \end{array}$	27. $\begin{array}{r} 4 \\ \times 7 \\ \hline \end{array}$	28. $\begin{array}{r} 9 \\ \times 1 \\ \hline \end{array}$	29. $\begin{array}{r} 8 \\ \times 5 \\ \hline \end{array}$	30. $\begin{array}{r} 7 \\ \times 1 \\ \hline \end{array}$

PROBLEM SOLVING

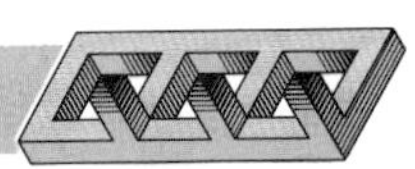

Make a multiplication table like this.
Fill in the pink squares.
Use related multiplication sentences to fill in the green squares.
Can you see a pattern?

Use squared paper.

×	1	2	3	4	5	6	7	8	9
1	1								
2		4							
3			9						
4				16					
5					25				
6									
7									
8									
9									

Multiplication: Groups of 6

Julie gave her little sister a counting book called Insects.

Julie wrote this song to the tune of The 12 Days of Christmas.

A. On the first page of Insects, Roberta showed to me...

a bee flying near a pear tree.

B. On the second page of Insects, Roberta showed to me...

2 dragonflies...

C. On the third page of Insects, Roberta showed to me...

3 chirping crickets...

D. On the fourth page of Insects, Roberta showed to me...

4 stinging wasps...

E. On the fifth page of Insects, Roberta showed to me...

5 butterflies...

F. On the sixth page of Insects, Roberta showed to me...

6 hornets humming...

G. On the seventh page of Insects, Roberta showed to me...

7 ants a-marching...

H. On the eighth page of Insects, Roberta showed to me...

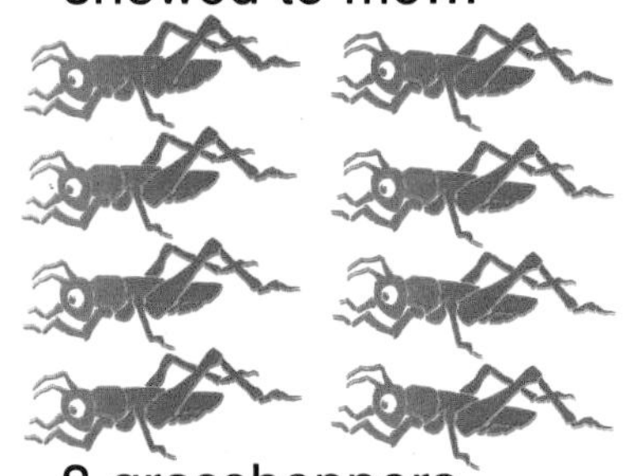

8 grasshoppers hopping...

I. On the ninth page of Insects, Roberta showed to me...

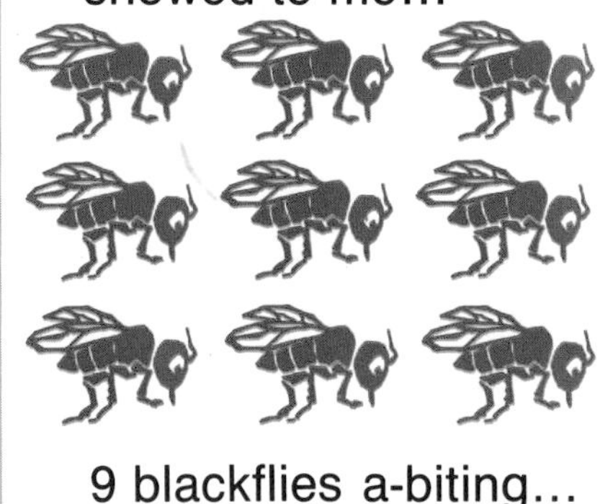

9 blackflies a-biting...

1. Every insect has 6 legs.
   Write a multiplication sentence to show how many legs are in each picture.

2. Copy and complete this table.

×	0	1	2	3	4	5	6	7	8	9
6	▒	▒	▒	▒	▒	▒	▒	▒	▒	▒

Copy and complete.

3. $\begin{array}{r} 6 \\ \times 4 \\ \hline \end{array}$

4. $\begin{array}{r} 6 \\ \times 6 \\ \hline \end{array}$

5. $\begin{array}{r} 1 \\ \times 6 \\ \hline \end{array}$

6. $\begin{array}{r} 3 \\ \times 6 \\ \hline \end{array}$

7. $\begin{array}{r} 5 \\ \times 6 \\ \hline \end{array}$

8. $\begin{array}{r} 6 \\ \times 0 \\ \hline \end{array}$

9. $\begin{array}{r} 8 \\ \times 6 \\ \hline \end{array}$

10. $\begin{array}{r} 6 \\ \times 7 \\ \hline \end{array}$

11. $\begin{array}{r} 9 \\ \times 6 \\ \hline \end{array}$

12. $\begin{array}{r} 5 \\ \times 6 \\ \hline \end{array}$

13. $\begin{array}{r} 6 \\ \times 8 \\ \hline \end{array}$

14. $\begin{array}{r} 7 \\ \times 6 \\ \hline \end{array}$

15. Miss Talbot works 6 days a week.
    How many days does she work in 8 weeks?

16. There are 6 apples in a bag.
    How many apples are in 7 bags?

17. It costs $8 for a guitar lesson.
    It costs $6 for a piano lesson.
    How much does it cost for 9 piano lessons?

18. There are 6 cans of pop in a carton.
    Tina has 3 cartons and Jack has 4 cartons.
    How many cans do they have in all?

Multiplication: Groups of 7

This graph shows how many weeks each child spent visiting grandparents.

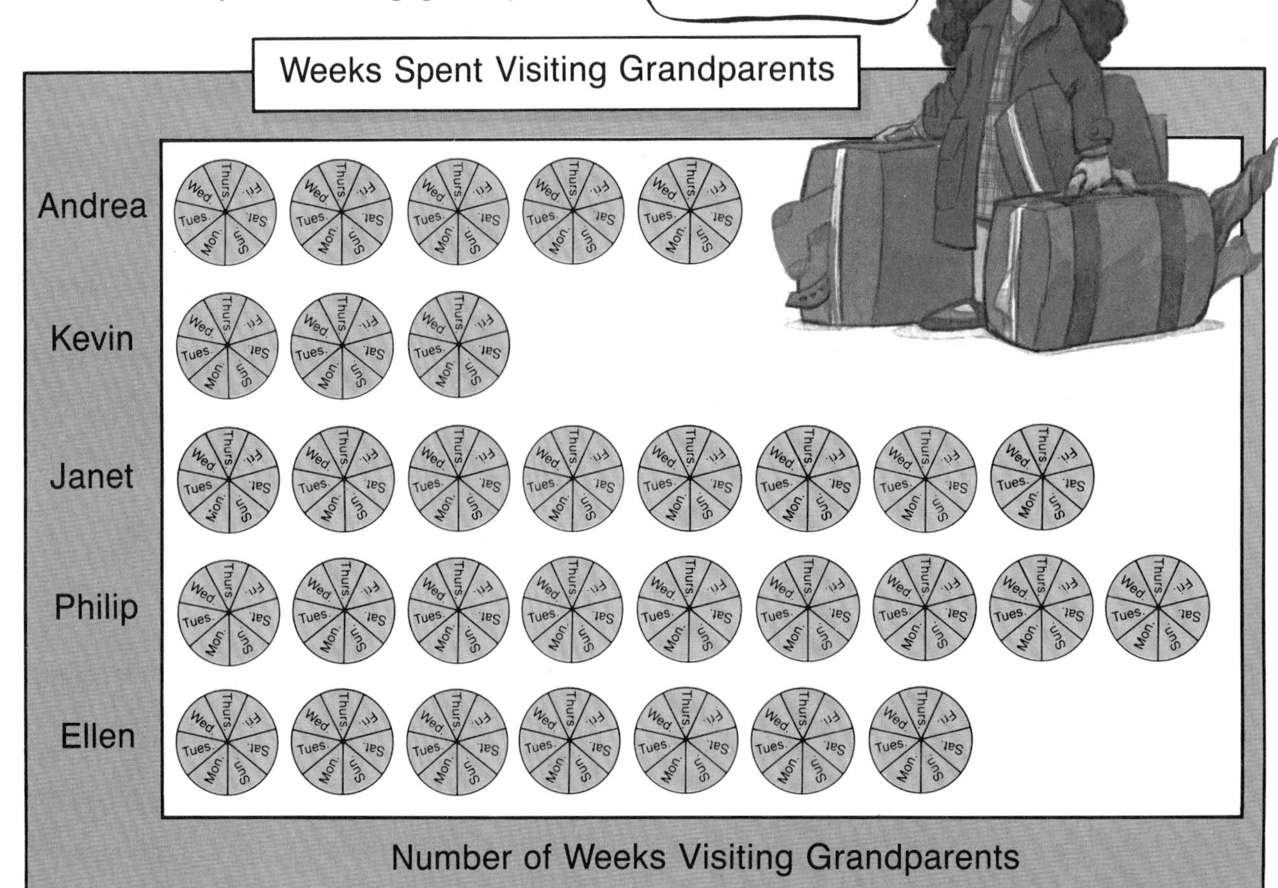

1. Make a table to show how many weeks and how many days each child spent visiting grandparents.

	Andrea	Kevin	Janet	Philip	Ellen
Number of Weeks					
Number of Days					

2. Copy and complete.

×	0	1	2	3	4	5	6	7	8	9
7										

Every seventh square on this 100-chart has been coloured.
Use the 100-chart to count by 7s to 98.

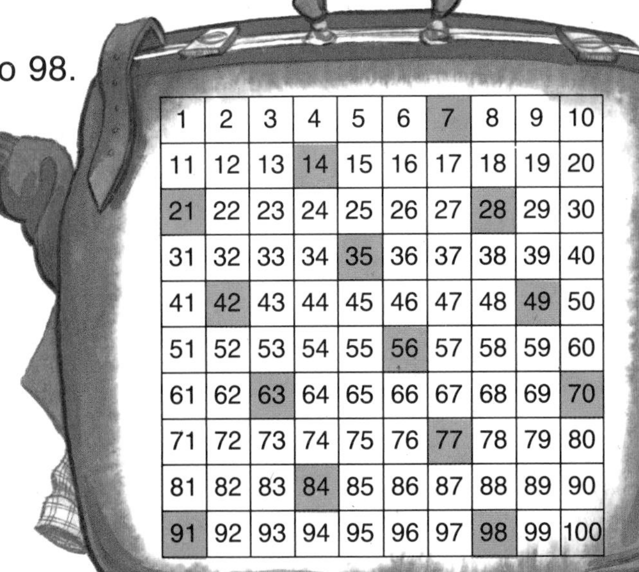

1	2	3	4	5	6	7	8	9	10
11	12	13	14	15	16	17	18	19	20
21	22	23	24	25	26	27	28	29	30
31	32	33	34	35	36	37	38	39	40
41	42	43	44	45	46	47	48	49	50
51	52	53	54	55	56	57	58	59	60
61	62	63	64	65	66	67	68	69	70
71	72	73	74	75	76	77	78	79	80
81	82	83	84	85	86	87	88	89	90
91	92	93	94	95	96	97	98	99	100

Complete each pattern.

1. 7, 14, 21, ■, ■, 42, ■
2. 56, 63, 70, ■, ■, 91, ■
3. 7, 21, 35, ■, ■, 77, ■
4. 14, 28, 42, ■, ■, 84, ■
5. 63, 56, 49, ■, 35, ■, ■

Copy and complete.

6. 7×2
7. 7×5
8. 0×7
9. 8×7
10. 7×6
11. 9×7
12. 5×7
13. 8×7
14. 3×7

15. How many days are there in 6 weeks?
16. There are 8 rows of stamps.
There are 7 stamps in each row.
How many stamps are there in all?
17. Rachel earns $5 per weekday.
She earns $9 on Saturday and $10 on Sunday.
How much will she earn working 7 days in a row?

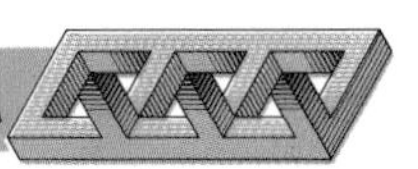

In the first 3 months of the year 2012:

How many weeks are there altogether?

How many days are there altogether?

JANUARY

S	M	T	W	T	F	S
1	2	3	4	5	6	7
8	9	10	11	12	13	14
15	16	17	18	19	20	21
22	23	24	25	26	27	28
29	30	31				

FEBRUARY

S	M	T	W	T	F	S
			1	2	3	4
5	6	7	8	9	10	11
12	13	14	15	16	17	18
19	20	21	22	23	24	25
26	27	28	29			

MARCH

S	M	T	W	T	F	S
				1	2	3
4	5	6	7	8	9	10
11	12	13	14	15	16	17
18	19	20	21	22	23	24
25	26	27	28	29	30	31

Multiplication: Groups of 8

Write a multiplication sentence for each picture.

1. How many wheels can be seen? $8 \times 4 = \square$

2. How many wheels are hidden? $\square \times \square = \square$

3. How many wheels are there in all? 

4. Each pizza has been divided into 8 slices.
   How many slices are there in all?

5. Write a multiplication sentence to show how many squares appear on a checkerboard.

Copy and complete each pattern.
Use the 100-chart to help you.

1	2	3	4	5	6	7	8	9	10
11	12	13	14	15	16	17	18	19	20
21	22	23	24	25	26	27	28	29	30
31	32	33	34	35	36	37	38	39	40
41	42	43	44	45	46	47	48	49	50
51	52	53	54	55	56	57	58	59	60
61	62	63	64	65	66	67	68	69	70
71	72	73	74	75	76	77	78	79	80
81	82	83	84	85	86	87	88	89	90
91	92	93	94	95	96	97	98	99	100

1. 8, 16, 24, ■, ■, 48, ■
2. 48, 56, 64, ■, ■, 88, ■
3. 16, 32, 48, ■, ■, ■
4. 8, 24, 40, ■, ■, 88, ■

Copy and complete.

5. 2×8
6. 4×8
7. 8×3
8. 5×8
9. 8×4
10. 8×7
11. 0×8
12. 6×8
13. 8×8
14. 8×1
15. 9×8
16. 8×6

Copy and complete.

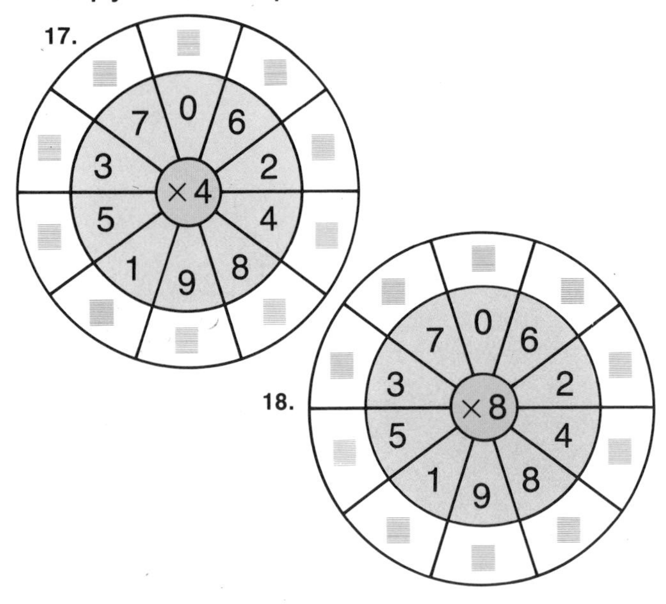

Copy and complete these tables.
What are the numbers
in the bottom row?

×	1	2	3	4	5	6	7	8	9
4	4	8	■	■	■	■	■	■	■
4	4	8	■	■	■	■	■	■	■

Add.

8	16	■	■	■	■	■	■	■

Complete:
When you add the 4 times table to itself you get the ■ times table.

Multiplication: Groups of 9

1. How many buttons are there on all the jackets?
   Write a multiplication sentence.

Copy and complete each pattern.
Use the 100-chart to help you.

1	2	3	4	5	6	7	8	9	10
11	12	13	14	15	16	17	18	19	20
21	22	23	24	25	26	27	28	29	30
31	32	33	34	35	36	37	38	39	40
41	42	43	44	45	46	47	48	49	50
51	52	53	54	55	56	57	58	59	60
61	62	63	64	65	66	67	68	69	70
71	72	73	74	75	76	77	78	79	80
81	82	83	84	85	86	87	88	89	90
91	92	93	94	95	96	97	98	99	100

2. 9, 18, 27, ▒ , ▒ , ▒

3. 36, 45, 54, ▒ , ▒ , ▒

4. 27, ▒ , 45, ▒ , 63, ▒

5. 18, 36, 54, ▒ , ▒

6. There are 6 bags.
   Each bag has 9 cookies.
   How many cookies are there in all?

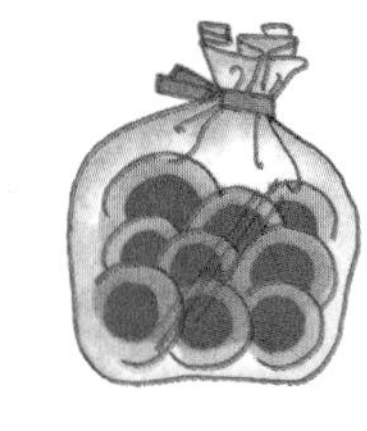

7. There are 7 boxes of pencils.
   Each box has 9 pencils.
   How many pencils are there in all?

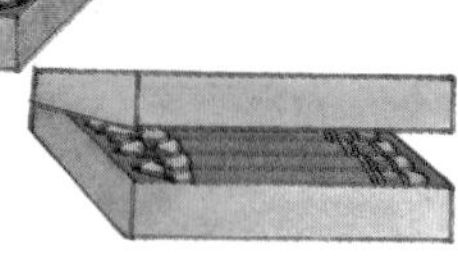

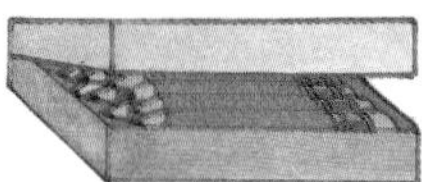
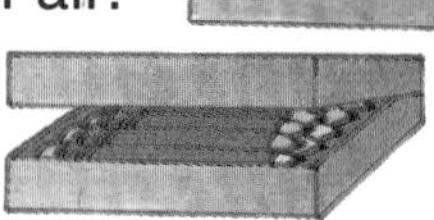

Copy and complete.

1. $4 \times 9 = \square$
   $9 \times 4 = \square$

2. $6 \times 9 = \square$
   $9 \times 6 = \square$

3. $7 \times 9 = \square$
   $9 \times 7 = \square$

4. $3 \times 9 = \square$
   $9 \times 3 = \square$

5. $2 \times 9 = \square$
   $9 \times 2 = \square$

6. $5 \times 9 = \square$
   $9 \times 5 = \square$

7. $8 \times 9 = \square$
   $9 \times 8 = \square$

8. $9 \times 1 = \square$
   $1 \times 9 = \square$

9. $\begin{array}{r} 3 \\ \times 9 \\ \hline \end{array}$
10. $\begin{array}{r} 5 \\ \times 9 \\ \hline \end{array}$
11. $\begin{array}{r} 0 \\ \times 9 \\ \hline \end{array}$
12. $\begin{array}{r} 2 \\ \times 9 \\ \hline \end{array}$
13. $\begin{array}{r} 8 \\ \times 9 \\ \hline \end{array}$
14. $\begin{array}{r} 9 \\ \times 7 \\ \hline \end{array}$

15. $\begin{array}{r} 9 \\ \times 8 \\ \hline \end{array}$
16. $\begin{array}{r} 4 \\ \times 9 \\ \hline \end{array}$
17. $\begin{array}{r} 6 \\ \times 9 \\ \hline \end{array}$
18. $\begin{array}{r} 9 \\ \times 9 \\ \hline \end{array}$
19. $\begin{array}{r} 9 \\ \times 6 \\ \hline \end{array}$
20. $\begin{array}{r} 9 \\ \times 5 \\ \hline \end{array}$

21.

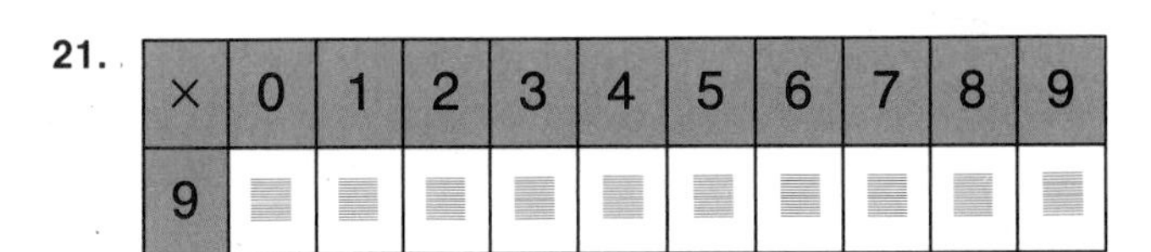

×	0	1	2	3	4	5	6	7	8	9
9	▒	▒	▒	▒	▒	▒	▒	▒	▒	▒

22.

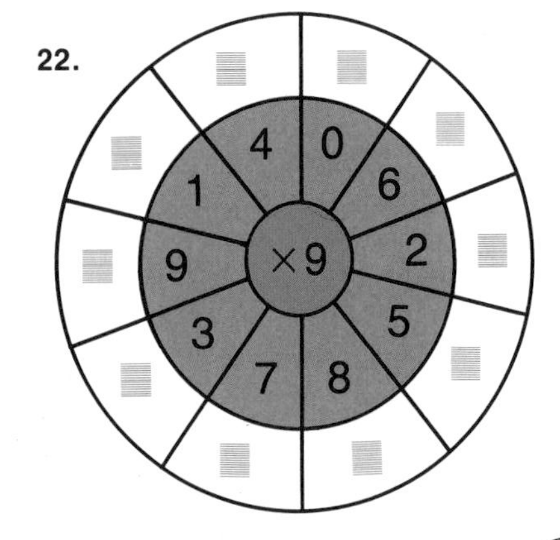

23. There are 52 cards in a fish deck.
    9 children are playing.
    Each child gets 5 cards.
    How many cards are left?

JUST FOR FUN

Each balloon has a partner.
The partners have the same product.
Write the letters for each partner.

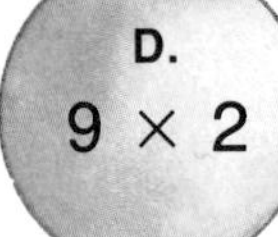

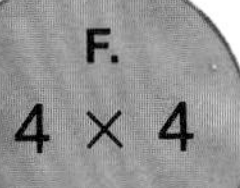

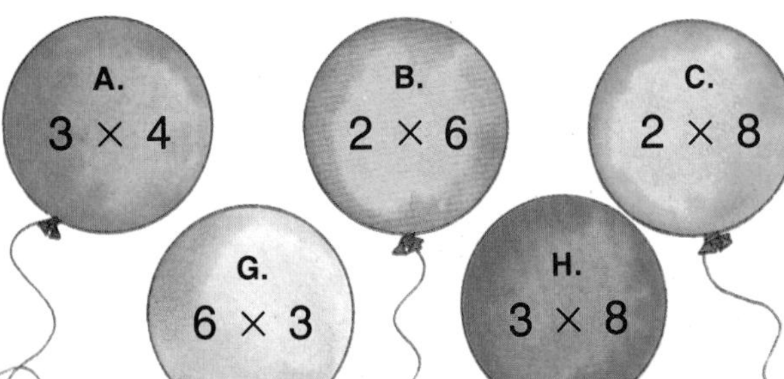

JUST FOR FUN

RIDDLE: What is a knight's favorite fish?

Multiply.

A. 3×6

B. 3×9

C. 6×0

D. 1×7

E. 4×5

F. 7×7

G. 9×5

H. 8×8

I. 7×9

J. 6×6

K. 5×9

L. 8×9

M. 7×5

N. 9×3

O. 9×6

P. 9×1

Q. 8×4

R. 6×7

S. 7×8

T. 0×8

U. 5×5

V. 5×7

W. 9×9

X. 2×9

Y. 9×4

Z. 9×8

To answer the riddle, write the letters of the questions with these answers.

   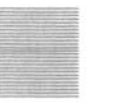 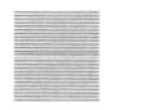   

56 81 54 42 7 49 63 56 64

RIDDLE: What time is most dangerous for knights?

Multiply.

A. 6×9 B. 9×5 C. 4×7

D. 8×4 E. 0×7 F. 7×7

G. 7×6 H. 9×9 I. 8×8 J. 8×6 K. 3×8

L. 3×7 M. 5×9 N. 8×7 O. 6×6 P. 7×5

Q. 5×5 R. 8×9 S. 8×5 T. 9×7 U. 5×8

V. 8×3 W. 9×8 X. 5×7 Y. 8×6 Z. 4×8

To answer the riddle, write the letters of the questions with these answers.

          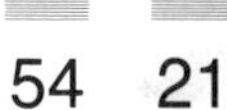  

56 64 42 81 63 49 54 21 21

Multiplication: Groups of 10

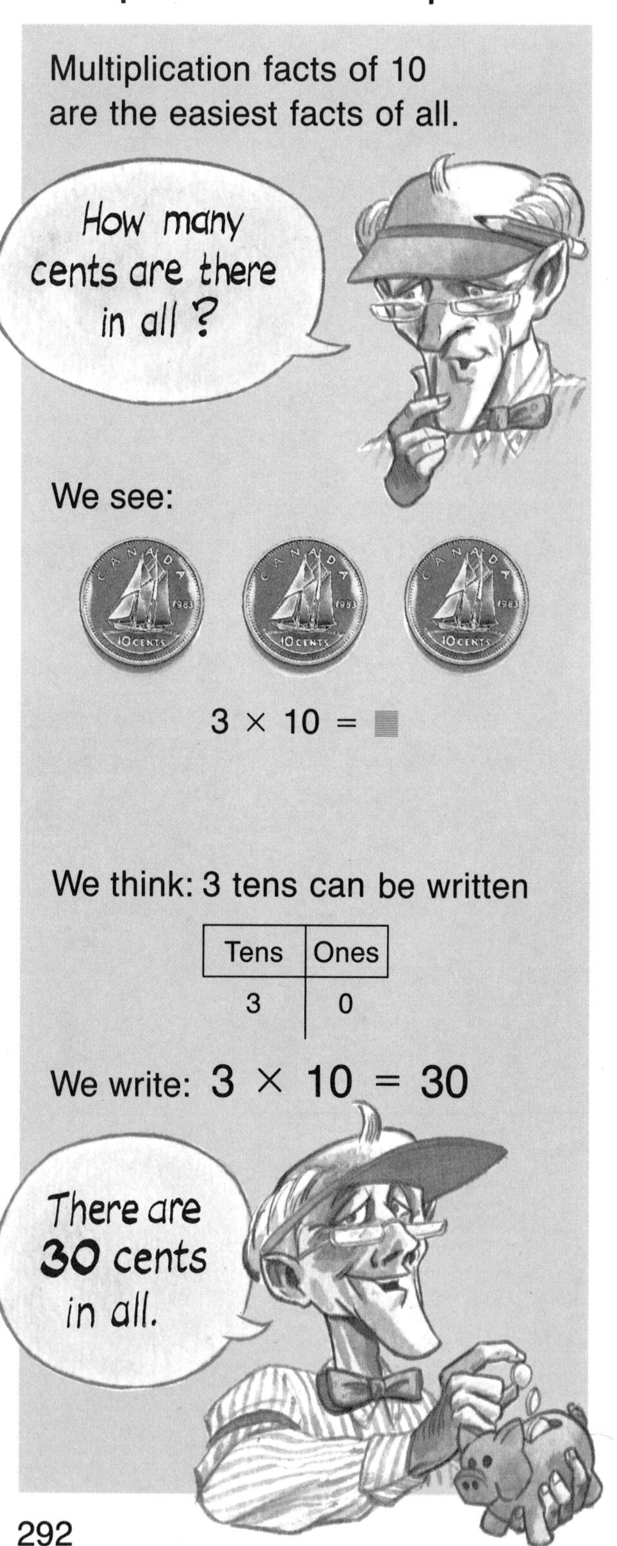

Multiplication facts of 10 are the easiest facts of all.

We see:

$3 \times 10 = \blacksquare$

We think: 3 tens can be written

Tens	Ones
3	0

We write: $3 \times 10 = 30$

Write a multiplication sentence to match each picture.

Copy and complete.

3. $\begin{array}{r} 2 \\ \times 10 \\ \hline \end{array}$

4. $\begin{array}{r} 5 \\ \times 10 \\ \hline \end{array}$

5. $\begin{array}{r} 4 \\ \times 10 \\ \hline \end{array}$

6. $\begin{array}{r} 8 \\ \times 10 \\ \hline \end{array}$

7. $\begin{array}{r} 0 \\ \times 10 \\ \hline \end{array}$

8. $\begin{array}{r} 9 \\ \times 10 \\ \hline \end{array}$

9. $\begin{array}{r} 10 \\ \times 1 \\ \hline \end{array}$

10. $\begin{array}{r} 10 \\ \times 6 \\ \hline \end{array}$

11. $\begin{array}{r} 10 \\ \times 3 \\ \hline \end{array}$

12. Use squared paper.

×	1	2	3	4	5	6	7	8	9	10
1										
2										
3										
4										
5										
6										
7										
8										
9										
10										

Multiplication: Groups of 100

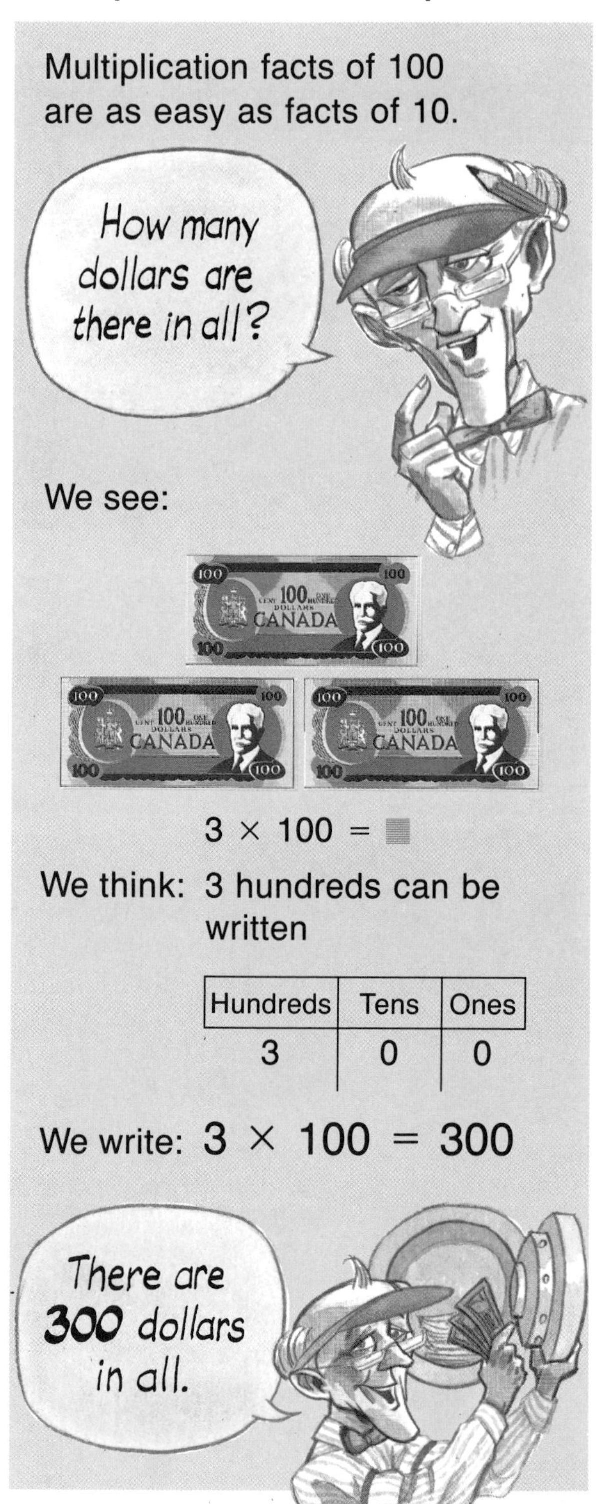

Write a multiplication sentence to match each picture.

1.

2.

Copy and complete.

3. $2 \times 100 = ■$
4. $6 \times 100 = ■$
5. $5 \times 100 = ■$
6. $100 \times 5 = ■$
7. $9 \times 100 = ■$
8. $100 \times 9 = ■$
9. 6×100
10. 4×100
11. 100×8
12. 7×100
13. 0×100
14. 100×2

Special Products

How many cents are there in all?

We think: 2 × 3 tens is 6 tens.

We write: $2 \times 30 = 60$

There are 60 cents in all.

Copy and complete.

1. $5 \times 30 = ▢$ [1. 5 x 30 = 150]
2. $3 \times 70 = ▢$
3. $4 \times 60 = ▢$
4. $2 \times 90 = ▢$
5. $5 \times 70 = ▢$
6. $2 \times 80 = ▢$
7. $3 \times 80 = ▢$
8. $3 \times 40 = ▢$
9. $5 \times 80 = ▢$
10. $6 \times 90 = ▢$
11. $7 \times 60 = ▢$

---

12. 60×3
13. 70×2
14. 90×4
15. 80×4
16. 40×5
17. 80×6
18. 20×7
19. 30×6
20. 60×5
21. 40×9
22. 70×8
23. 40×7

---

24. A bottle of pop costs 30¢.
How much does it cost for 3 bottles?

25. A bus holds 60 people.
How many people can 4 buses hold?

26. It takes the moon about 30 days for one orbit around the Earth.
How many days does it take for 9 orbits?

27. An orchard has 7 rows of pear trees.
It has 9 rows of apple trees.
There are 90 trees in each row.
How many apple trees are there in all?

How many dollars are there in all?

We see:

$$3 \times 400 = \square$$

We think: 3×4 hundreds is 12 hundreds.

We write: $3 \times 400 = 1200$

There are 1200 dollars in all.

Copy and complete.

1. $2 \times 300 = \square$	2. $3 \times 500 = \square$	3. $2 \times 900 = \square$	4. $3 \times 700 = \square$
5. $4 \times 600 = \square$	6. $4 \times 800 = \square$	7. $6 \times 700 = \square$	8. $6 \times 300 = \square$
9. $5 \times 200 = \square$	10. $7 \times 300 = \square$	11. $8 \times 800 = \square$	12. $9 \times 400 = \square$

---

13. $\begin{array}{r} 200 \\ \times\ 9 \\ \hline \end{array}$	14. $\begin{array}{r} 500 \\ \times\ 7 \\ \hline \end{array}$	15. $\begin{array}{r} 300 \\ \times\ 3 \\ \hline \end{array}$	16. $\begin{array}{r} 600 \\ \times\ 5 \\ \hline \end{array}$	17. $\begin{array}{r} 300 \\ \times\ 8 \\ \hline \end{array}$	18. $\begin{array}{r} 700 \\ \times\ 9 \\ \hline \end{array}$
19. $\begin{array}{r} 800 \\ \times\ 5 \\ \hline \end{array}$	20. $\begin{array}{r} 600 \\ \times\ 8 \\ \hline \end{array}$	21. $\begin{array}{r} 900 \\ \times\ 3 \\ \hline \end{array}$	22. $\begin{array}{r} 800 \\ \times\ 6 \\ \hline \end{array}$	23. $\begin{array}{r} 700 \\ \times\ 8 \\ \hline \end{array}$	24. $\begin{array}{r} 900 \\ \times\ 6 \\ \hline \end{array}$

Multiplying Tens and Ones

Four children have 23¢ each. How much do they have altogether?

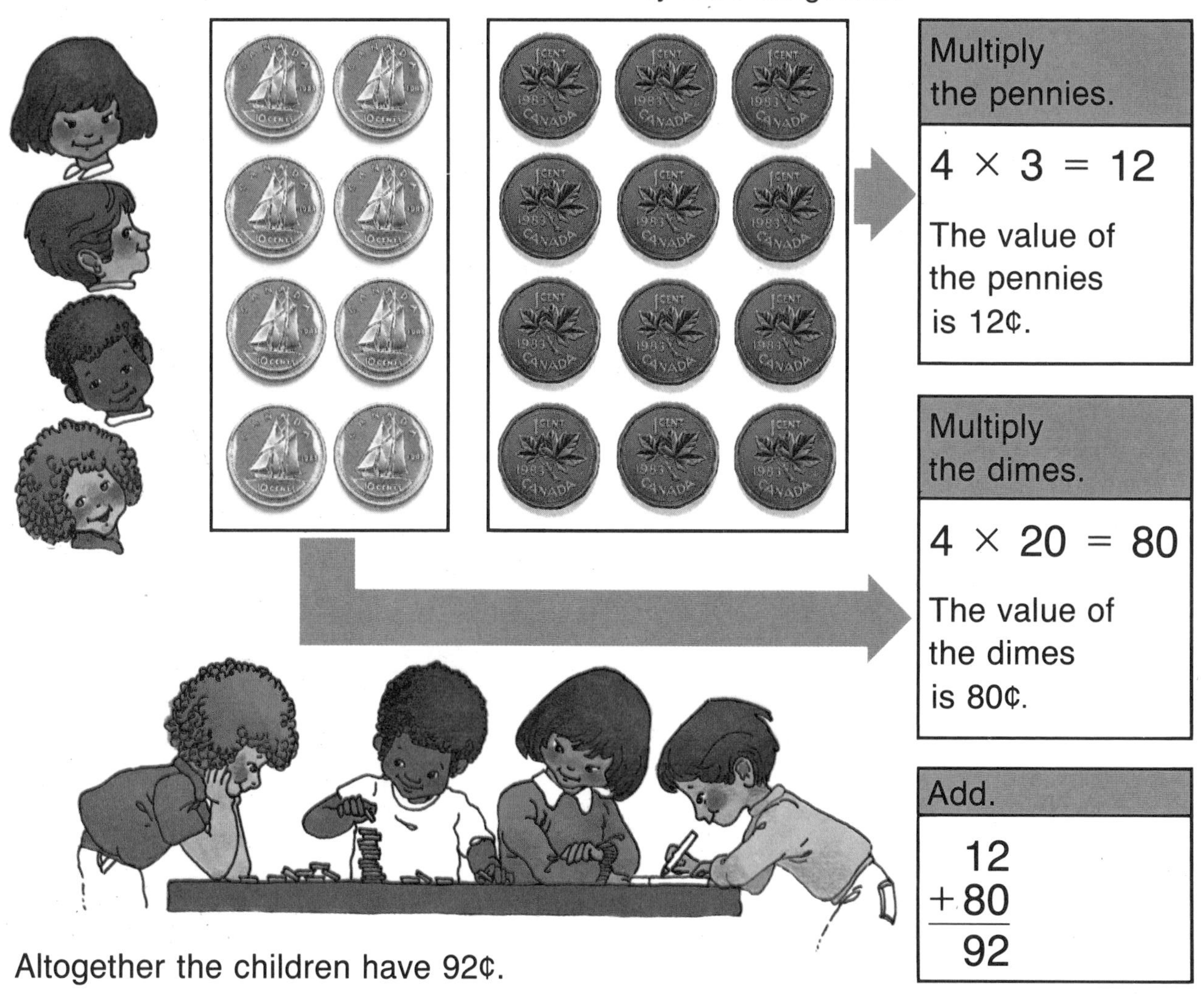

Altogether the children have 92¢.

Here is a shorter way to write multiplication.

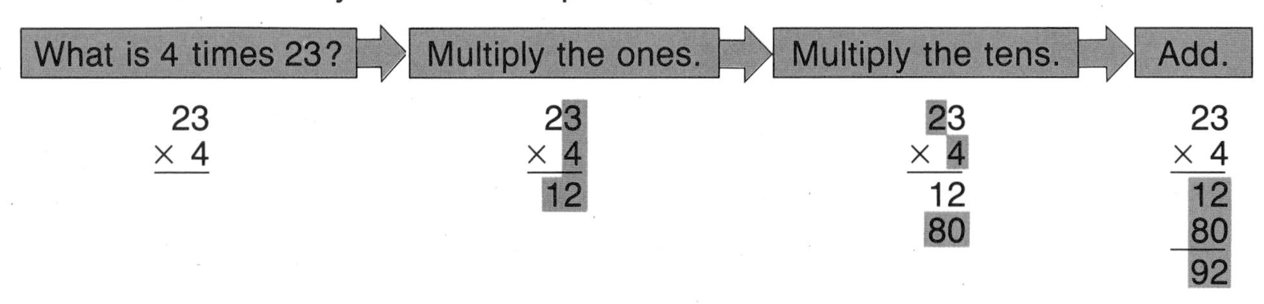

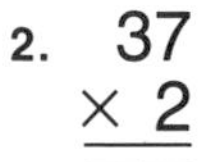

1. $\begin{array}{r} 46 \\ \times\ 2 \\ \hline \square\square \\ 8\square \\ \hline \square\square \end{array}$	2. $\begin{array}{r} 37 \\ \times\ 2 \\ \hline \square\square \\ \square\square \\ \hline \square\square \end{array}$	3. $\begin{array}{r} 48 \\ \times\ 2 \\ \hline \square\square \\ \square\square \\ \hline \square\square \end{array}$	4. $\begin{array}{r} 69 \\ \times\ 3 \\ \hline \square\square \\ 180 \\ \hline \square\square\square \end{array}$	5. $\begin{array}{r} 56 \\ \times\ 4 \\ \hline \square\square \\ \square\square\square \\ \hline \square\square\square \end{array}$	6. $\begin{array}{r} 29 \\ \times\ 5 \\ \hline \square\square \\ \square\square\square \\ \hline \square\square\square \end{array}$
7. $\begin{array}{r} 29 \\ \times\ 3 \\ \hline \end{array}$	8. $\begin{array}{r} 32 \\ \times\ 5 \\ \hline \end{array}$	9. $\begin{array}{r} 63 \\ \times\ 4 \\ \hline \end{array}$	10. $\begin{array}{r} 56 \\ \times\ 6 \\ \hline \end{array}$	11. $\begin{array}{r} 23 \\ \times\ 7 \\ \hline \end{array}$	12. $\begin{array}{r} 52 \\ \times\ 3 \\ \hline \end{array}$
13. $\begin{array}{r} 25 \\ \times\ 5 \\ \hline \end{array}$	14. $\begin{array}{r} 44 \\ \times\ 7 \\ \hline \end{array}$	15. $\begin{array}{r} 60 \\ \times\ 9 \\ \hline \end{array}$	16. $\begin{array}{r} 76 \\ \times\ 3 \\ \hline \end{array}$	17. $\begin{array}{r} 57 \\ \times\ 8 \\ \hline \end{array}$	18. $\begin{array}{r} 82 \\ \times\ 8 \\ \hline \end{array}$
19. $\begin{array}{r} 68 \\ \times\ 5 \\ \hline \end{array}$	20. $\begin{array}{r} 27 \\ \times\ 9 \\ \hline \end{array}$	21. $\begin{array}{r} 38 \\ \times\ 3 \\ \hline \end{array}$	22. $\begin{array}{r} 72 \\ \times\ 8 \\ \hline \end{array}$	23. $\begin{array}{r} 63 \\ \times\ 7 \\ \hline \end{array}$	24. $\begin{array}{r} 52 \\ \times\ 5 \\ \hline \end{array}$

25. There are 12 muffins in a dozen. How many muffins are there in 4 dozen?

26. Teresa delivers 56 newspapers each day. How many newspapers does she deliver in 6 days?

ESTIMATING

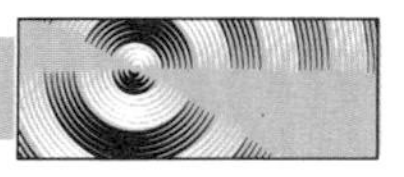

Estimate: How many squares do you think there are in the quilt?

Making Maple Syrup

The Minuks live on a farm.
Part of the farm is a bush
with hundreds of maple trees.

Every spring they drill small holes
in these trees to collect sap.
This is called tapping the trees.

1. In the north bush they tapped 374 trees.
   In the south bush they tapped 319 trees.
   How many trees were tapped this year?

2. Last year they tapped 497 trees.
   How many more trees did they tap
   this year?

The sap runs out of the trees.
It runs through plastic tubes into
large tanks.

3. One tank collected 380 L of sap.
   Another tank collected 493 L of sap.
   A third tank collected 307 L of sap.
   How much sap was in all 3 tanks?

Mr. Minuk boils the sap
until syrup begins to appear.

4. It takes 10 L of sap to make one jug of maple syrup. The Minuks made 90 jugs of maple syrup this year. How much sap did they collect?

5. The Minuks kept 20 jugs for themselves. They sold the rest. How many jugs did they sell?

6. Each jug sold for $4. How much did the Minuks receive for the jugs they sold?

7. Last year the Minuks collected $248 for their syrup. How much more did they collect this year?

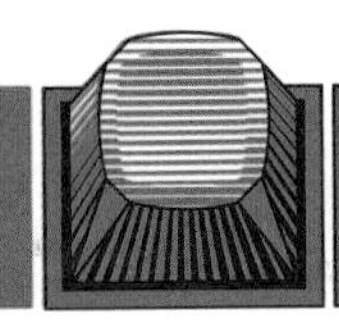 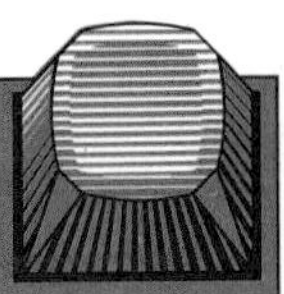 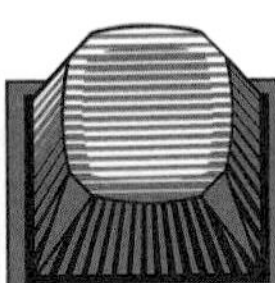

BITS AND BYTES

When the turtle sees this computer command . . . → it moves forward 110 units.

110 units

When the turtle sees this computer command . . . → it turns right.

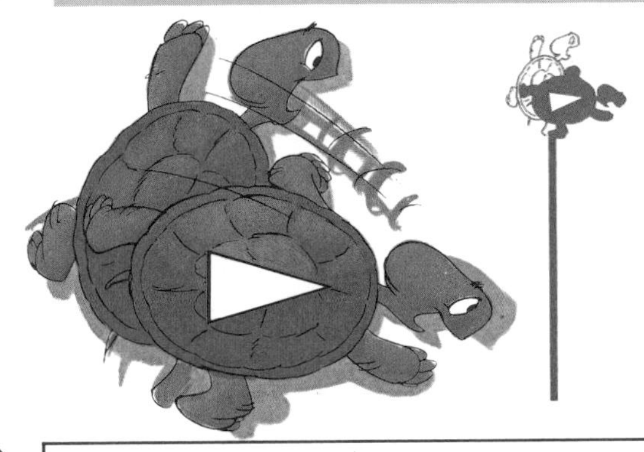

When the turtle sees this computer program . . . → it moves forward 110 units and turns right. It repeats this 4 times to make a square.

How many units forward did the turtle walk to make this square?

RIGHT 90 FORWARD 110 RIGHT 90

FORWARD 110 FORWARD 110

RIGHT 90 FORWARD 110 RIGHT 90

1. The turtle used these programs to make squares. How far is it around each square?

A.

B.

C.

D.

REPEAT 4
FORWARD 72
RIGHT 90

E.

F.

2. How far did the turtle walk altogether in question 1?

3. Write a program which makes a square with each side 45 units long.

CHECK-UP

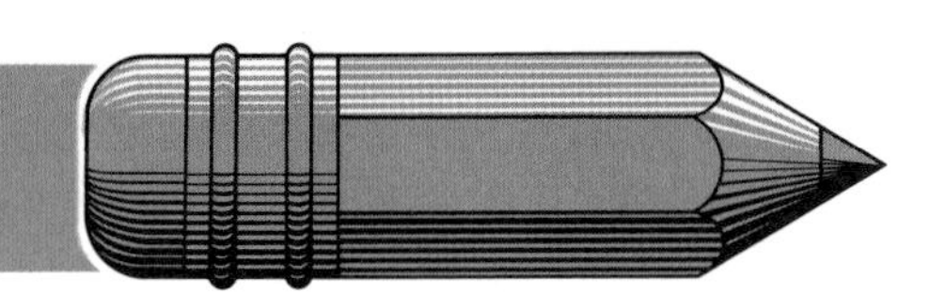

Write a multiplication sentence to answer each question.

1. How many leaves are there?

2. How many shells are there?

Multiply.

3. $\begin{array}{r} 3 \\ \times 7 \\ \hline \end{array}$ 4. $\begin{array}{r} 5 \\ \times 8 \\ \hline \end{array}$ 5. $\begin{array}{r} 4 \\ \times 6 \\ \hline \end{array}$ 6. $\begin{array}{r} 3 \\ \times 9 \\ \hline \end{array}$ 7. $\begin{array}{r} 9 \\ \times 7 \\ \hline \end{array}$ 8. $\begin{array}{r} 8 \\ \times 6 \\ \hline \end{array}$

9. $\begin{array}{r} 9 \\ \times 8 \\ \hline \end{array}$ 10. $\begin{array}{r} 4 \\ \times 9 \\ \hline \end{array}$ 11. $\begin{array}{r} 0 \\ \times 7 \\ \hline \end{array}$ 12. $\begin{array}{r} 7 \\ \times 6 \\ \hline \end{array}$ 13. $\begin{array}{r} 4 \\ \times 8 \\ \hline \end{array}$ 14. $\begin{array}{r} 5 \\ \times 9 \\ \hline \end{array}$

15. Copy and complete the table.

×	6	9	7	8
5				
7				
2				

16. There are 6 legs on an insect. How many legs are there on 9 insects?

Copy and complete.

17. $8 \times 10 = \blacksquare$ 18. $9 \times 100 = \blacksquare$ 19. $2 \times 80 = \blacksquare$

20. $\begin{array}{r} 40 \\ \times\ 5 \\ \hline \end{array}$ 21. $\begin{array}{r} 63 \\ \times\ 3 \\ \hline \end{array}$ 22. $\begin{array}{r} 41 \\ \times\ 8 \\ \hline \end{array}$ 23. $\begin{array}{r} 67 \\ \times\ 5 \\ \hline \end{array}$ 24. $\begin{array}{r} 28 \\ \times\ 3 \\ \hline \end{array}$ 25. $\begin{array}{r} 35 \\ \times\ 2 \\ \hline \end{array}$

MORE DIVISION

FOG

The fog comes
on little cat feet.

It sits looking
over harbor and city
on silent haunches
and then moves on.

—Carl Sandburg

Division Sentences

There are 15 seals.
How many groups of 5 are there?

We see:

We think: 15 divided into groups of 5 equals 3.

We write: $15 \div 5 = 3$

There are 3 groups of seals.

Write a division sentence to answer each question.

1. How many pairs of gloves are there?

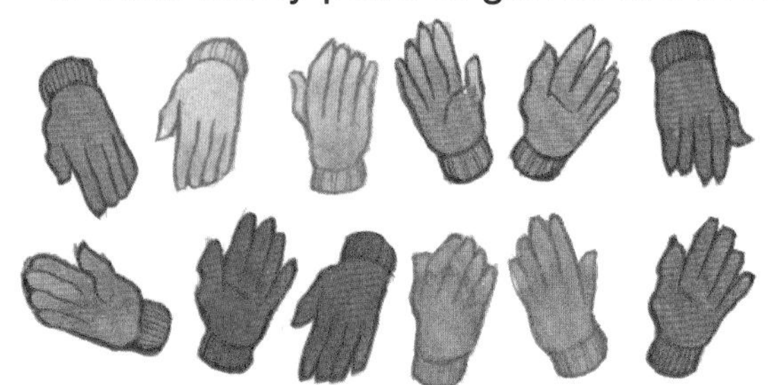

2. There are 20 books.
   How many piles of 4 can be made?

Draw a picture, then write a division sentence to answer each question.

3. John had 45¢ in nickels.
   How many nickels did he have?
4. 36 flowers are planted in 4 equal rows.
   How many are in each row?
5. 18 children are divided into teams of 3.
   How many teams are there?
6. There are 12 cookies.
   4 are placed in each bag.
   How many bags are needed?

There are 12 balloons.
They are shared equally among 4 clowns.
How many does each clown get?

We see:

We think: 12 balloons shared among 4 clowns is 3 balloons each.

We write: $12 \div 4 = 3$

Each clown gets 3 balloons.

Draw a picture, then write a division sentence.

1. How many balloons does each child get?

2. 28 cards are given out to 4 children. How many cards does each child get?

3. 12 pencils are shared by 4 children. How many does each child get?

PROBLEM SOLVING

A turtle walked a distance of 36 cm around a square.

Using a Table to Divide

We can use a multiplication table to divide.

To divide $4\overline{)20}$…

Step 1

Find the number 20 in row 4.

Step 2

Write the column in which the number 20 appears.

$$\text{row } 4\overline{)20}^{\;5 \text{ column}}$$

×	1	2	3	4	5
1	1	2	3	4	5
2	2	4	6	8	10
3	3	6	9	12	15
4	4	8	12	16	20
5	5	10	15	20	25

Use the table to divide.

1. $2\overline{)4}$
2. $2\overline{)6}$
3. $3\overline{)12}$
4. $3\overline{)15}$
5. $3\overline{)6}$
6. $2\overline{)2}$
7. $4\overline{)8}$
8. $2\overline{)8}$
9. $3\overline{)3}$
10. $4\overline{)12}$
11. $3\overline{)9}$
12. $4\overline{)16}$
13. $1\overline{)2}$
14. $5\overline{)5}$
15. $1\overline{)1}$
16. $5\overline{)10}$
17. $2\overline{)10}$
18. $1\overline{)4}$
19. $5\overline{)15}$
20. $1\overline{)5}$

To divide $8\overline{)48}$...

Step 1

Find the number 48 in row 8.

Step 2

Write the column in which the number 48 appears.

$$\text{row } 8\overline{)48}\ \ \overset{6}{} \text{ column}$$

×	1	2	3	4	5	6	7	8	9
1	1	2	3	4	5	6	7	8	9
2	2	4	6	8	10	12	14	16	18
3	3	6	9	12	15	18	21	24	27
4	4	8	12	16	20	24	28	32	36
5	5	10	15	20	25	30	35	40	45
6	6	12	18	24	30	36	42	48	54
7	7	14	21	28	35	42	49	[illegible]	[illegible]
8	8	16	24	32	40	48	56	[illegible]	[illegible]
9	9	18	27	36	45	54	63	[illegible]	[illegible]

Use the table to divide.

1. $6\overline{)42}$
2. $7\overline{)28}$
3. $9\overline{)36}$
4. $9\overline{)45}$
5. $8\overline{)24}$
6. $7\overline{)35}$
7. $6\overline{)48}$
8. $8\overline{)32}$
9. $9\overline{)54}$
10. $8\overline{)40}$
11. $7\overline{)49}$
12. $9\overline{)63}$
13. $7\overline{)42}$
14. $6\overline{)54}$
15. $8\overline{)16}$
16. $7\overline{)21}$
17. $9\overline{)18}$
18. $6\overline{)36}$
19. $8\overline{)56}$
20. $9\overline{)72}$

Division: Groups of 6

There are 6 cans of juice in one carton. How many cartons will hold 54 cans?

The number of groups of 6 in 54 is

$$6\overline{)54}$$

From the the table, we see $\begin{array}{r}9\\6\overline{)54}\end{array}$.

9 cartons hold 54 cans.

Row 6 of the table shows the division facts of 6.

×	1	2	3	4	5	6	7	8	9
1	1	2	3	4	5	6	7	8	9
2	2	4	6	8	10	12	14	16	18
3	3	6	9	12	15	18	21	24	27
4	4	8	12	16	20	24	28	32	36
5	5	10	15	20	25	30	35	40	45
6	6	12	18	24	30	36	42	48	54
7	7	14	21	28	35	42	49	56	63
8	8	16	24	32	40	48	56	64	72
9	9	18	27	36	45	54	63	72	81

Divide.

1. $6\overline{)30}$
2. $6\overline{)18}$
3. $6\overline{)36}$
4. $6\overline{)12}$
5. $6\overline{)42}$
6. $6\overline{)24}$
7. $6\overline{)54}$
8. $6\overline{)48}$
9. 18 muffins are shared by 6 people. How many does each person get?
10. 6 tickets cost $42. How much does each ticket cost?
11. There are 30 children in teams of 6. There are 4 referees. How many teams are there?
12. 48 chocolate bars are wrapped in packages of 6. How many packages are there?

Division: Groups of 7

Susan went to visit her aunt for 21 days. How many weeks was she away?

The number of weeks in 21 days is

$$7\overline{)21}$$

From the the table, we see $\begin{array}{r} 3 \\ 7\overline{)21} \end{array}$.

21 days make 3 weeks.

Row 7 of the table shows the division facts of 7.

×	1	2	3	4	5	6	7	8	9
1	1	2	3	4	5	6	7	8	9
2	2	4	6	8	10	12	14	16	18
3	3	6	9	12	15	18	21	24	27
4	4	8	12	16	20	24	28	32	36
5	5	10	15	20	25	30	35	40	45
6	6	12	18	24	30	36	42	48	54
7	7	14	21	28	35	42	49	56	63
8	8	16	24	32	40	48	56	64	72
9	9	18	27	36	45	54	63	72	81

Divide.

1. $7\overline{)7}$
2. $7\overline{)28}$
3. $7\overline{)14}$
4. $7\overline{)42}$
5. $7\overline{)35}$
6. $7\overline{)63}$
7. $7\overline{)49}$
8. $7\overline{)56}$

9. Gordon planted 63 trees in rows of 7. How many rows are there?

10. 35 cards are dealt to 7 people. How many cards does each person get?

11. How many weeks are there in 42 days?

12. How many days are there in 21 weeks?

Division: Groups of 8

There are 56 tires needed.
Each tanker needs 8 tires.
How many tankers are there?

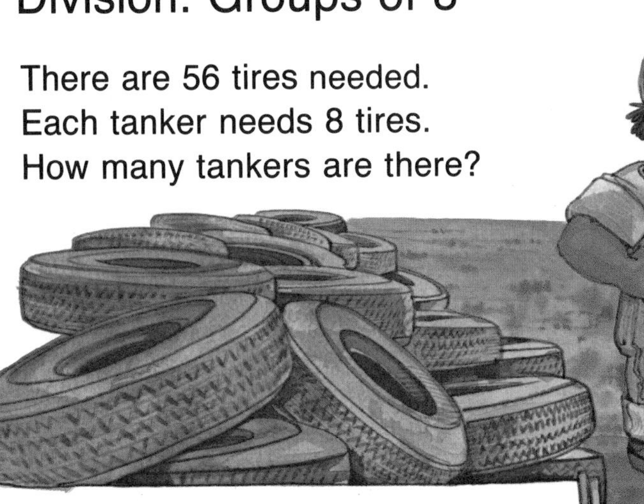

The number of groups of 8 in 56 is

$$8\overline{)56}$$

From the the table, we see $\begin{array}{r} 7 \\ 8\overline{)56} \end{array}$.

There are 7 tankers.

Row 8 of the table shows the division facts of 8.

×	1	2	3	4	5	6	7	8	9
1	1	2	3	4	5	6	7	8	9
2	2	4	6	8	10	12	14	16	18
3	3	6	9	12	15	18	21	24	27
4	4	8	12	16	20	24	28	32	36
5	5	10	15	20	25	30	35	40	45
6	6	12	18	24	30	36	42	48	54
7	7	14	21	28	35	42	49	56	63
8	8	16	24	32	40	48	56	64	72
9	9	18	27	36	45	54	63	72	81

Divide.

1. $8\overline{)16}$
2. $8\overline{)40}$
3. $8\overline{)8}$
4. $8\overline{)32}$
5. $8\overline{)56}$
6. $8\overline{)72}$
7. $8\overline{)48}$
8. $8\overline{)64}$

9. A checkerboard has 64 squares in 8 rows.
   How many squares are there in each row?

10. Grandpa Kelly is 72 years old.
    He is 8 times older than his grandson Kevin.
    How old is Kevin?

11. 8 people share a prize of $40.
    How much does each person receive?

12. 6 people want 4 slices of pizza each.
    How many 8-slice pizzas are needed?

Division: Groups of 9

72 witches are at a broom party.
How many groups of 9 are there?

×	1	2	3	4	5	6	7	8	9
1	1	2	3	4	5	6	7	8	9
2	2	4	6	8	10	12	14	16	18
3	3	6	9	12	15	18	21	24	27
4	4	8	12	16	20	24	28	32	36
5	5	10	15	20	25	30	35	40	45
6	6	12	18	24	30	36	42	48	54
7	7	14	21	28	35	42	49	56	63
8	8	16	24	32	40	48	56	64	72
9	9	18	27	36	45	54	63	72	81

Row 9 of the table shows the division facts of 9.

From the the table, we see $9\overline{)72}$ with quotient 8.

There are 8 groups of 9 witches.

Divide.

1. $9\overline{)27}$
2. $9\overline{)9}$
3. $9\overline{)18}$
4. $9\overline{)45}$
5. $9\overline{)36}$
6. $9\overline{)54}$
7. $9\overline{)81}$
8. $9\overline{)63}$

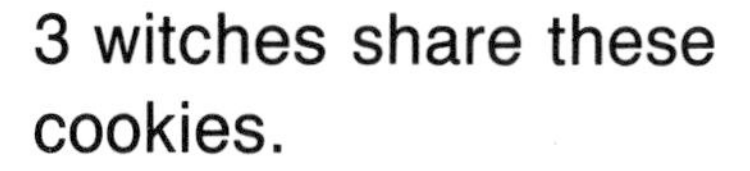

ESTIMATING

3 witches share these cookies.

Estimate: About how many cookies are there for each witch?

Leftovers

A. 13 was the favorite number of Winnifred Witch.

B. When 2 witches tried to share 13 cupcakes…1 cupcake was left over.

$\begin{array}{r}6\\2\overline{)13}\end{array}$ and 1 leftover

C. Then the witches fought for the leftover cupcake.

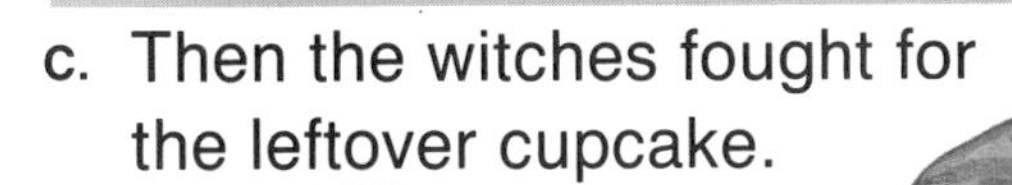

D. When 3 witches tried to share 13 cupcakes…1 cupcake was left over.

$\begin{array}{r}4\\3\overline{)13}\end{array}$ and 1 leftover

E. Again the witches fought for the leftover cupcake.

F. When 4 witches tried to share 13 cupcakes…1 cupcake was left over.

$\begin{array}{r}3\\4\overline{)13}\end{array}$ and 1 leftover

G. Once more the witches fought for the leftover cupcake.

H. Finally, the witches solved the problem.

I. They agreed to give the leftovers to the cat.

J. Now everyone is happy...

Draw pictures to show sharing.

1. Divide 15 pencils among 7 children.
How many are there for each child?
How many are left over?

2. Divide 14 balls among 6 children.
How many are there for each child?
How many are left over?

Division with Leftovers

Write a division sentence to match each picture.

1.

2.

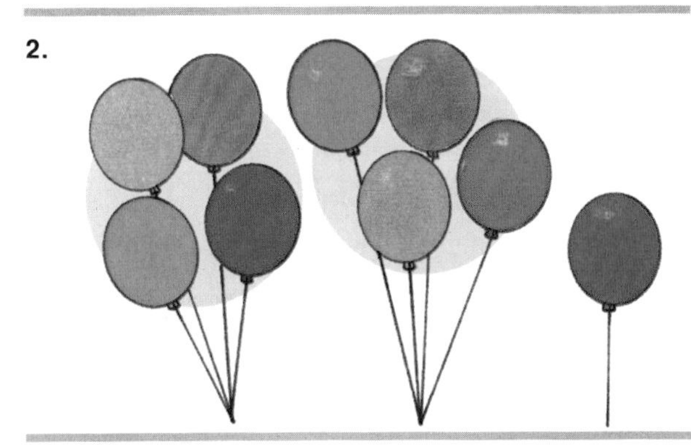

3.

4.

We see: $3\overline{)22}$

We think:

22 divided into 3 shares
is 7 and 1 leftover.

We write: $\begin{array}{r} 7 \text{ and } 1 \text{ leftover} \\ 3\overline{)22} \end{array}$

Divide.

1. $2\overline{)17}$
2. $4\overline{)15}$
3. $3\overline{)20}$
4. $2\overline{)19}$
5. $4\overline{)9}$
6. $3\overline{)29}$
7. $2\overline{)15}$
8. $4\overline{)21}$
9. $2\overline{)9}$
10. $4\overline{)27}$
11. $3\overline{)24}$
12. $4\overline{)39}$
13. $5\overline{)33}$
14. $5\overline{)45}$
15. $4\overline{)34}$

---

16. 12 hamburgers are shared equally by 5 people.
    How many are left over?
17. Tickets cost $4 each.
    How many can be bought with $22?
18. 30 stamps are arranged in rows of 7.
    How many are left over?

PROBLEM SOLVING

Each child at the party ate 3 hot dogs.
There were 24 hot dogs in all.
Rex ate 3 that were left over.
How many children were at the party?

The Magnifying GLASS

Things look bigger when seen through Maria's magnifying glass.

This is the real size of a thumb.

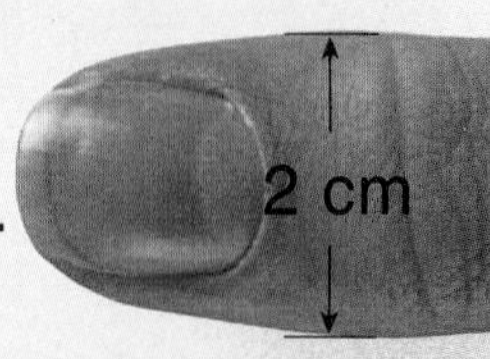

This is how it looks through a magnifying glass.

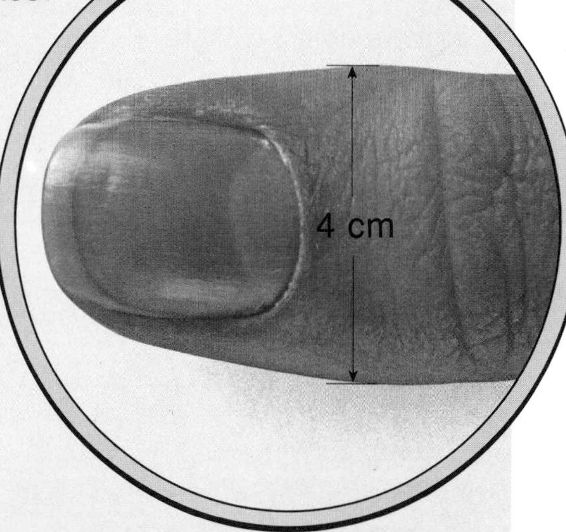

The magnifying glass makes the thumb look twice as big.

Look at each picture.
Then answer the question below.

1. This is the real size of a fly.

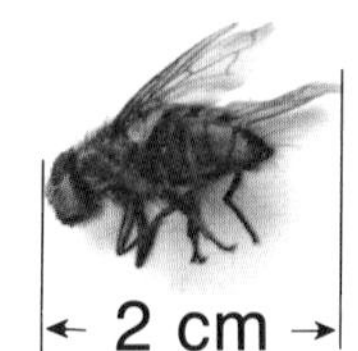

This is how it looks through a magnifying glass.

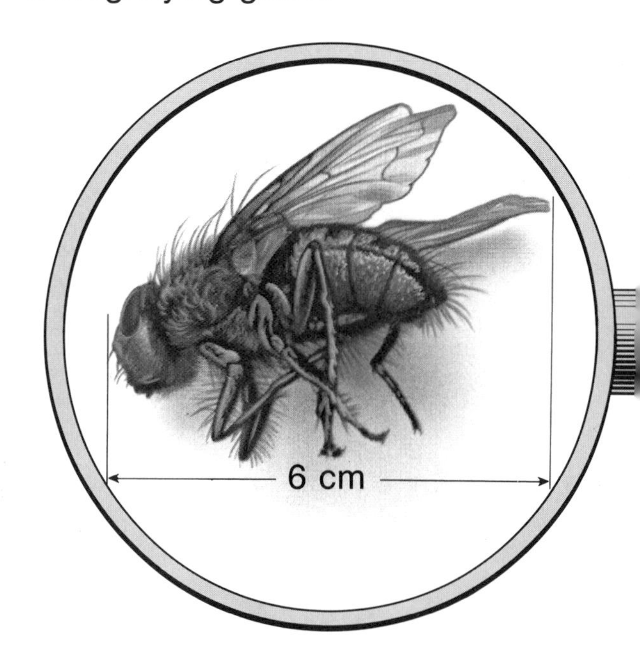

How many times as big does it look?

Applications

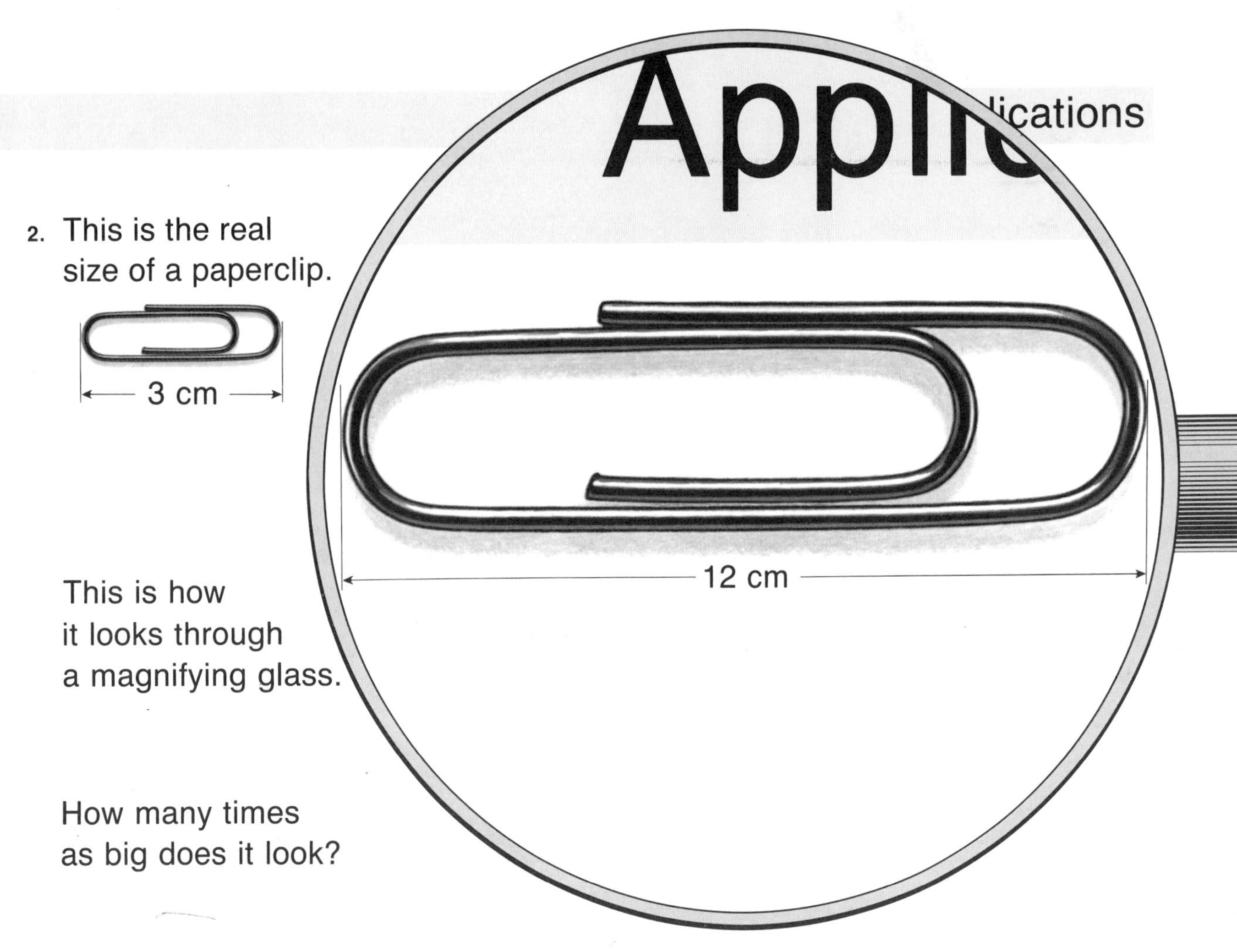

2. This is the real size of a paperclip.

   This is how it looks through a magnifying glass.

   How many times as big does it look?

3. A coin is 9 units wide.
   The magnifying glass makes it look 8 times as big.
   How wide would the coin look?

4. A leaf is 9 cm wide.
   It looks 36 cm wide through a magnifying glass.
   How many times as big does it look?

5. A fingernail is 1 cm wide.
   It looks 7 times as big through a magnifying glass.
   How wide does it look?

6. A pencil is 8 cm long.
   Through a magnifying glass it looks 40 cm long.
   How many times as big does it look?

7. A bug looks 56 units wide through a magnifying glass.
   The magnifying glass makes things look 7 times as big.
   What is the real width of the bug?

8. A flea is 7 units wide.
   The magnifying glass makes it look 4 times as big.
   How wide does it look through a magnifying glass?

PROBLEM SOLVING

Obtaining Information from a Picture

Miss Melba's class went to the museum. They stopped at Frank's Fast Foods for lunch.

Use the menu on the wall to answer these questions.

1. How much does it cost for 27 hamburgers?
2. How many small drinks can they buy for $1?
3. How many small drinks can they buy for $6?
4. How many cheeseburgers would they get for $18?
5. Which costs more: 27 cheeseburgers or 35 hamburgers?
6. 9 boys bought one cheeseburger and one milkshake each. How much did this cost?

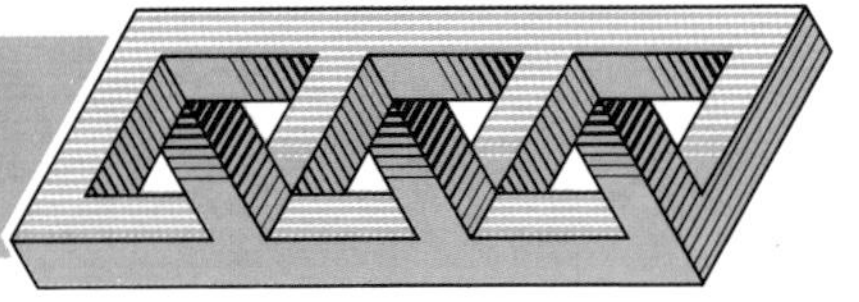

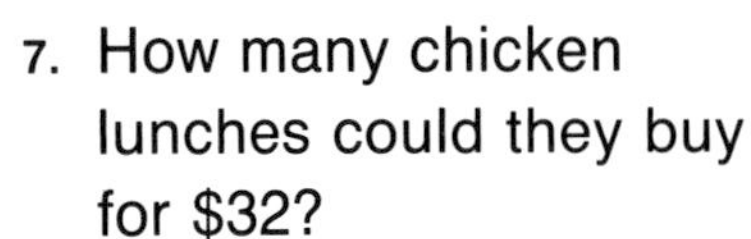

7. How many chicken lunches could they buy for $32?

8. 18 girls each bought a hamburger and a large drink. How much did this cost?

9. What is the total cost of: 7 chicken lunches, 3 cheeseburgers, 11 milkshakes and 1 hamburger?

10. Write your own problem about the picture.

PROBLEM SOLVING

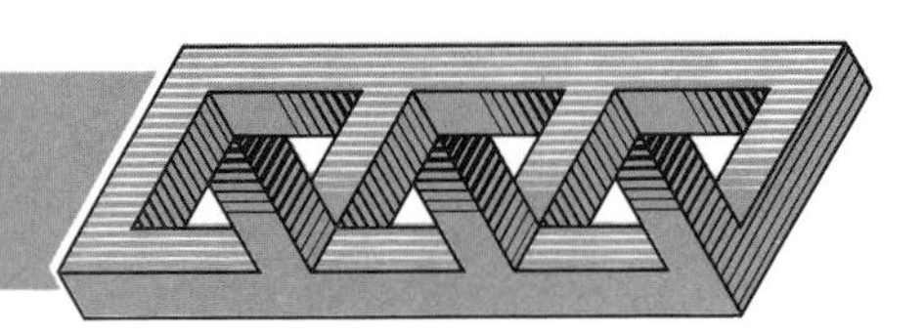

Multi-step Problems

A chessboard has 64 squares.
At the start of a game,
Kim placed 16 white pieces
on the board.
Pat placed 16 black pieces
on the board.
How many squares did not
have pieces on them?

We think: Number of pieces in all

$$\begin{array}{r} 16 \\ +16 \\ \hline 32 \end{array}$$

Number of squares without pieces

$$\begin{array}{r} 64 \\ -32 \\ \hline 32 \end{array}$$

There are 32 squares without pieces.

1. At one point in the game,
Kim had 12 pieces left.
Pat had 15 pieces left.
How many blank squares
were there then?

2. A little later,
Kim had 5 pawns and 4 other pieces.
Pat had 6 pawns and 8 other pieces.
How many more pieces did Pat
have than Kim?

3. Still later,
there were 47 blank squares.
Kim had 6 pieces on the board.
How many pieces did Pat have then?

BUYING PACKAGED GOODS

Some items are sold in packages. Often we must buy more than we need to get the number we want.

Hot dog buns come in packages of 8.

Kim needs 30 hot dogs for her party. How many packages must she buy?

We think: 3 packages of 8 buns is

$$3 \times 8 = 24$$

24 buns are not enough.

4 packages of 8 buns is

$$4 \times 8 = 32$$

32 buns are enough.

We write: Kim must buy 4 packages.

1. Hot dog buns come in packages of 8. 70 buns are needed for a wiener roast. How many packages are needed?

2. Tennis balls are sold in tin cans. Each can holds 3 balls. A tennis class needs 25 balls. How many cans must they buy?

3. Glass mirrors are made from square glass tiles like this.

   These tiles are sold in boxes of 6. How many boxes of 6 are needed to make the mirrors shown?

A.

B.

Read the sign to help you answer these questions.

1. How many geraniums are there in 7 boxes?
2. What is the cost of 7 boxes of geraniums?
3. Mary bought 40 marigolds. How many boxes did she buy?
4. Gary bought 18 petunias. How many boxes did he buy?
5. How much would it cost for 3 boxes of geraniums and 2 boxes of petunias?
6. How many petunias can be bought for $16?

7. What is the total cost of all these flowers?

8. How much more would it cost for 5 boxes of geraniums than for 5 boxes of marigolds?

9. Terri paid for 3 boxes of marigolds with a 10 dollar bill. How much change will she receive?

10. Nancy's father paid $45 for geraniums. How many geraniums did he buy?

11. Which costs more: 40 marigolds or 16 geraniums?

12. What is the total cost of 12 geraniums and 12 petunias?

13. Susan's mother bought 56 marigolds. How much did she pay?

14. How many marigolds can you buy for $27?

15. Emma bought 8 flowers for $10. What kind of flower did she buy?

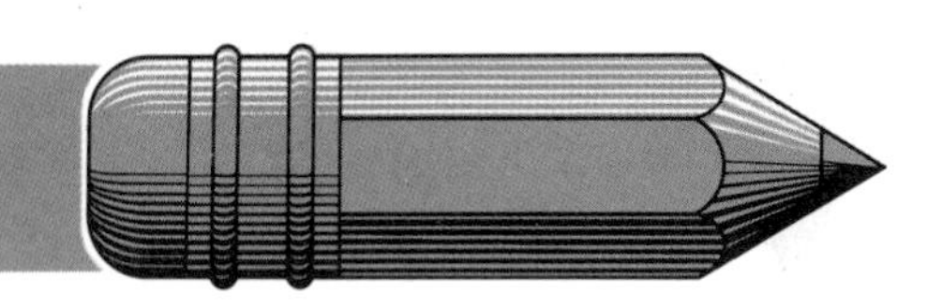

CHECK-UP

Divide.

1. $6\overline{)54}$ 2. $8\overline{)72}$ 3. $9\overline{)36}$ 4. $7\overline{)56}$

5. $6\overline{)30}$ 6. $8\overline{)64}$ 7. $9\overline{)45}$ 8. $7\overline{)49}$

9. 24 muffins are shared by 8 people. How many muffins are there for each?

10. 35 trees are planted in rows of 7. How many rows are there?

11. Divide 13 crayons among 6 children. How many are there for each child? How many are left over?

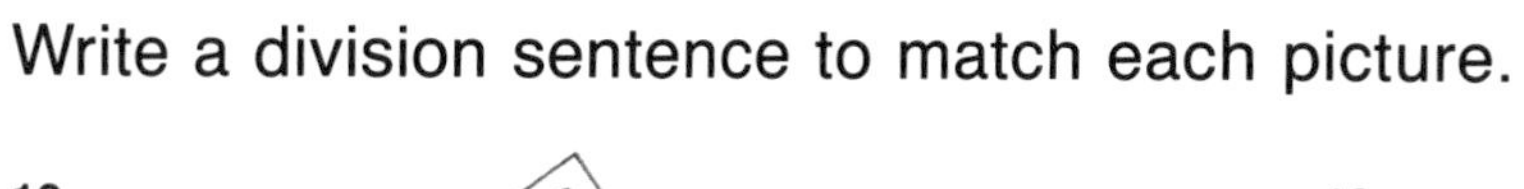

Write a division sentence to match each picture.

12.

13.

Divide.

14. $2\overline{)11}$ 15. $5\overline{)43}$ 16. $4\overline{)25}$ 17. $4\overline{)37}$

FRACTIONS AND DECIMALS

Inuit Chant

There is joy in
Feeling the warmth
Come to the great world
And seeing the sun
Follow its old footprints
In the summer night.

There is fear in
Feeling the cold
Come to the great world
And seeing the moon
— Now new moon, now full moon —
Follow its old footprints
In the winter night.

—translated by Knud Rasmussen

Equal Parts

We can fold paper different ways to show equal parts.

Both of these show 4 equal parts.

Take 4 pieces of paper. Fold each a different way to make 8 equal parts.

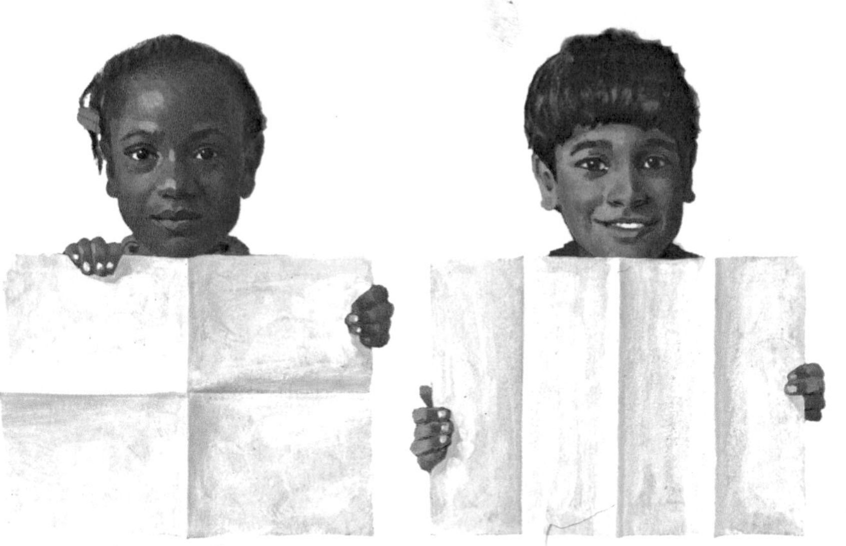

Which figures are divided into equal parts?

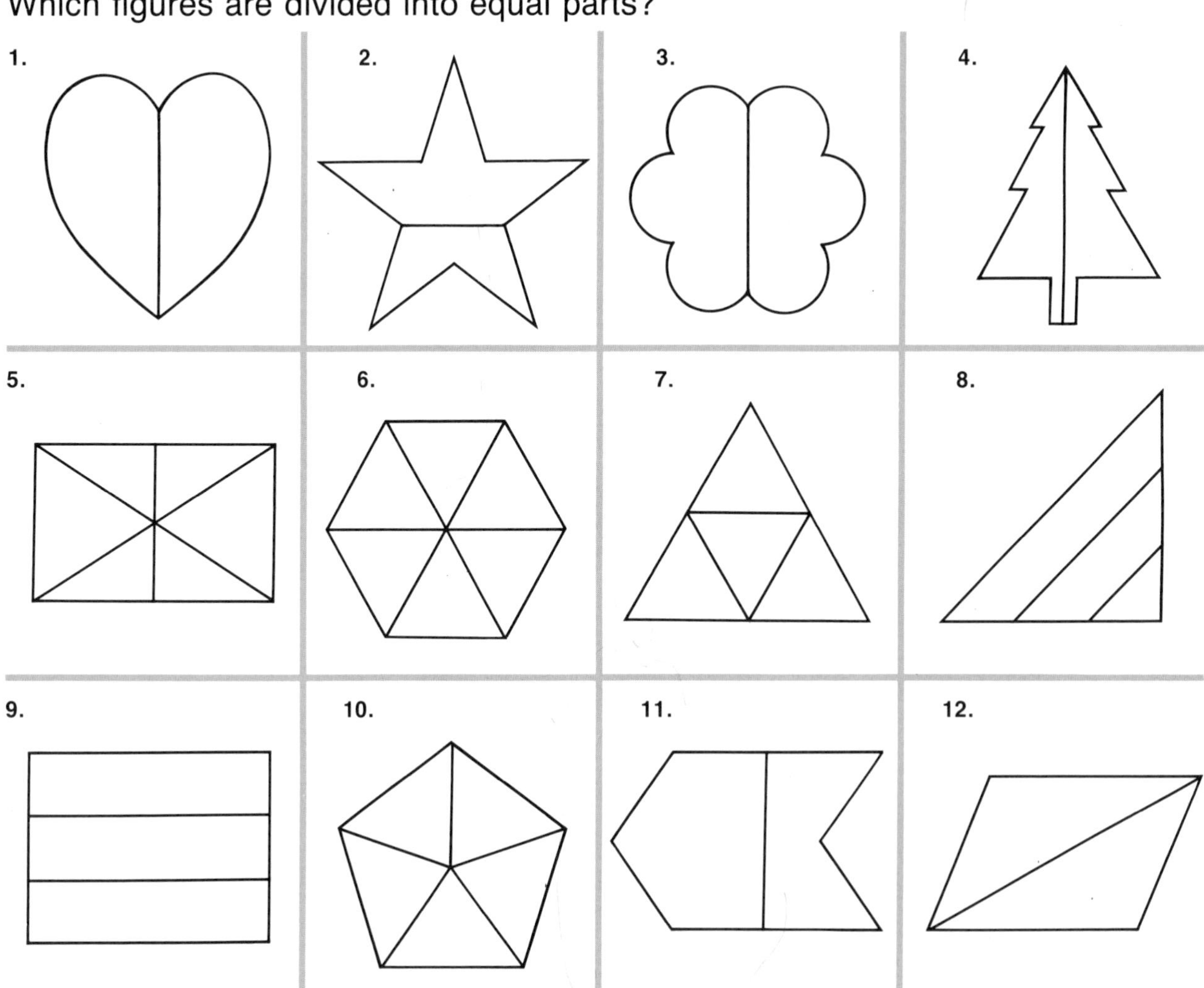

Writing Fractions

We see:	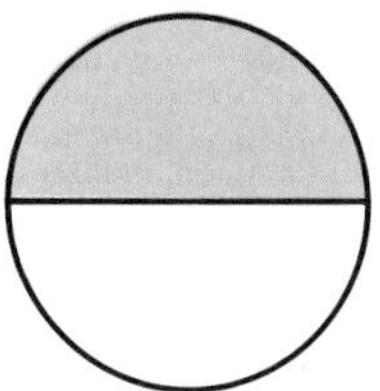	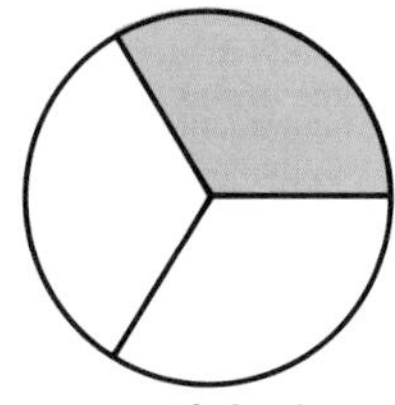	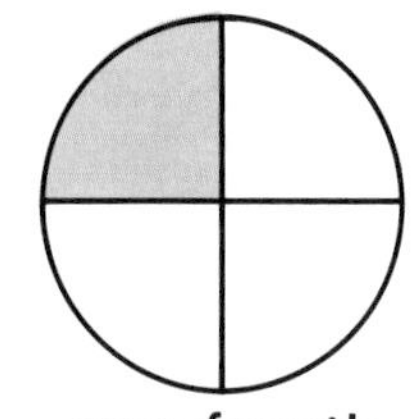	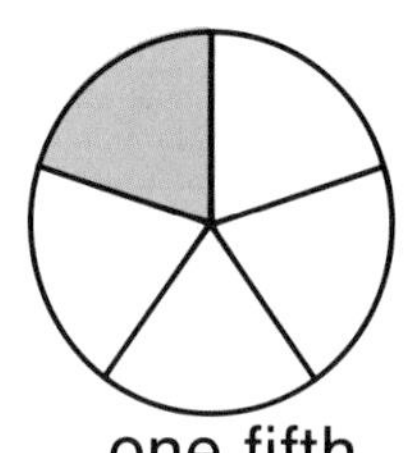
We say:	one half	one third	one fourth or one quarter	one fifth
We write:	$\frac{1}{2}$	$\frac{1}{3}$	$\frac{1}{4}$	$\frac{1}{5}$

Write a fraction for the coloured part of each figure.

1. 

2. 

3. 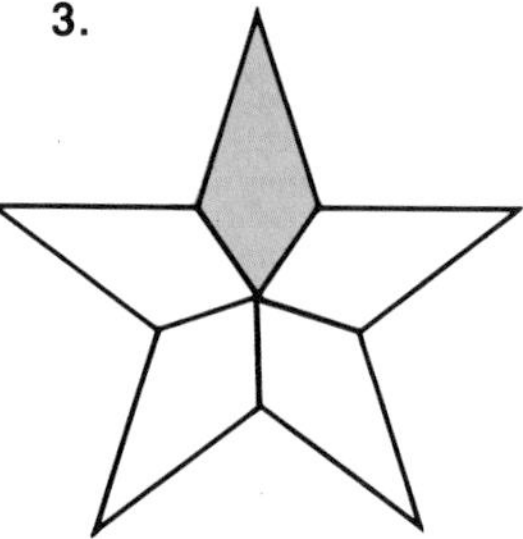

4. 

5. 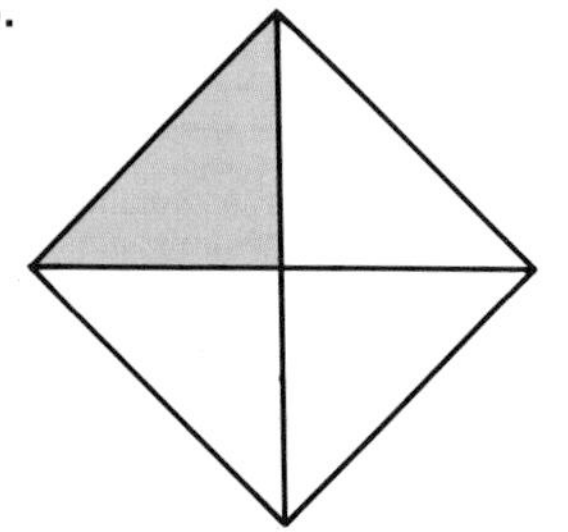

6. 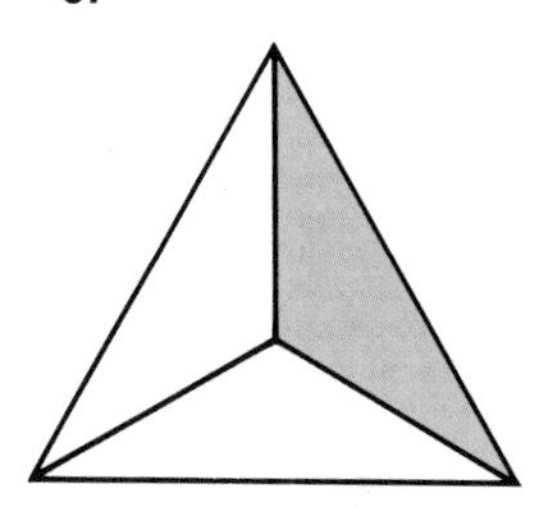

7. 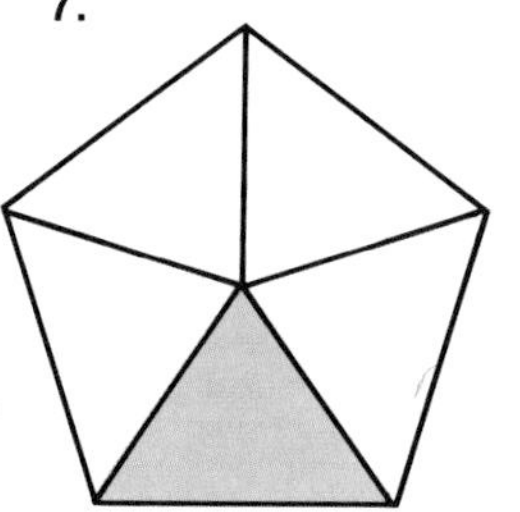

8. 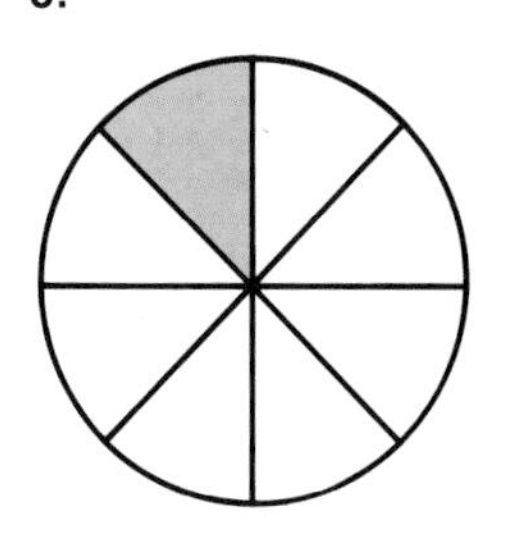

9. 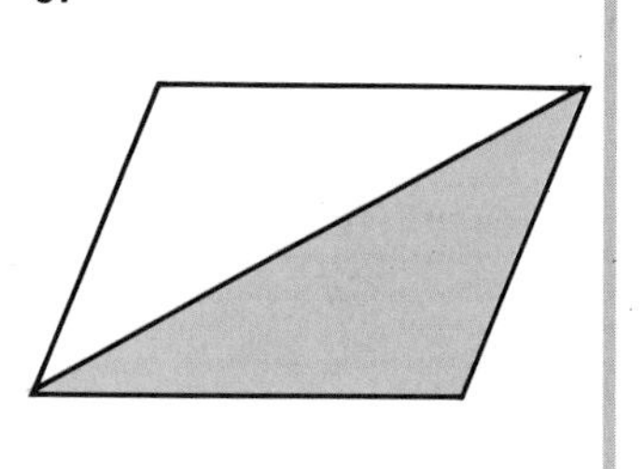

10. 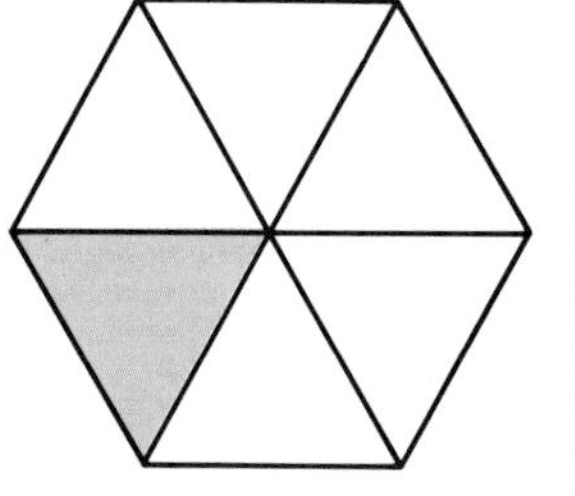

11. 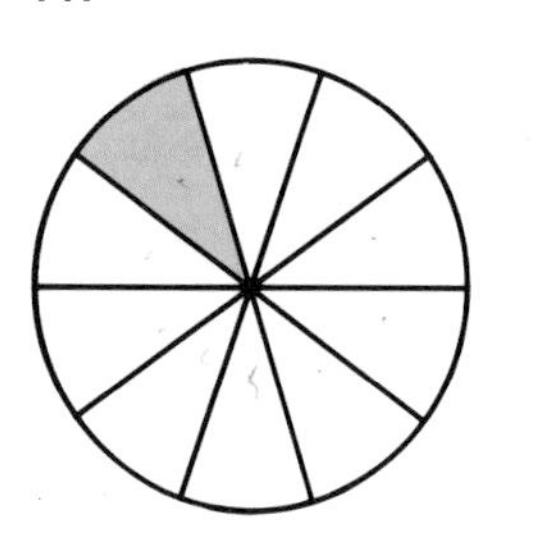

12. 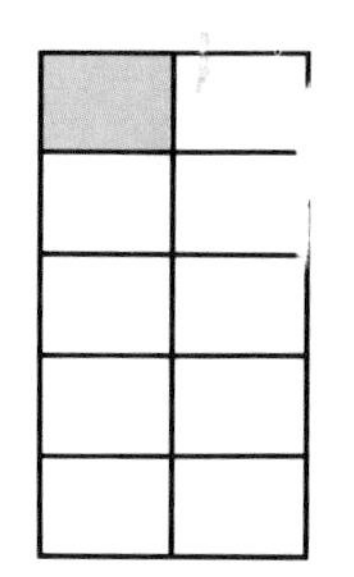

Writing Fractions

We see: 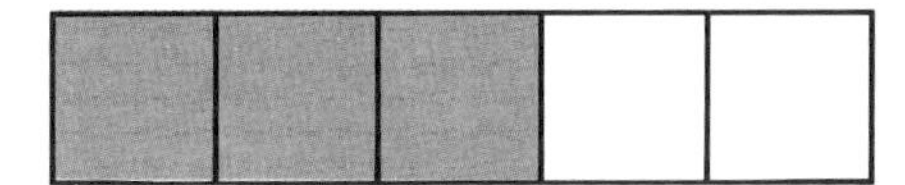

We say: 3 fifths are coloured.

We write: $\frac{3}{5}$ ← number of parts coloured / ← number of parts in all

What fraction of each figure is coloured?

1. 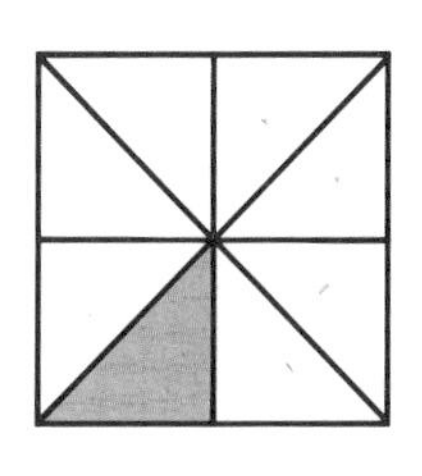

2. 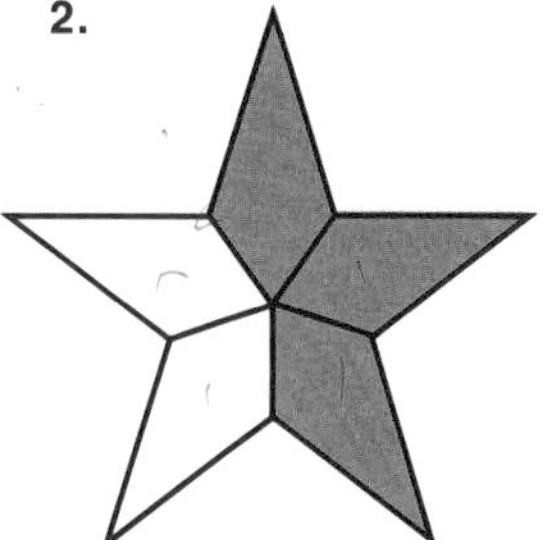

3. 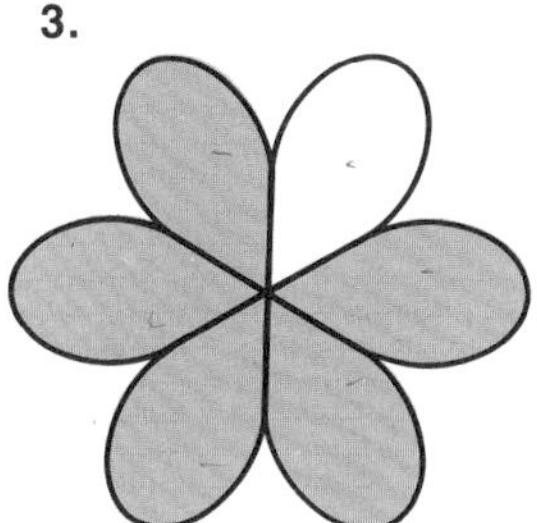

4. 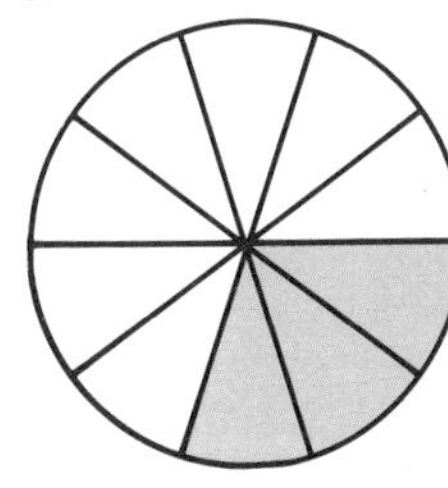

5. 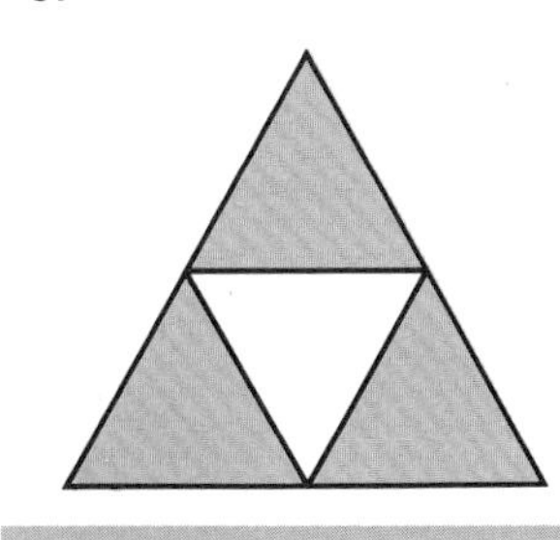

6. 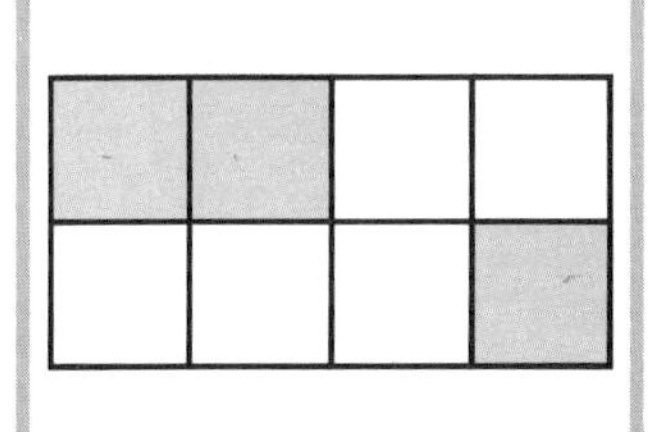

7. 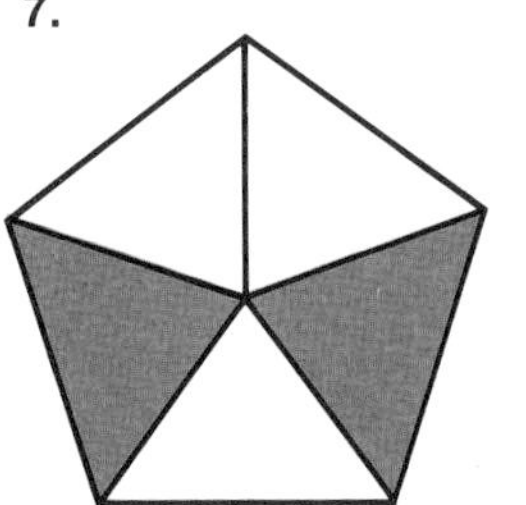

8. 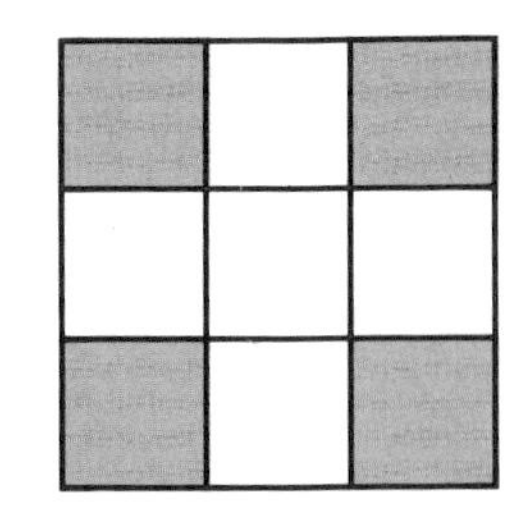

Cut out these figures from squared paper.
Colour the fraction named.

9. $\frac{5}{8}$ 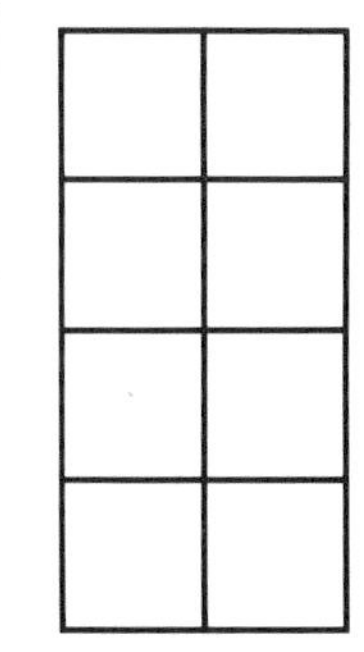

10. $\frac{2}{9}$ 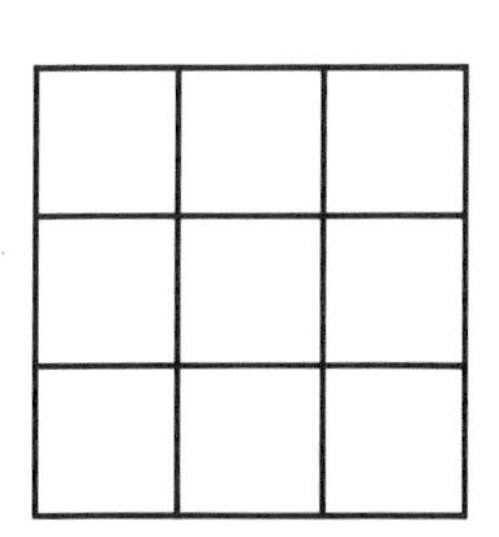

11. $\frac{3}{10}$ 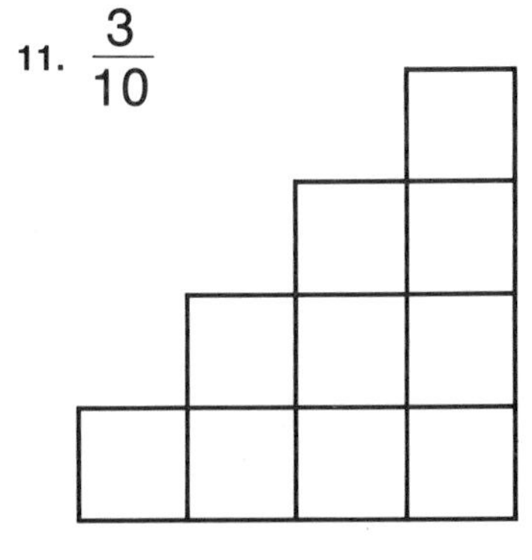

12. Use squared paper.
Cut out a figure.
Colour part of it.
Name your fraction.

Comparing Fractions

Which fraction is greater?

$\frac{2}{3}$ or $\frac{1}{4}$

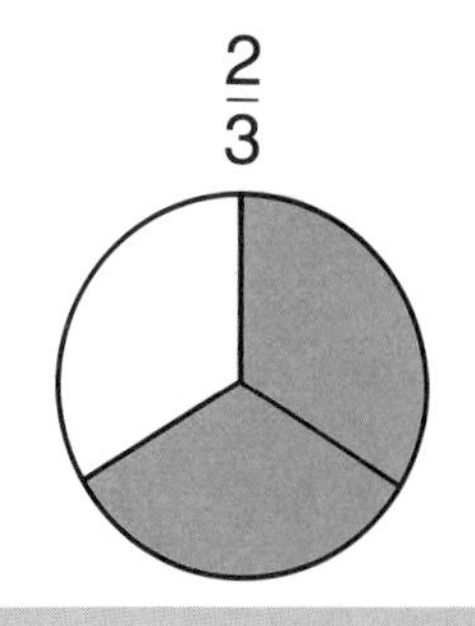

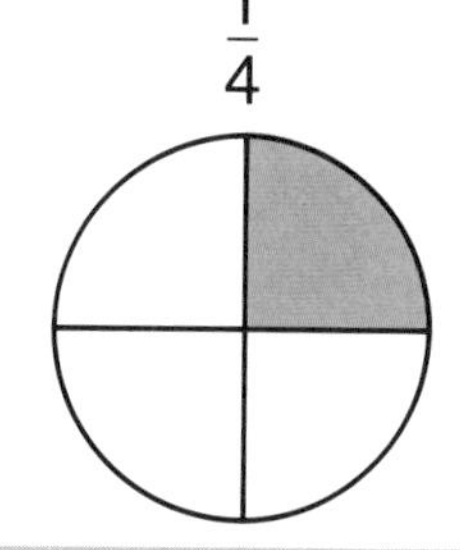

By putting one on top of the other, we see:

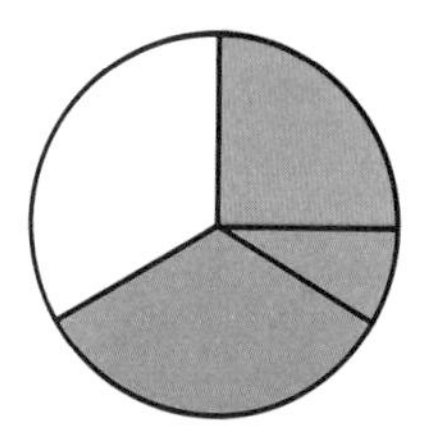

$\frac{2}{3}$ is greater than $\frac{1}{4}$

Which is greater?

1. 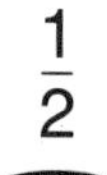 $\frac{1}{2}$ or $\frac{3}{4}$

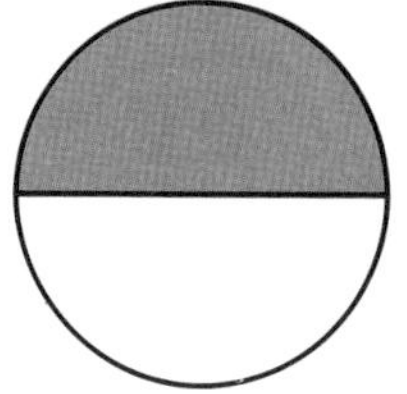

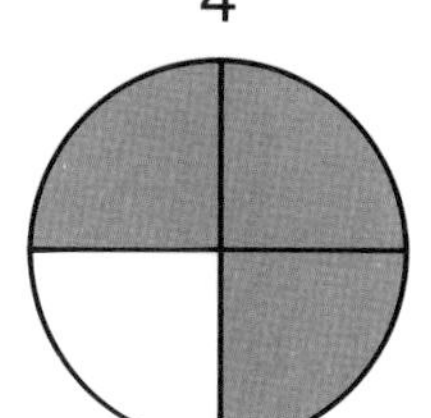

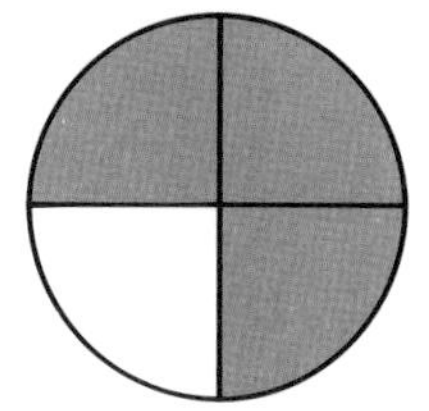

2. $\frac{3}{5}$ or $\frac{2}{5}$

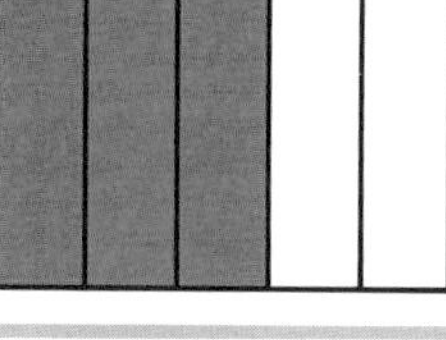

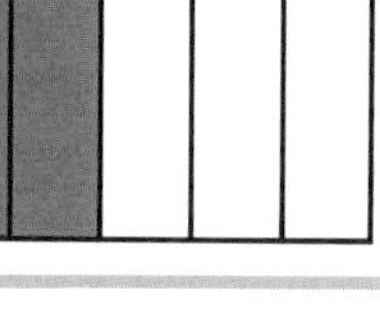

3.  $\frac{2}{3}$ or $\frac{5}{6}$

4. $\frac{1}{2}$ or 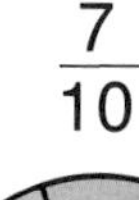 $\frac{7}{10}$

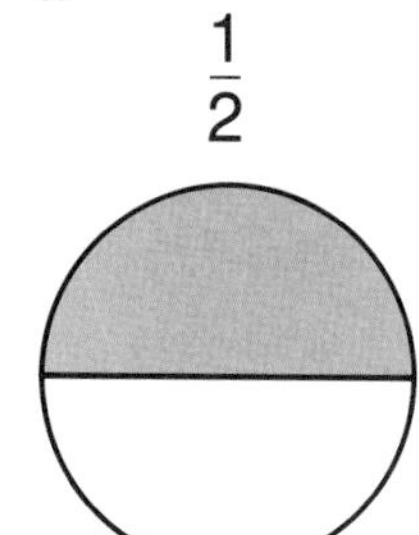

5. $\frac{1}{4}$ or $\frac{1}{5}$

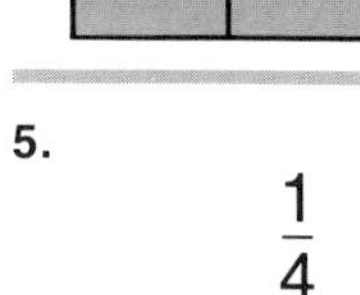

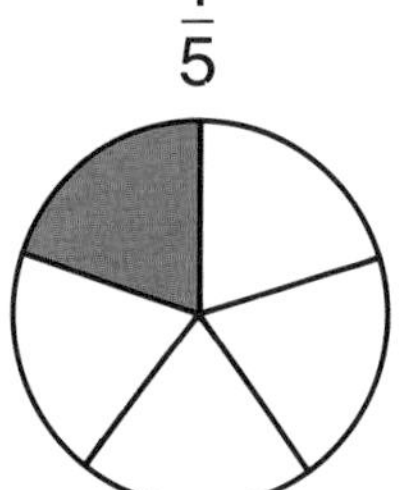

6. $\frac{3}{4}$ or 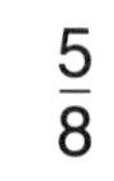 $\frac{5}{8}$

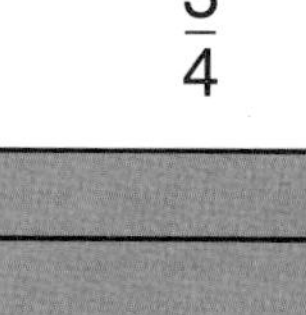

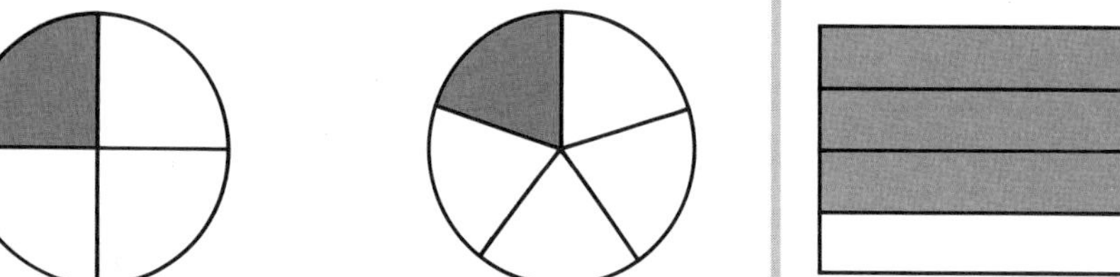

Tenths

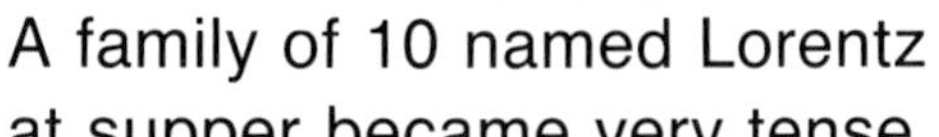

A family of 10 named Lorentz
at supper became very tense.

"Unfair" they would cry
when they shared pizza pie,

'Til the pie was divided in tenths.

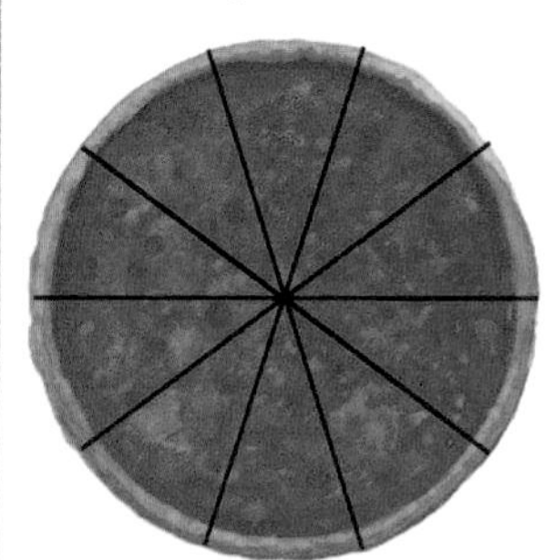

Each piece is $\frac{1}{10}$ of the pie.

We can write 0.1 for one tenth

We see:	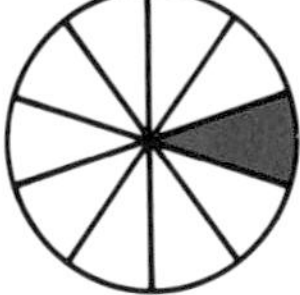		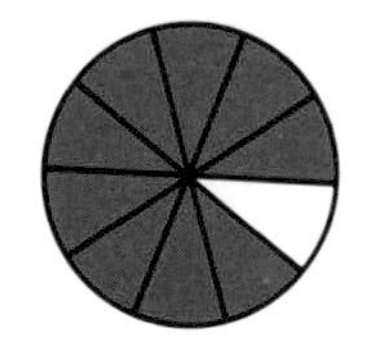	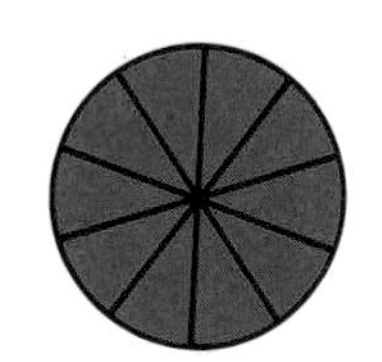
We say:	one tenth	eight tenths	nine tenths	ten tenths or one
We write:	$\frac{1}{10}$ or 0.1	$\frac{8}{10}$ or 0.8	$\frac{9}{10}$ or 0.9	$\frac{10}{10}$ or 1

Write a fraction and a decimal number for the coloured part of each figure.

1. 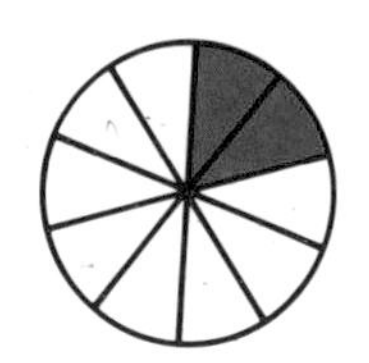
2. 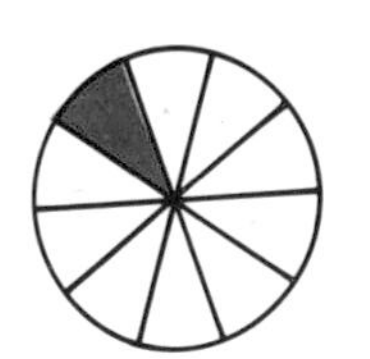
3. 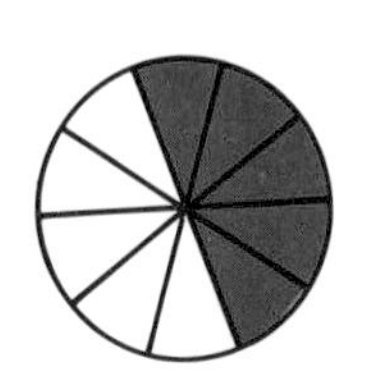
4. 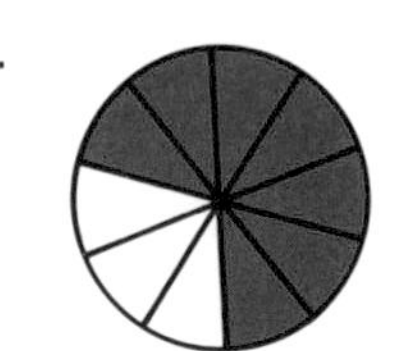
5. 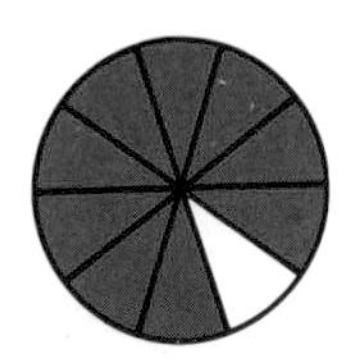

Write a fraction and a decimal number for the coloured part of each figure.

1.

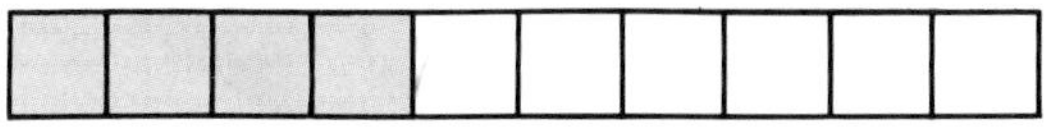

2.

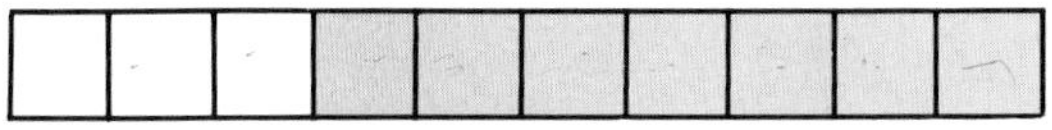

3.

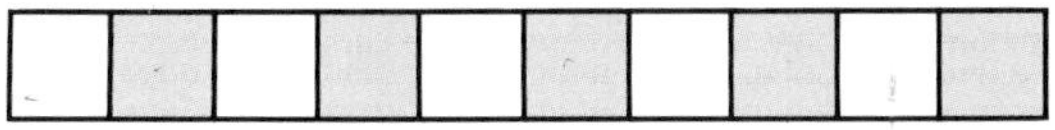

4.

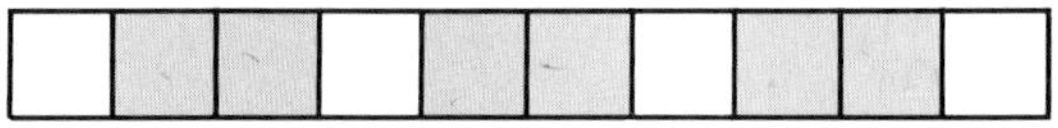

5.

6.

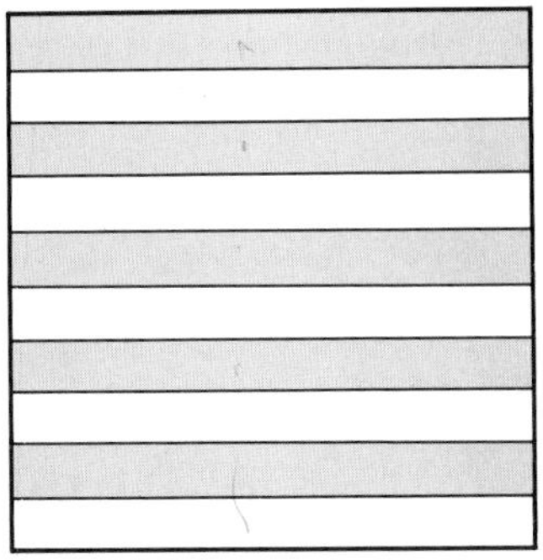

7.

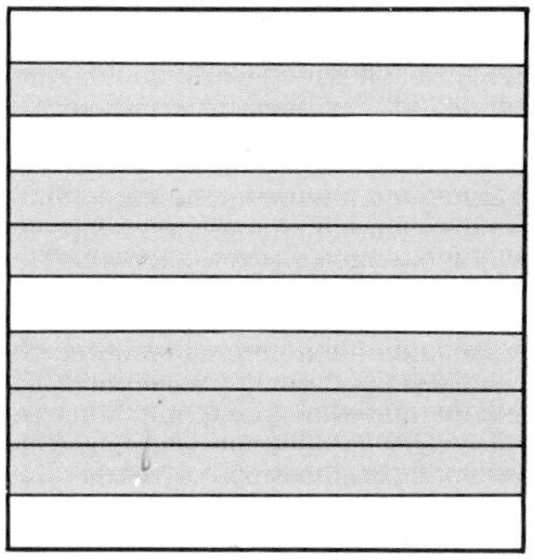

8.

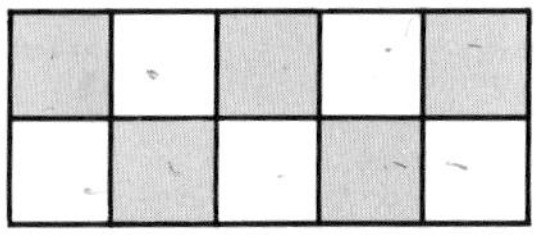

9.

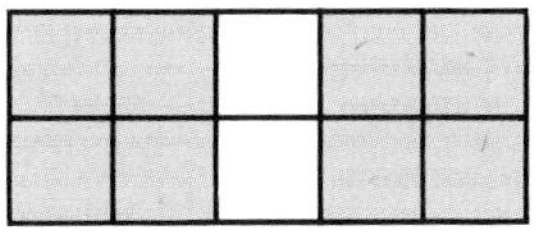

10.

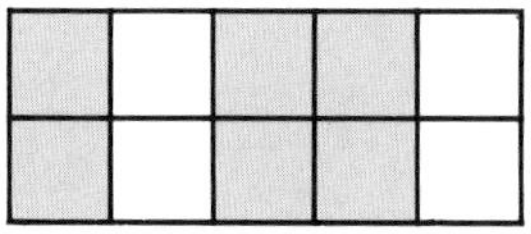

Use strips of paper divided into tenths.
Colour them to show each fraction.

11. 0.1

12. 0.6

13. 0.9

14. three tenths

15. seven tenths

16. ten tenths

Ones and Tenths

We see:

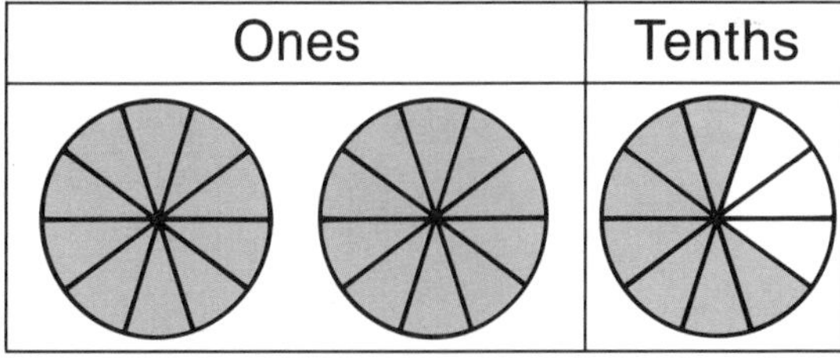

We say: two and seven tenths

We write: 2.7

Write a decimal number to tell how much is coloured.

1.

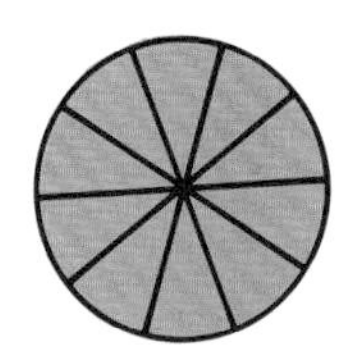 

2.

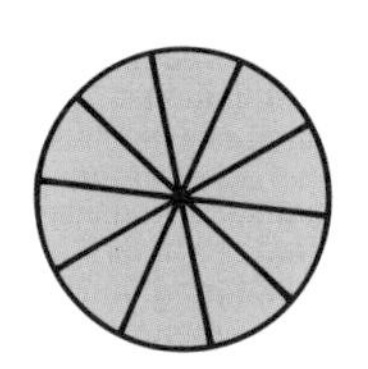  

3.

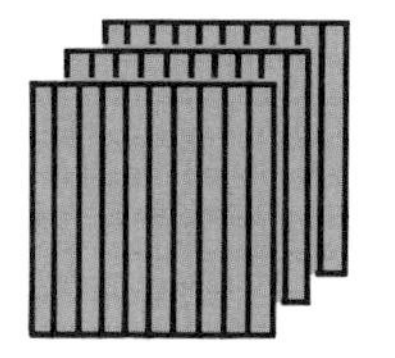 

4.

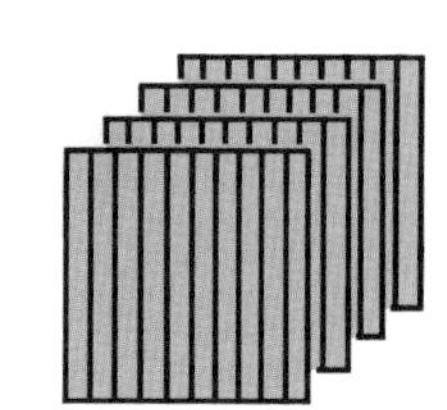 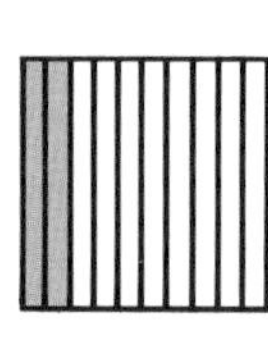

5.

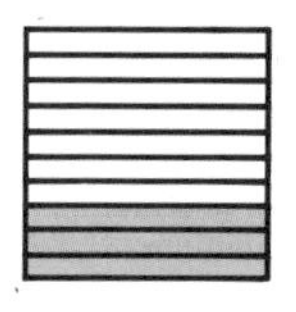 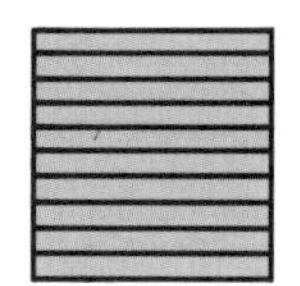

6.

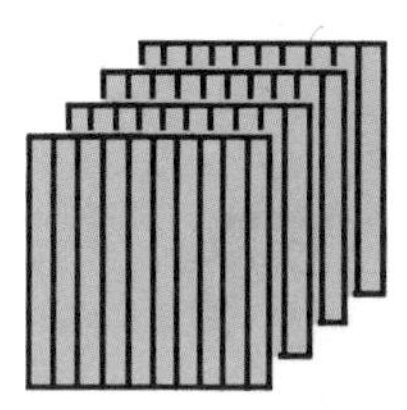

7.

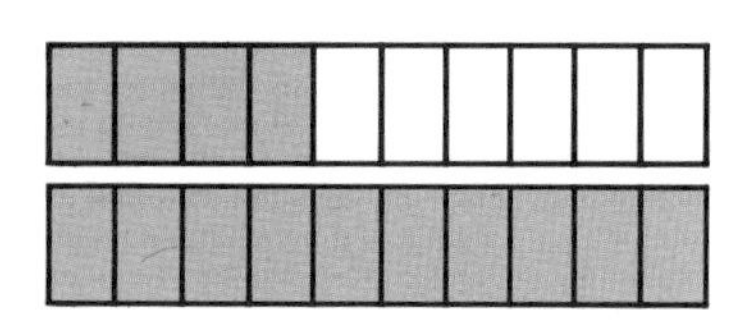

8.

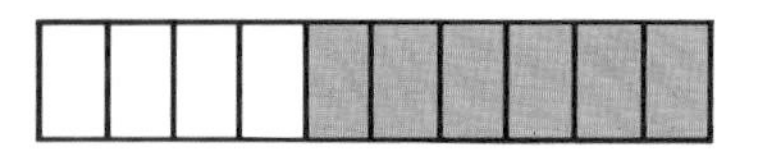

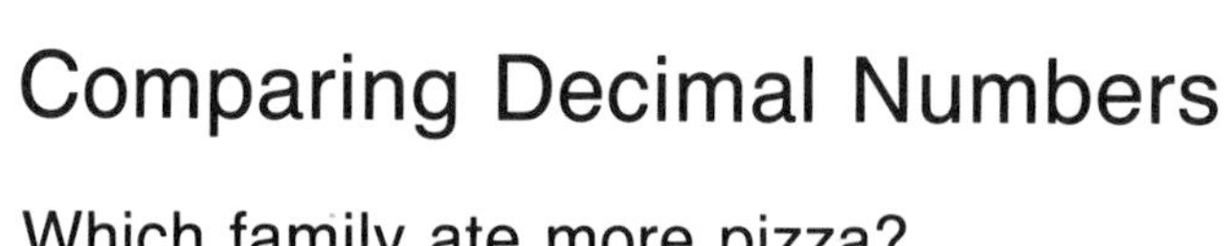

Comparing Decimal Numbers

Which family ate more pizza?

We think: 2.9 is two and nine tenths — 3.1 is three and one tenth

Ones	Tenths

Ones	Tenths

We say: 3.1 is greater than 2.9

We write: $3.1 > 2.9$

The Howard family ate more pizza.

Write the greater number.
Use materials if you wish.

1. 1.6, 1.4
2. 2.3, 2.7
3. 0.8, 0.7
4. 1.2, 0.9
5. 3.4, 4.3
6. 5.8, 3.9
7. 0.9, 1.8
8. 12.7, 13.0

Adding Decimal Numbers

How much pie is there in all?

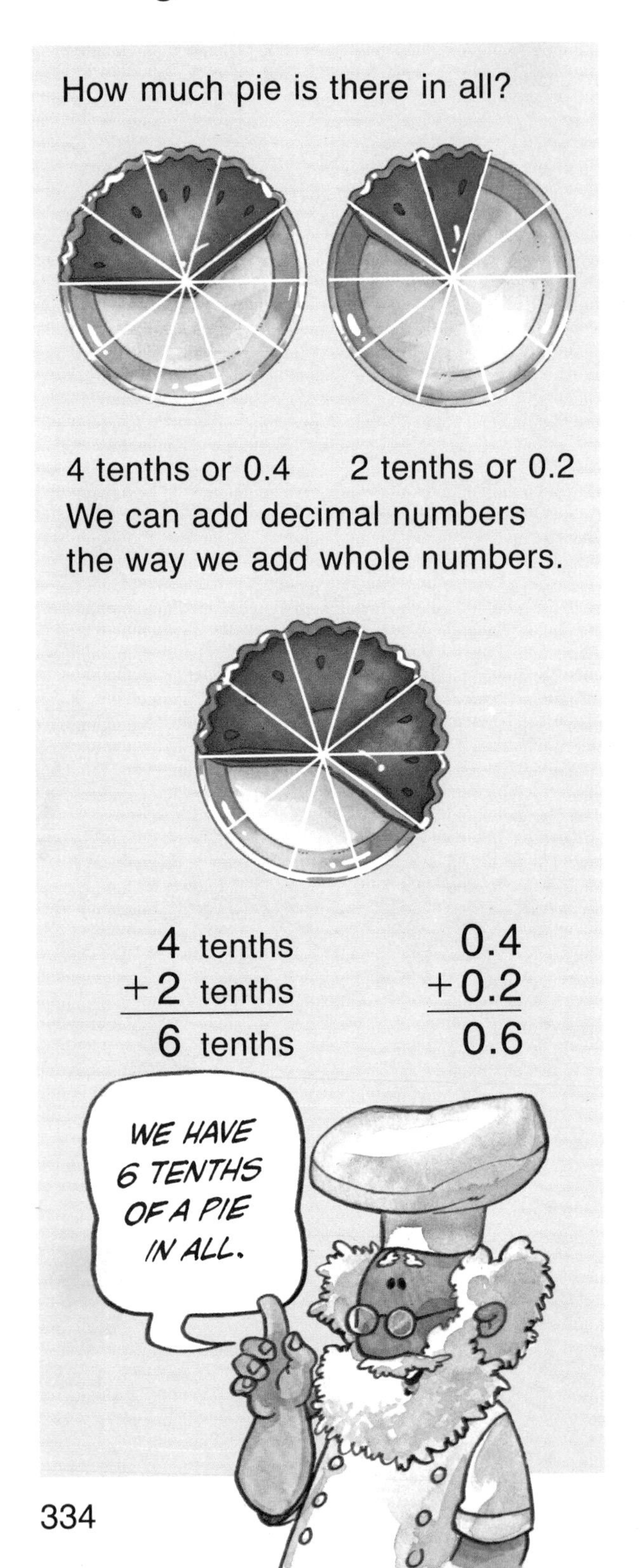

4 tenths or 0.4   2 tenths or 0.2

We can add decimal numbers the way we add whole numbers.

4 tenths	0.4
+2 tenths	+0.2
6 tenths	0.6

Write an addition sentence to match each picture.

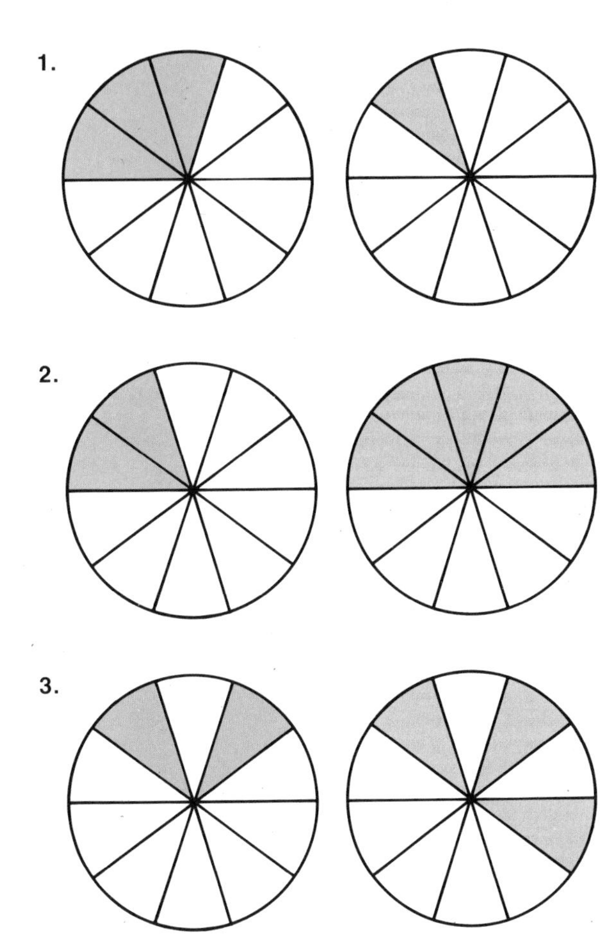

1.

2.

3.

Add.
Use materials if you wish.

4. $0.1 + 0.3$

5. $0.5 + 0.2$

6. $0.6 + 0.3$

7. $0.4 + 0.4$

8. $0.2 + 0.5$

9. $0.3 + 0.8$

Adding Decimal Numbers

How many are there in all?

1.4 red pies

2.8 blue pies

We can add decimal numbers the way we add whole numbers.

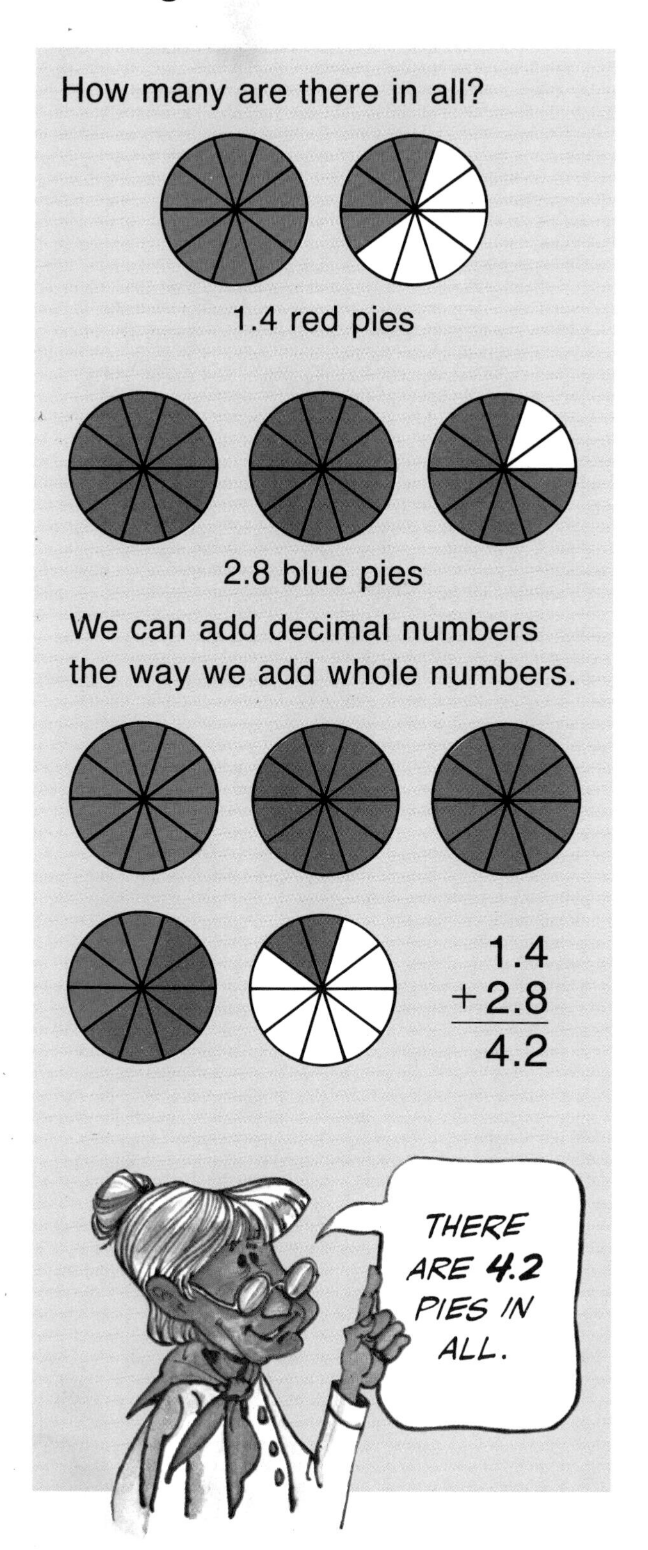

$$\begin{array}{r} 1.4 \\ +2.8 \\ \hline 4.2 \end{array}$$

Write an addition sentence to match each picture.

1.

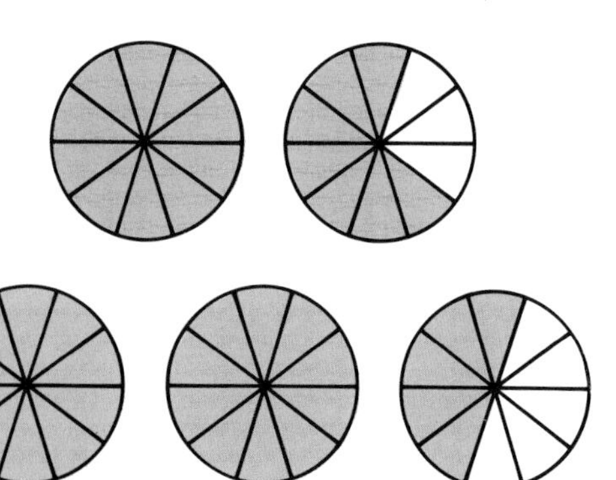

2.

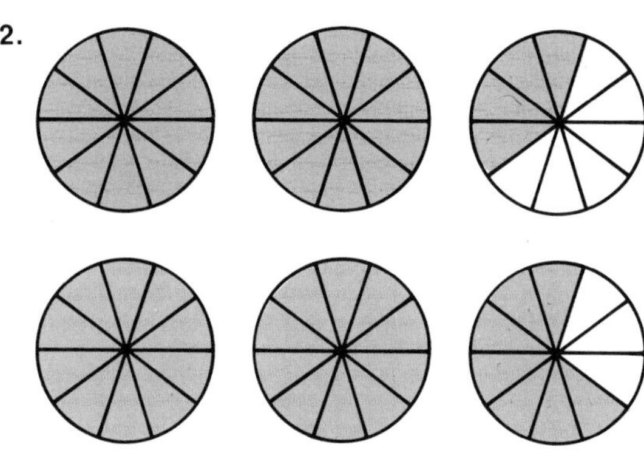

Use materials to add.

3. $\begin{array}{r} 1.3 \\ +2.2 \\ \hline \end{array}$

4. $\begin{array}{r} 3.1 \\ +1.8 \\ \hline \end{array}$

5. $\begin{array}{r} 2.6 \\ +0.5 \\ \hline \end{array}$

6. $\begin{array}{r} 3.7 \\ +1.4 \\ \hline \end{array}$

7. $\begin{array}{r} 1.6 \\ +2.8 \\ \hline \end{array}$

8. $\begin{array}{r} 0.6 \\ +0.9 \\ \hline \end{array}$

Subtracting Decimal Numbers

Sam saw 6 tenths of a pie.
He ate 2 tenths.
How much was left?

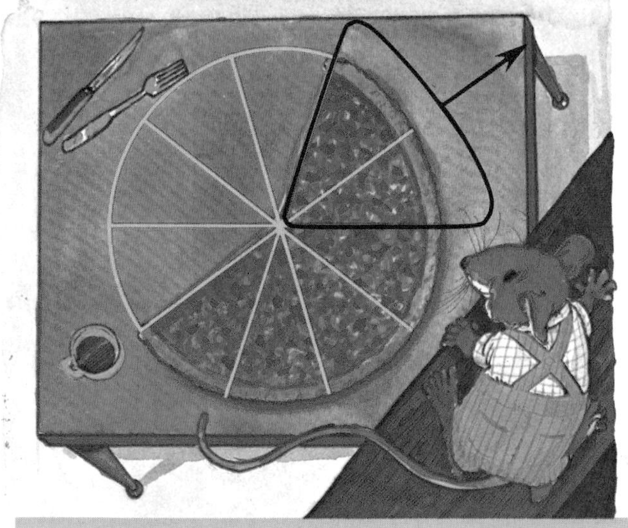

We can subtract decimal numbers the way we subtract whole numbers.

$$\begin{array}{r} 6 \text{ tenths} \\ -\,2 \text{ tenths} \\ \hline 4 \text{ tenths} \end{array} \qquad \begin{array}{r} 0.6 \\ -\,0.2 \\ \hline 0.4 \end{array}$$

4 tenths of a pie was left.

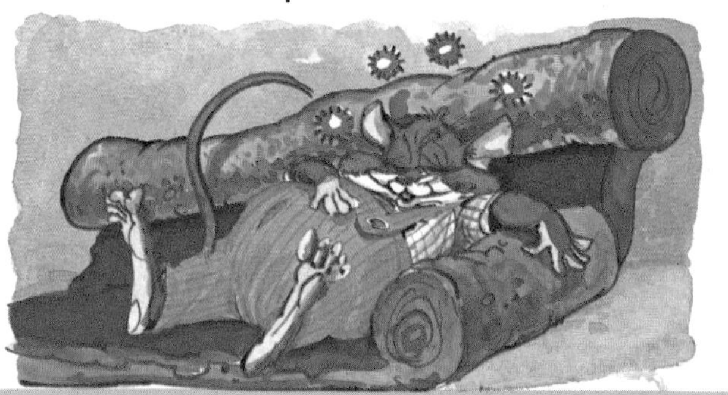

Write a subtraction sentence to match each picture.

1\.

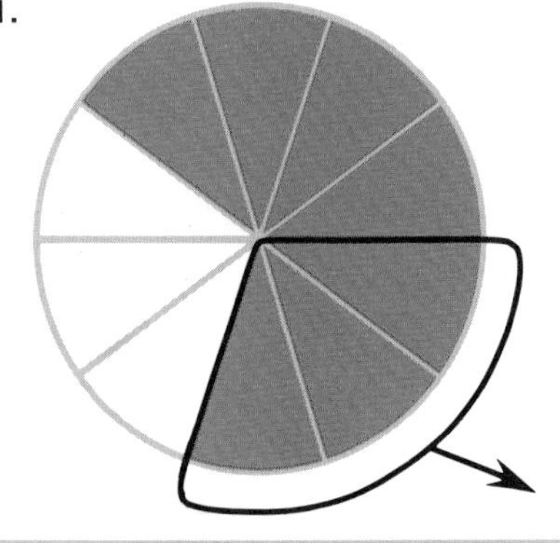

2\.

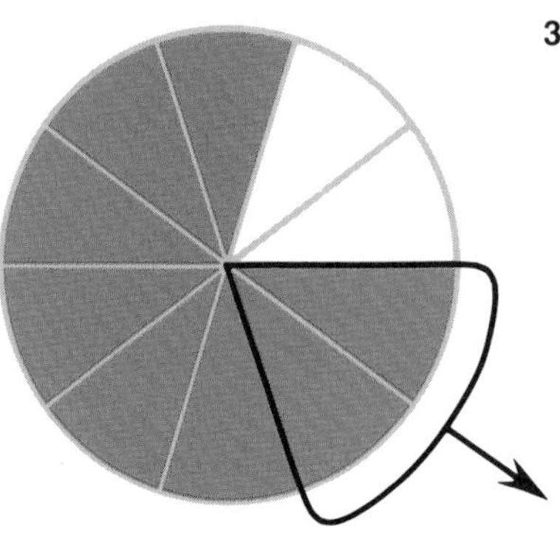

3\.

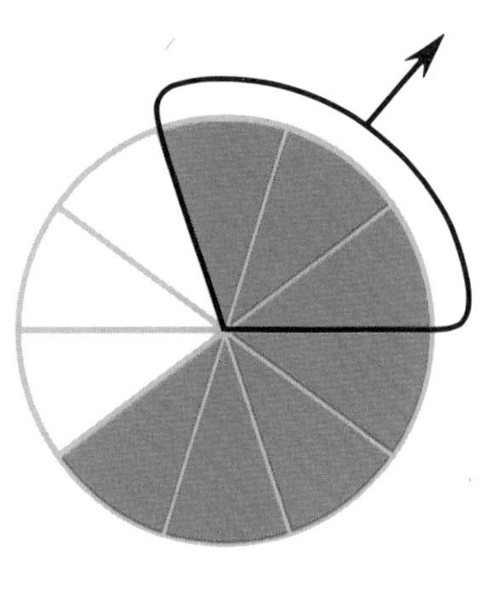

Subtract.
Use materials if you wish.

4\. $\begin{array}{r} 0.8 \\ -\,0.2 \\ \hline \end{array}$ 5. $\begin{array}{r} 0.3 \\ -\,0.1 \\ \hline \end{array}$ 6. $\begin{array}{r} 0.5 \\ -\,0.4 \\ \hline \end{array}$ 7. $\begin{array}{r} 0.9 \\ -\,0.4 \\ \hline \end{array}$ 8. $\begin{array}{r} 0.7 \\ -\,0.6 \\ \hline \end{array}$ 9. $\begin{array}{r} 0.8 \\ -\,0.5 \\ \hline \end{array}$

Subtracting Decimal Numbers

There were 3.2 pies.
Lynne took away 1.7 pies.
How much was left?

We can subtract decimal numbers the way we subtract whole numbers.

$$\begin{array}{r} 3.2 \\ -1.7 \\ \hline 1.5 \end{array}$$

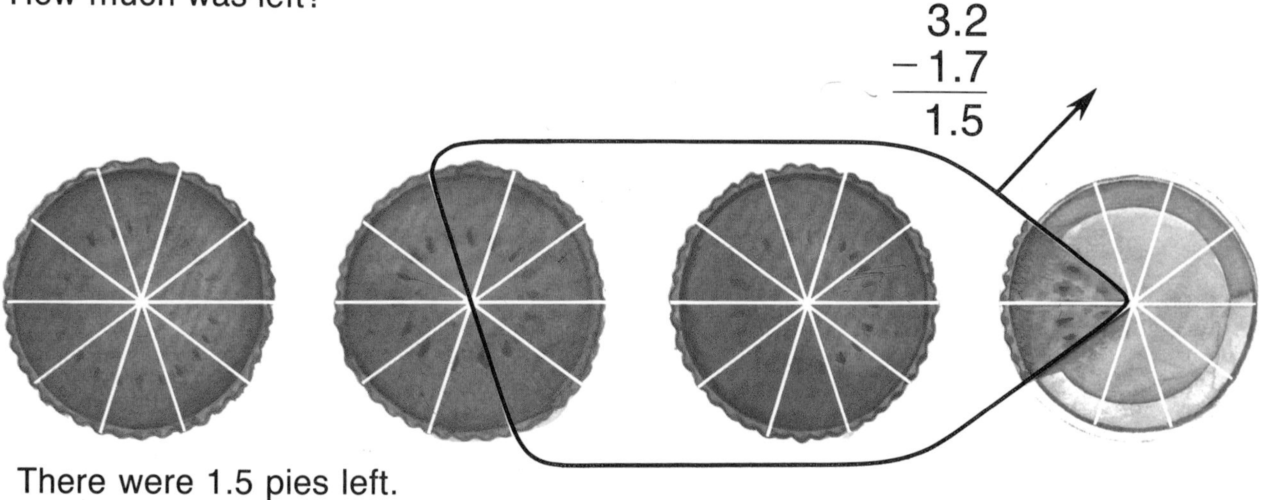

There were 1.5 pies left.

Write a subtraction sentence to match each picture.

1.

2.

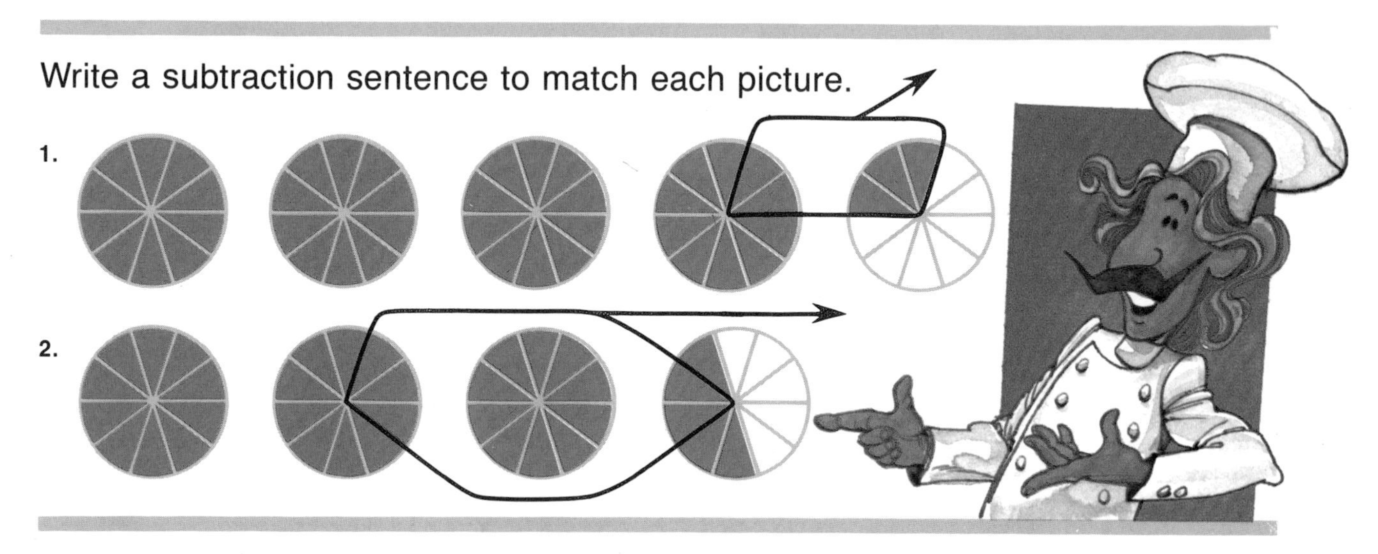

Use materials to subtract.

3. $\begin{array}{r} 3.5 \\ -1.6 \\ \hline \end{array}$

4. $\begin{array}{r} 1.5 \\ -0.9 \\ \hline \end{array}$

5. $\begin{array}{r} 2.4 \\ -1.6 \\ \hline \end{array}$

6. $\begin{array}{r} 4.2 \\ -2.4 \\ \hline \end{array}$

7. $\begin{array}{r} 3.2 \\ -2.6 \\ \hline \end{array}$

8. $\begin{array}{r} 2.3 \\ -1.9 \\ \hline \end{array}$

DOLLAR$ and ¢ENTS

Dollars and Cents

We use decimal numbers to write dollars and cents.

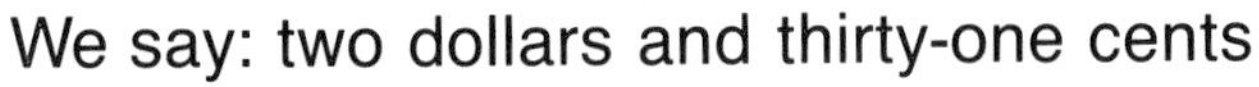

We say: two dollars and thirty-one cents

We write: $2.31

How much money is shown?

1\.

2\.

3\.

4\.

How much money is shown?

Write these amounts using decimal numbers.

5. Four dollars and seventy-five cents
6. Three dollars and forty-nine cents
7. Six dollars and nine cents
8. Eight dollars and eighty cents
9. Seventy-four cents
10. Ninety-nine cents
11. One dollar and sixty-eight cents
12. One dollar and seven cents

Copy and complete.

	Dollars	Dimes	Pennies	Amount
	3	7	8	$3.78
13.	5	8	2	
14.	3	0	9	
15.	0	0	7	
16.	5		2	$5.72

ESTIMATING

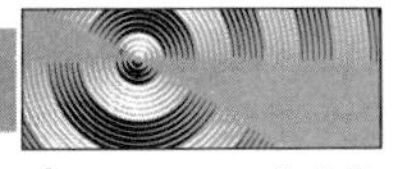

Estimate: How many pennies would it take to cover your MathQuest 3 book?

Adding Dollars and Cents

Monster Mash costs $7.93.
The batteries cost $1.69.
How much do they cost together?

Mario adds cents only.

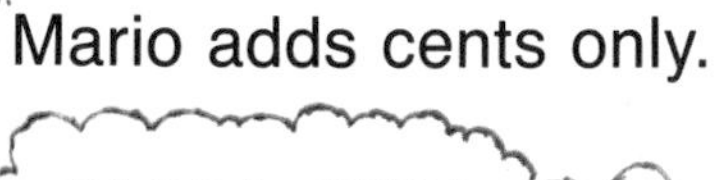

962¢ is $9.62

```
  793¢
+ 169¢
  962¢
```

Leah adds dollars and cents.

```
  $7.93
+  1.69
  $9.62
```

The total cost is $9.62.

Add.

1. $1.26 + 2.43
2. $3.07 + 1.31
3. $2.52 + 6.35
4. $5.21 + 3.69
5. $4.62 + 0.21
6. $6.28 + 1.03
7. $7.25 + 1.25
8. $2.81 + 0.17
9. $2.81 + 5.83
10. $6.94 + 0.03
11. $5.42 + 1.71
12. $8.48 + 0.61
13. $7.47 + 1.65
14. $6.72 + 1.18
15. $3.76 + 4.34

Subtracting Dollars and Cents

The calculator costs $7.98.
A customer gave
the cashier $9.00.
How much change
should he get?

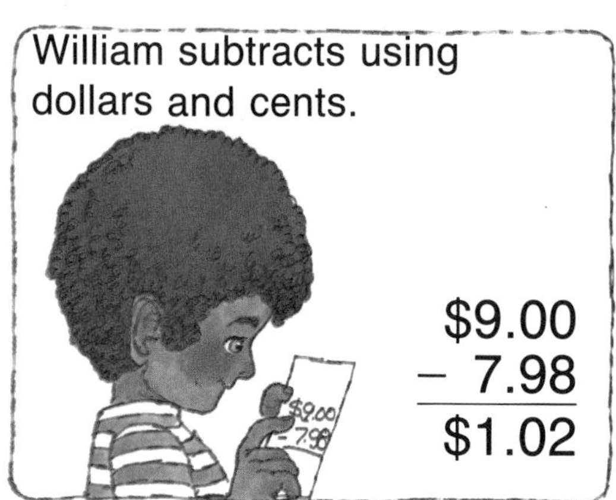

The customer gets $1.02.

Subtract.

1. $6.78 − 3.26	2. $9.47 − 5.26	3. $5.93 − 2.02	4. $7.87 − 3.51	5. $9.26 − 7.17
6. $7.47 − 0.85	7. $5.25 − 3.83	8. $6.73 − 4.82	9. $9.27 − 6.39	10. $8.48 − 3.29
11. $8.80 − 5.20	12. $7.30 − 4.60	13. $8.00 − 4.29	14. $7.00 − 3.88	15. $9.00 − 4.63

Catalog
Shopping

nifty
notes
$1.30
Nifty Note Pad
50 sheets

Invisible
INK
$1.75
For your secret
messages.

Box of
24
COLOURED PENCILS
$4.50

ERASERS
Any 3 for $1.50

Ye Olde Pen
Use with invisible ink
or regular ink.
$2.20

Space Pen
$1.60
Writes 5 different
colours!

Gold
Writer
Only
$3.98
Write
glittery letters!

Copy and complete the bills.

1.

1 Gold Writer:	▒
1 Nifty Note Pad	▒
Total:	▒

2.

1 Ye Olde Pen:	▒
1 Invisible Ink:	▒
Total:	▒

3.

1 Box Coloured Pencils:	▒
3 Erasers:	▒
Total:	▒

4. Jenny bought a Gold Writer pen. She gave the clerk $5.00. What change did she get?

5. Bill spent exactly $3.05. What 2 things did he buy?

6. What was Bill's change from $10.00?

7. How much more does the Gold Writer cost than the Space Pen?

8. How much would 6 erasers cost?

PROBLEM SOLVING

You have $10.00.
What items would you buy?
Find the total cost.

BITS AND BYTES

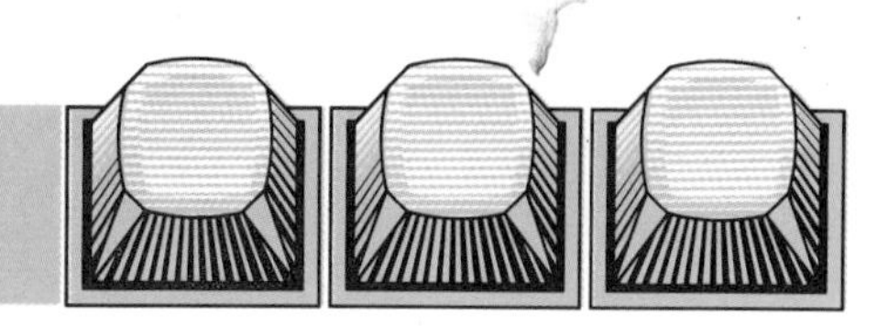

Calculators in stores add dollars and cents quickly.

To find the total for this bill on a calculator, press these keys in order.

3 . 8 7 + 4 . 2 9 =

Your answer should be 8.16.

Use your calculator to find the total for each bill.

1.	2.	3.	4.	5.
$3.97	$4.83	$7.16	$4.36	$4.82
$2.53	$3.72	$4.19	$2.37	$5.33
$6.47	$4.28	$2.81	$3.63	$3.56
		$2.22	$1.91	$1.44

To find the change for $8.16 from $10, press these keys.

1 0 . 0 0 − 8 . 1 6 =

The change is $1.84.

Use your calculator to calculate the change.

6. Money given: $8.00
   Total cost: $7.56

7. Money given: $10.00
   Total cost: $ 7.23

8. Money given: $10.03
   Total cost: $ 9.78

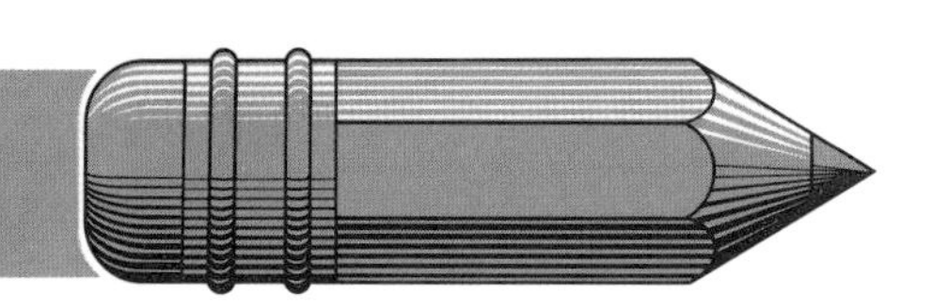

CHECK-UP

Write a fraction for the coloured part of each figure.

1. 

2. 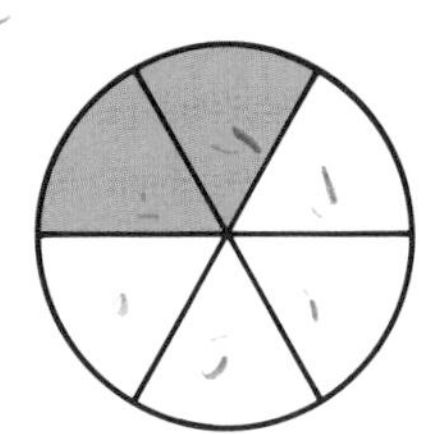

3. 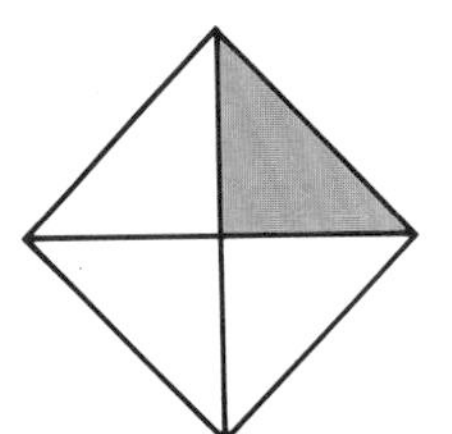

Write a fraction and a decimal number for the coloured part of each picture.

4. 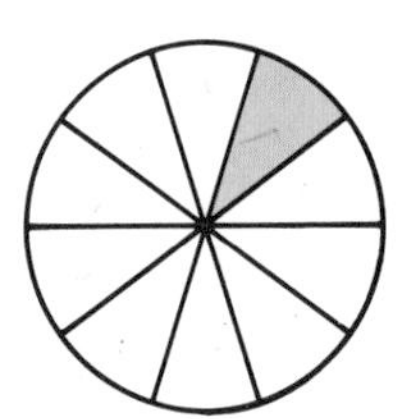

5. 

6. 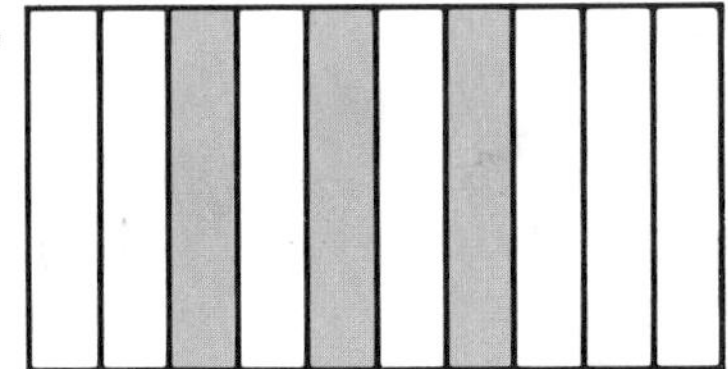

Write a decimal number to tell how much is coloured.

7. 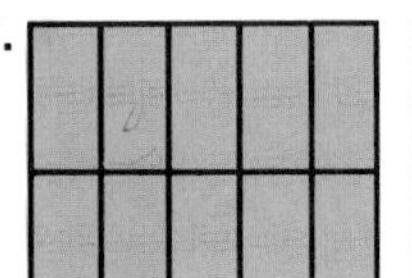

8. 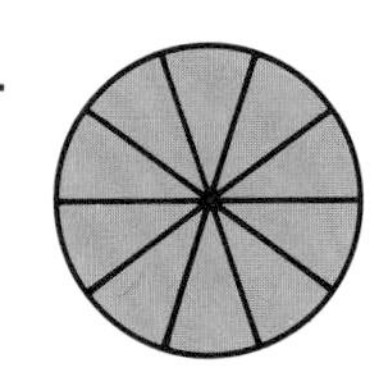

9. 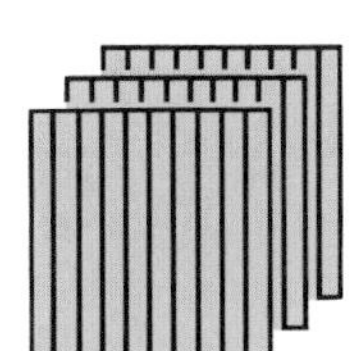

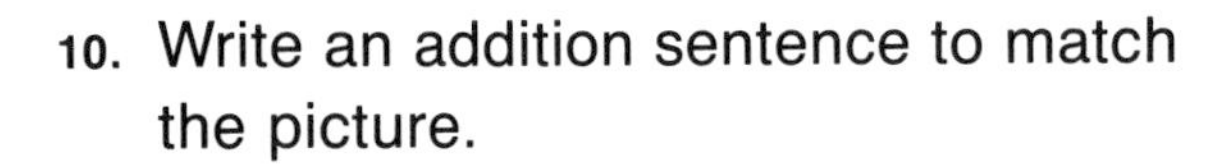

10. Write an addition sentence to match the picture.

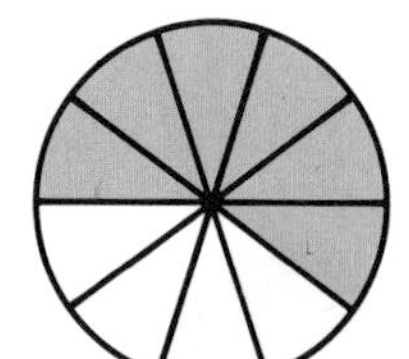

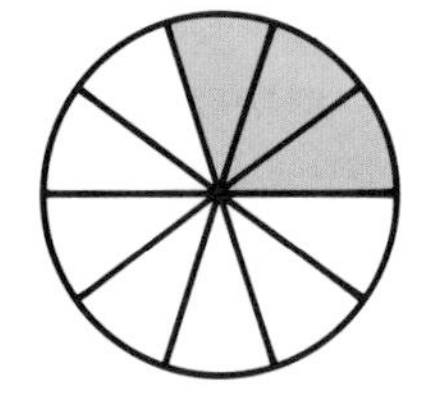

11. Write a subtraction sentence to match the picture.

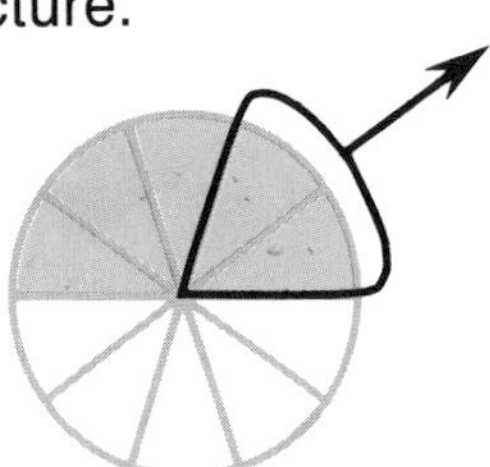

CUMULATIVE REVIEW

Write 2 multiplication sentences for each picture.

1. 

2. 

Copy and complete each sentence.

3. $3 \times 6 = \blacksquare$
   $6 \times 3 = \blacksquare$

4. $1 \times 8 = \blacksquare$
   $8 \times 1 = \blacksquare$

5. $4 \times 7 = \blacksquare$
   $7 \times 4 = \blacksquare$

6. $3 \times 9 = \blacksquare$
   $9 \times 3 = \blacksquare$

Multiply.

7. 4×6
8. 2×7
9. 6×8
10. 3×8
11. 7×7
12. 6×9
13. 8×8
14. 9×6
15. 5×9
16. 2×6
17. 8×7
18. 5×8
19. 10×3
20. 40×5
21. 100×6
22. 300×2
23. 26×2
24. 63×7

Divide.

25. $6\overline{)54}$
26. $9\overline{)27}$
27. $7\overline{)28}$
28. $6\overline{)30}$
29. $9\overline{)36}$
30. $8\overline{)24}$
31. $7\overline{)14}$
32. $6\overline{)36}$
33. $8\overline{)32}$
34. $9\overline{)45}$
35. $7\overline{)21}$
36. $8\overline{)16}$
37. $4\overline{)17}$
38. $2\overline{)19}$
39. $3\overline{)29}$
40. $5\overline{)18}$
41. $4\overline{)26}$
42. $3\overline{)16}$

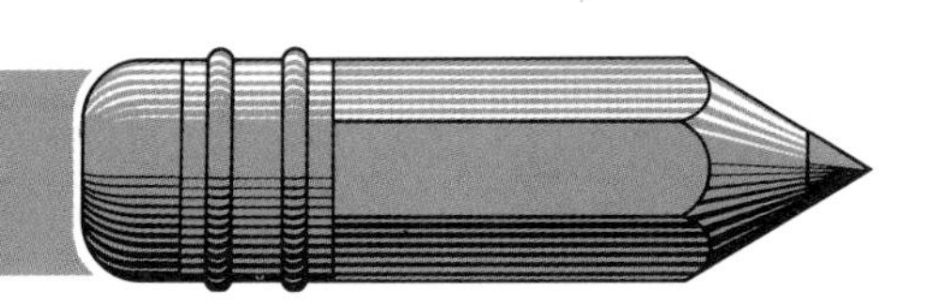

43. A family of 4 shared
12 hot dogs equally.
How many did each person get?

44. There are 7 days in a week.
How many days are there in 8 weeks?

45. Your heart beats 72 times each minute.
How many times does it beat in 5 minutes?

46. 20 tickets are shared equally among 3 children.
How many tickets are left over?

Write the greater number.

47. 1.5, 1.7

48. 2.9, 3.1

49. 4.1, 4.0

Tell whether each figure is divided into equal parts.

50. 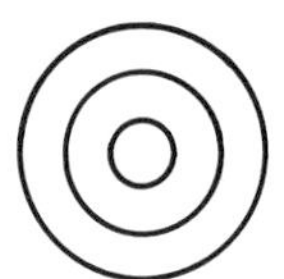

51. 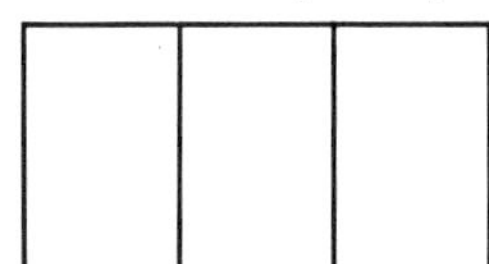

52. 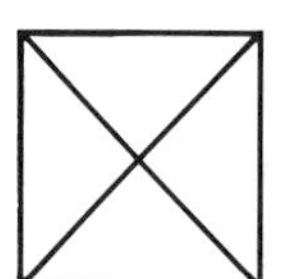

Write a fraction and a decimal number for the coloured part of each picture.

53. 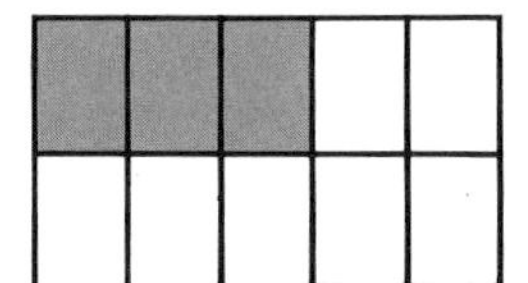

54. 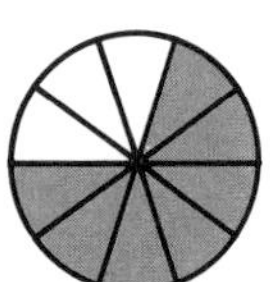

55. 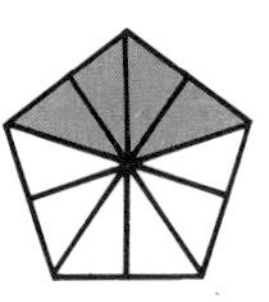

Add.

56. $\begin{array}{r} 0.2 \\ +0.4 \\ \hline \end{array}$

57. $\begin{array}{r} 0.3 \\ +0.6 \\ \hline \end{array}$

58. $\begin{array}{r} 1.4 \\ +2.3 \\ \hline \end{array}$

59. $\begin{array}{r} 2.5 \\ +3.6 \\ \hline \end{array}$

60. $\begin{array}{r} \$1.25 \\ +2.40 \\ \hline \end{array}$

61. $\begin{array}{r} \$2.55 \\ +3.38 \\ \hline \end{array}$

Subtract.

62. $\begin{array}{r} 0.5 \\ -0.2 \\ \hline \end{array}$

63. $\begin{array}{r} 0.7 \\ -0.3 \\ \hline \end{array}$

64. $\begin{array}{r} 2.8 \\ -0.7 \\ \hline \end{array}$

65. $\begin{array}{r} 4.2 \\ -1.6 \\ \hline \end{array}$

66. $\begin{array}{r} \$5.49 \\ -3.25 \\ \hline \end{array}$

67. $\begin{array}{r} \$4.30 \\ -2.70 \\ \hline \end{array}$

SKILLS CHECK-UP

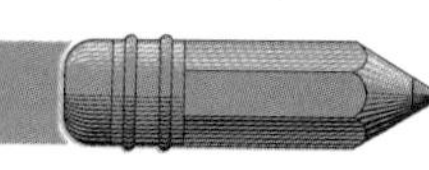

Add.

1. 3 + 5	2. 5 + 6	3. 2 + 8	4. 9 + 3	5. 4 + 2	6. 4 + 8
7. 4 + 3	8. 5 + 4	9. 0 + 5	10. 5 + 9	11. 3 + 0	12. 5 + 7
13. 6 + 4	14. 6 + 6	15. 4 + 3	16. 9 + 7	17. 5 + 2	18. 8 + 5
19. 5 + 3	20. 7 + 8	21. 6 + 1	22. 4 + 9	23. 7 + 1	24. 3 + 8
25. 4 + 6	26. 6 + 7	27. 7 + 2	28. 9 + 6	29. 9 + 0	30. 4 + 7

Add.

31. 2 + 7	32. 8 + 7	33. 9 + 1	34. 2 + 9	35. 6 + 2	36. 8 + 6
37. 3 + 6	38. 6 + 5	39. 2 + 0	40. 9 + 4	41. 3 + 1	42. 8 + 4
43. 4 + 5	44. 9 + 9	45. 7 + 3	46. 9 + 8	47. 8 + 1	48. 7 + 5
49. 6 + 0	50. 4 + 4	51. 5 + 1	52. 7 + 9	53. 3 + 7	54. 7 + 4
55. 3 + 2	56. 7 + 6	57. 8 + 2	58. 5 + 9	59. 1 + 4	60. 6 + 8

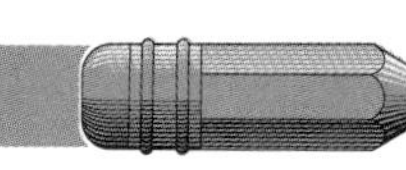

SKILLS CHECK-UP

Subtract.

1. $4 - 2$
2. $6 - 3$
3. $9 - 7$
4. $13 - 9$
5. $7 - 0$
6. $12 - 8$
7. $10 - 2$
8. $10 - 6$
9. $7 - 1$
10. $15 - 6$
11. $8 - 6$
12. $12 - 6$
13. $8 - 7$
14. $14 - 7$
15. $3 - 2$
16. $11 - 2$
17. $10 - 7$
18. $11 - 3$
19. $9 - 4$
20. $2 - 2$
21. $7 - 4$
22. $16 - 8$
23. $6 - 2$
24. $14 - 6$
25. $8 - 3$
26. $8 - 4$
27. $10 - 9$
28. $14 - 9$
29. $10 - 5$
30. $11 - 7$

Subtract.

31. $5 - 1$
32. $2 - 1$
33. $6 - 5$
34. $12 - 3$
35. $8 - 5$
36. $13 - 5$
37. $9 - 3$
38. $12 - 5$
39. $9 - 1$
40. $17 - 8$
41. $10 - 3$
42. $12 - 7$
43. $7 - 5$
44. $9 - 0$
45. $8 - 6$
46. $15 - 7$
47. $9 - 2$
48. $14 - 8$
49. $8 - 0$
50. $16 - 9$
51. $4 - 4$
52. $13 - 6$
53. $7 - 3$
54. $11 - 8$
55. $5 - 3$
56. $18 - 9$
57. $4 - 3$
58. $11 - 6$
59. $0 - 0$
60. $13 - 8$

SKILLS CHECK-UP

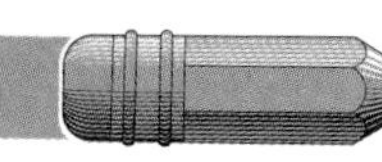

Add.

1. 23 + 2	2. 6 + 37	3. 96 + 22	4. 20 + 77	5. 57 + 24	6. 53 + 69
7. 16 + 31	8. 17 + 19	9. 59 + 60	10. 3 + 93	11. 46 + 16	12. 45 + 66
13. 55 + 40	14. 14 + 46	15. 48 + 71	16. 22 + 27	17. 59 + 33	18. 74 + 68
19. 17 + 41	20. 55 + 5	21. 82 + 57	22. 39 + 50	23. 48 + 47	24. 99 + 21
25. 63 + 5	26. 76 + 15	27. 65 + 43	28. 60 + 30	29. 39 + 29	30. 86 + 57

Add.

31. 36 + 33	32. 67 + 17	33. 98 + 31	34. 57 + 2	35. 26 + 15	36. 83 + 77
37. 59 + 10	38. 9 + 89	39. 27 + 81	40. 47 + 22	41. 68 + 22	42. 86 + 35
43. 25 + 42	44. 39 + 33	45. 63 + 76	46. 16 + 13	47. 24 + 38	48. 42 + 98
49. 11 + 17	50. 69 + 11	51. 87 + 81	52. 80 + 9	53. 88 + 8	54. 39 + 79
55. 5 + 81	56. 28 + 35	57. 70 + 90	58. 73 + 14	59. 78 + 5	60. 97 + 44

SKILLS CHECK-UP

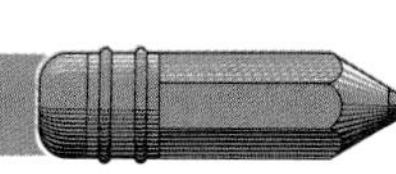

Subtract.

1. $50 - 20$	2. $41 - 25$	3. $58 - 49$	4. $50 - 11$	5. $27 - 22$	6. $71 - 45$
7. $85 - 13$	8. $60 - 12$	9. $76 - 58$	10. $73 - 17$	11. $87 - 70$	12. $33 - 28$
13. $41 - 10$	14. $45 - 39$	15. $92 - 25$	16. $42 - 23$	17. $85 - 21$	18. $48 - 19$
19. $54 - 42$	20. $52 - 13$	21. $74 - 29$	22. $62 - 35$	23. $66 - 6$	24. $55 - 46$
25. $27 - 26$	26. $82 - 66$	27. $81 - 49$	28. $87 - 68$	29. $97 - 15$	30. $36 - 17$

Subtract.

31. $38 - 14$	32. $78 - 59$	33. $32 - 16$	34. $51 - 36$	35. $16 - 12$	36. $64 - 56$
37. $19 - 7$	38. $34 - 8$	39. $40 - 37$	40. $53 - 14$	41. $88 - 61$	42. $24 - 7$
43. $96 - 55$	44. $70 - 11$	45. $61 - 33$	46. $94 - 75$	47. $43 - 30$	48. $92 - 13$
49. $63 - 32$	50. $22 - 6$	51. $84 - 37$	52. $95 - 87$	53. $59 - 37$	54. $67 - 28$
55. $72 - 31$	56. $93 - 84$	57. $68 - 49$	58. $90 - 52$	59. $74 - 61$	60. $80 - 16$

SKILLS CHECK-UP

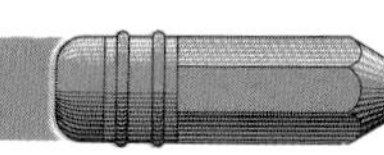

Add.

1. 126 + 831	2. 461 + 109	3. 588 + 291	4. 127 + 696	5. 885 + 257	6. 721 + 557
7. 756 + 243	8. 112 + 118	9. 365 + 554	10. 235 + 589	11. 756 + 986	12. 601 + 888
13. 101 + 368	14. 119 + 677	15. 244 + 382	16. 358 + 173	17. 688 + 432	18. 940 + 930
19. 409 + 310	20. 405 + 516	21. 493 + 416	22. 279 + 284	23. 741 + 299	24. 562 + 511
25. 582 + 17	26. 843 + 37	27. 618 + 190	28. 352 + 499	29. 675 + 398	30. 815 + 250

Add.

31. 663 + 233	32. 715 + 148	33. 175 + 740	34. 367 + 68	35. 333 + 769	36. 100 + 972
37. 727 + 152	38. 927 + 36	39. 787 + 180	40. 225 + 196	41. 136 + 897	42. 411 + 778
43. 464 + 324	44. 225 + 369	45. 822 + 95	46. 867 + 53	47. 661 + 839	48. 759 + 430
49. 553 + 401	50. 555 + 238	51. 247 + 361	52. 499 + 252	53. 638 + 992	54. 221 + 851
55. 516 + 281	56. 379 + 314	57. 33 + 574	58. 193 + 188	59. 783 + 718	60. 569 + 830

SKILLS CHECK-UP

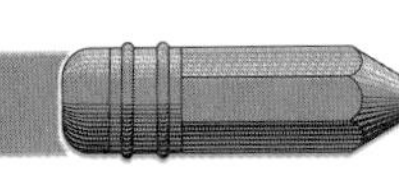

Subtract.

1. 554 − 231	2. 823 − 408	3. 188 − 92	4. 828 − 189	5. 201 − 94	6. 314 − 99
7. 758 − 151	8. 651 − 234	9. 634 − 152	10. 623 − 474	11. 303 − 136	12. 724 − 399
13. 939 − 18	14. 378 − 229	15. 819 − 637	16. 331 − 192	17. 608 − 259	18. 741 − 572
19. 195 − 122	20. 327 − 108	21. 279 − 185	22. 716 − 587	23. 802 − 311	24. 347 − 267
25. 362 − 140	26. 581 − 227	27. 586 − 293	28. 474 − 377	29. 704 − 595	30. 198 − 119

Subtract.

31. 236 − 16	32. 725 − 119	33. 444 − 150	34. 721 − 188	35. 106 − 87	36. 488 − 77
37. 407 − 305	38. 694 − 75	39. 382 − 291	40. 563 − 378	41. 308 − 249	42. 903 − 28
43. 811 − 611	44. 433 − 116	45. 765 − 574	46. 382 − 283	47. 507 − 128	48. 772 − 149
49. 543 − 322	50. 936 − 707	51. 915 − 745	52. 977 − 689	53. 900 − 297	54. 378 − 164
55. 668 − 535	56. 292 − 114	57. 288 − 191	58. 121 − 74	59. 405 − 186	60. 896 − 527

SKILLS CHECK-UP

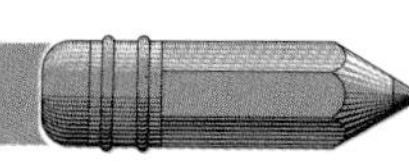

Complete.

1. $\begin{array}{r}38\\-12\\\hline\end{array}$	2. $\begin{array}{r}75\\+18\\\hline\end{array}$	3. $\begin{array}{r}287\\+655\\\hline\end{array}$	4. $\begin{array}{r}172\\-91\\\hline\end{array}$	5. $\begin{array}{r}13\\+26\\\hline\end{array}$	6. $\begin{array}{r}832\\-264\\\hline\end{array}$
7. $\begin{array}{r}16\\-12\\\hline\end{array}$	8. $\begin{array}{r}237\\+26\\\hline\end{array}$	9. $\begin{array}{r}97\\+83\\\hline\end{array}$	10. $\begin{array}{r}670\\-439\\\hline\end{array}$	11. $\begin{array}{r}483\\+15\\\hline\end{array}$	12. $\begin{array}{r}300\\-172\\\hline\end{array}$
13. $\begin{array}{r}683\\-61\\\hline\end{array}$	14. $\begin{array}{r}871\\+84\\\hline\end{array}$	15. $\begin{array}{r}195\\+36\\\hline\end{array}$	16. $\begin{array}{r}852\\-18\\\hline\end{array}$	17. $\begin{array}{r}11\\+8\\\hline\end{array}$	18. $\begin{array}{r}411\\-137\\\hline\end{array}$
19. $\begin{array}{r}359\\-126\\\hline\end{array}$	20. $\begin{array}{r}68\\+50\\\hline\end{array}$	21. $\begin{array}{r}78\\+277\\\hline\end{array}$	22. $\begin{array}{r}95\\-68\\\hline\end{array}$	23. $\begin{array}{r}62\\+24\\\hline\end{array}$	24. $\begin{array}{r}746\\-69\\\hline\end{array}$
25. $\begin{array}{r}96\\-4\\\hline\end{array}$	26. $\begin{array}{r}614\\+139\\\hline\end{array}$	27. $\begin{array}{r}682\\+598\\\hline\end{array}$	28. $\begin{array}{r}73\\-58\\\hline\end{array}$	29. $\begin{array}{r}222\\+175\\\hline\end{array}$	30. $\begin{array}{r}682\\-95\\\hline\end{array}$

Complete.

31. $\begin{array}{r}592\\-371\\\hline\end{array}$	32. $\begin{array}{r}463\\+292\\\hline\end{array}$	33. $\begin{array}{r}59\\+43\\\hline\end{array}$	34. $\begin{array}{r}205\\-131\\\hline\end{array}$	35. $\begin{array}{r}830\\+139\\\hline\end{array}$	36. $\begin{array}{r}526\\-468\\\hline\end{array}$
37. $\begin{array}{r}98\\-17\\\hline\end{array}$	38. $\begin{array}{r}32\\+81\\\hline\end{array}$	39. $\begin{array}{r}99\\+25\\\hline\end{array}$	40. $\begin{array}{r}613\\-350\\\hline\end{array}$	41. $\begin{array}{r}127\\+111\\\hline\end{array}$	42. $\begin{array}{r}200\\-87\\\hline\end{array}$
43. $\begin{array}{r}67\\-3\\\hline\end{array}$	44. $\begin{array}{r}80\\+65\\\hline\end{array}$	45. $\begin{array}{r}117\\+894\\\hline\end{array}$	46. $\begin{array}{r}38\\-9\\\hline\end{array}$	47. $\begin{array}{r}37\\+40\\\hline\end{array}$	48. $\begin{array}{r}122\\-63\\\hline\end{array}$
49. $\begin{array}{r}444\\-21\\\hline\end{array}$	50. $\begin{array}{r}52\\+9\\\hline\end{array}$	51. $\begin{array}{r}356\\+486\\\hline\end{array}$	52. $\begin{array}{r}80\\-42\\\hline\end{array}$	53. $\begin{array}{r}28\\+60\\\hline\end{array}$	54. $\begin{array}{r}903\\-387\\\hline\end{array}$
55. $\begin{array}{r}896\\-280\\\hline\end{array}$	56. $\begin{array}{r}67\\+6\\\hline\end{array}$	57. $\begin{array}{r}257\\+568\\\hline\end{array}$	58. $\begin{array}{r}70\\-6\\\hline\end{array}$	59. $\begin{array}{r}97\\+2\\\hline\end{array}$	60. $\begin{array}{r}840\\-385\\\hline\end{array}$

SKILLS CHECK-UP

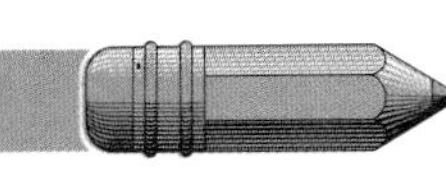

Multiply.

1. $\begin{array}{r} 1 \\ \times 3 \\ \hline \end{array}$	2. $\begin{array}{r} 7 \\ \times 3 \\ \hline \end{array}$	3. $\begin{array}{r} 7 \\ \times 0 \\ \hline \end{array}$	4. $\begin{array}{r} 8 \\ \times 2 \\ \hline \end{array}$	5. $\begin{array}{r} 5 \\ \times 5 \\ \hline \end{array}$	6. $\begin{array}{r} 3 \\ \times 9 \\ \hline \end{array}$
7. $\begin{array}{r} 0 \\ \times 4 \\ \hline \end{array}$	8. $\begin{array}{r} 3 \\ \times 6 \\ \hline \end{array}$	9. $\begin{array}{r} 7 \\ \times 7 \\ \hline \end{array}$	10. $\begin{array}{r} 8 \\ \times 4 \\ \hline \end{array}$	11. $\begin{array}{r} 5 \\ \times 4 \\ \hline \end{array}$	12. $\begin{array}{r} 9 \\ \times 4 \\ \hline \end{array}$
13. $\begin{array}{r} 5 \\ \times 1 \\ \hline \end{array}$	14. $\begin{array}{r} 9 \\ \times 3 \\ \hline \end{array}$	15. $\begin{array}{r} 2 \\ \times 7 \\ \hline \end{array}$	16. $\begin{array}{r} 6 \\ \times 8 \\ \hline \end{array}$	17. $\begin{array}{r} 8 \\ \times 5 \\ \hline \end{array}$	18. $\begin{array}{r} 9 \\ \times 7 \\ \hline \end{array}$
19. $\begin{array}{r} 6 \\ \times 1 \\ \hline \end{array}$	20. $\begin{array}{r} 3 \\ \times 8 \\ \hline \end{array}$	21. $\begin{array}{r} 7 \\ \times 9 \\ \hline \end{array}$	22. $\begin{array}{r} 8 \\ \times 9 \\ \hline \end{array}$	23. $\begin{array}{r} 5 \\ \times 7 \\ \hline \end{array}$	24. $\begin{array}{r} 6 \\ \times 9 \\ \hline \end{array}$
25. $\begin{array}{r} 6 \\ \times 0 \\ \hline \end{array}$	26. $\begin{array}{r} 4 \\ \times 3 \\ \hline \end{array}$	27. $\begin{array}{r} 1 \\ \times 7 \\ \hline \end{array}$	28. $\begin{array}{r} 1 \\ \times 8 \\ \hline \end{array}$	29. $\begin{array}{r} 6 \\ \times 5 \\ \hline \end{array}$	30. $\begin{array}{r} 9 \\ \times 8 \\ \hline \end{array}$

Multiply.

31. $\begin{array}{r} 4 \\ \times 2 \\ \hline \end{array}$	32. $\begin{array}{r} 6 \\ \times 6 \\ \hline \end{array}$	33. $\begin{array}{r} 6 \\ \times 7 \\ \hline \end{array}$	34. $\begin{array}{r} 8 \\ \times 3 \\ \hline \end{array}$	35. $\begin{array}{r} 4 \\ \times 5 \\ \hline \end{array}$	36. $\begin{array}{r} 9 \\ \times 2 \\ \hline \end{array}$
37. $\begin{array}{r} 7 \\ \times 2 \\ \hline \end{array}$	38. $\begin{array}{r} 7 \\ \times 6 \\ \hline \end{array}$	39. $\begin{array}{r} 8 \\ \times 7 \\ \hline \end{array}$	40. $\begin{array}{r} 8 \\ \times 8 \\ \hline \end{array}$	41. $\begin{array}{r} 4 \\ \times 8 \\ \hline \end{array}$	42. $\begin{array}{r} 4 \\ \times 9 \\ \hline \end{array}$
43. $\begin{array}{r} 2 \\ \times 8 \\ \hline \end{array}$	44. $\begin{array}{r} 8 \\ \times 6 \\ \hline \end{array}$	45. $\begin{array}{r} 4 \\ \times 7 \\ \hline \end{array}$	46. $\begin{array}{r} 5 \\ \times 8 \\ \hline \end{array}$	47. $\begin{array}{r} 4 \\ \times 6 \\ \hline \end{array}$	48. $\begin{array}{r} 9 \\ \times 6 \\ \hline \end{array}$
49. $\begin{array}{r} 2 \\ \times 3 \\ \hline \end{array}$	50. $\begin{array}{r} 6 \\ \times 4 \\ \hline \end{array}$	51. $\begin{array}{r} 7 \\ \times 5 \\ \hline \end{array}$	52. $\begin{array}{r} 0 \\ \times 8 \\ \hline \end{array}$	53. $\begin{array}{r} 7 \\ \times 4 \\ \hline \end{array}$	54. $\begin{array}{r} 0 \\ \times 9 \\ \hline \end{array}$
55. $\begin{array}{r} 2 \\ \times 6 \\ \hline \end{array}$	56. $\begin{array}{r} 5 \\ \times 6 \\ \hline \end{array}$	57. $\begin{array}{r} 3 \\ \times 7 \\ \hline \end{array}$	58. $\begin{array}{r} 7 \\ \times 8 \\ \hline \end{array}$	59. $\begin{array}{r} 3 \\ \times 4 \\ \hline \end{array}$	60. $\begin{array}{r} 9 \\ \times 1 \\ \hline \end{array}$

SKILLS CHECK-UP

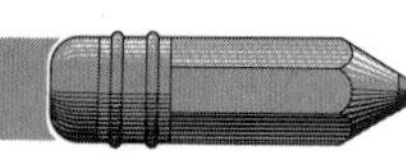

Divide.

1. $3\overline{)3}$	2. $4\overline{)32}$	3. $5\overline{)30}$	4. $3\overline{)15}$	5. $8\overline{)32}$	6. $2\overline{)18}$
7. $1\overline{)4}$	8. $3\overline{)12}$	9. $6\overline{)18}$	10. $3\overline{)27}$	11. $8\overline{)56}$	12. $9\overline{)27}$
13. $1\overline{)7}$	14. $9\overline{)36}$	15. $6\overline{)24}$	16. $3\overline{)24}$	17. $8\overline{)8}$	18. $6\overline{)54}$
19. $6\overline{)6}$	20. $4\overline{)16}$	21. $6\overline{)12}$	22. $3\overline{)9}$	23. $8\overline{)24}$	24. $9\overline{)72}$
25. $2\overline{)2}$	26. $4\overline{)28}$	27. $8\overline{)48}$	28. $7\overline{)21}$	29. $5\overline{)40}$	30. $4\overline{)36}$

Divide.

31. $6\overline{)12}$	32. $8\overline{)56}$	33. $3\overline{)18}$	34. $2\overline{)10}$	35. $8\overline{)64}$	36. $9\overline{)63}$
37. $2\overline{)8}$	38. $3\overline{)21}$	39. $6\overline{)36}$	40. $5\overline{)5}$	41. $7\overline{)56}$	42. $9\overline{)9}$
43. $3\overline{)6}$	44. $5\overline{)35}$	45. $6\overline{)42}$	46. $8\overline{)40}$	47. $6\overline{)48}$	48. $9\overline{)45}$
49. $2\overline{)14}$	50. $7\overline{)28}$	51. $4\overline{)24}$	52. $5\overline{)35}$	53. $8\overline{)16}$	54. $9\overline{)81}$
55. $2\overline{)16}$	56. $7\overline{)42}$	57. $9\overline{)54}$	58. $5\overline{)45}$	59. $8\overline{)72}$	60. $7\overline{)63}$

SKILLS CHECK-UP

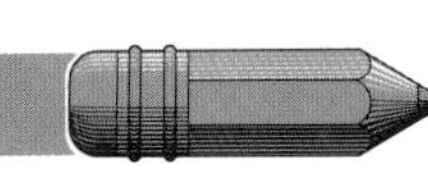

Complete.

1. 3×8
2. $3\overline{)9}$
3. 4×7
4. 6×6
5. $4\overline{)36}$
6. 6×8
7. $5\overline{)40}$
8. 9×5
9. $5\overline{)25}$
10. $7\overline{)42}$
11. $9\overline{)81}$
12. 5×5
13. 4×3
14. 8×8
15. 2×8
16. $5\overline{)15}$
17. 5×7
18. $6\overline{)24}$
19. $7\overline{)35}$
20. 0×9
21. 7×7
22. $4\overline{)20}$
23. 6×3
24. $6\overline{)48}$
25. 9×9
26. $2\overline{)14}$
27. 6×7
28. $7\overline{)49}$
29. 4×2
30. $3\overline{)27}$

Complete.

31. $4\overline{)28}$
32. $9\overline{)63}$
33. $7\overline{)56}$
34. 7×9
35. 8×4
36. 7×3
37. $3\overline{)24}$
38. 8×7
39. $6\overline{)18}$
40. 8×9
41. 4×4
42. $8\overline{)72}$
43. 6×9
44. $4\overline{)16}$
45. 6×5
46. $8\overline{)64}$
47. 1×8
48. 3×5
49. $4\overline{)12}$
50. 8×5
51. $6\overline{)6}$
52. 9×4
53. $9\overline{)45}$
54. $8\overline{)32}$
55. 5×4
56. 4×6
57. 9×3
58. $5\overline{)30}$
59. $5\overline{)0}$
60. $6\overline{)54}$

INDEX